DATE DUE

*f*P

ALSO BY CHARLES FISHMAN

THE WAL-MART EFFECT
How the World's Most Powerful Company Really Works—
and How It's Transforming the American Economy

THE
BIG
THIRST

≈≈≈≈≈

*The Secret Life
and Turbulent Future of
Water*

CHARLES FISHMAN

FREE PRESS

New York ■ London ■ Toronto ■ Sydney

FREE PRESS

A Division of Simon & Schuster, Inc.
1230 Avenue of the Americas
New York, NY 10020

First Free Press hardcover edition April 2011

FREE PRESS and colophon are trademarks of Simon & Schuster, Inc.

For information about special discounts for bulk purchases, please contact Simon & Schuster Special Sales at 1-866-506-1949 or business@simonandschuster.com.

The Simon & Schuster Speakers Bureau can bring authors to your live event. For more information or to book an event contact the Simon & Schuster Speakers Bureau at 1-866-248-3049 or visit our website at www.simonspeakers.com.

Designed by Katy Riegel

Excerpt from Paige Taylor and Victoria Laurie, "How a Simple Flat Tyre Killed Artist and Bushman," *The Australian*, Jan. 16, 2007, reprinted by permission of Paige Taylor and Victoria Laurie.

Manufactured in the United States of America

1   3   5   7   9   10   8   6   4   2

Library of Congress Cataloging-in-Publication Data
Fishman, Charles.
The big thirst: the secret life and turbulent future of water /
Charles Fishman.
        p.   cm.
1. Water resources development.   2. Water supply.   3. Water use.
        4. Water consumption.   I. Title.
HD1691.F55 2011
333.91—dc22   2010033989

ISBN 978-1-4391-0207-7
ISBN 978-1-4391-2493-2 (ebook)

For Trish,
like water, beautiful, mysterious, and essential

CONTENTS

1. The Revenge of Water — *1*

2. The Secret Life of Water — *25*

3. Dolphins in the Desert — *51*

4. Water Under Water — *88*

5. The Money in the Pipes — *112*

6. The Yuck Factor — *145*

7. Who Stopped the Rain? — *182*

8. Where Water Is Worshipped, but Gets No Respect — *218*

9. It's Water. Of Course It's Free — *265*

10. The Fate of Water — *293*

Notes — *315*

Water Measurements of All Kinds — *371*

Acknowledgments — *372*

Index — *375*

1

The Revenge of Water

Water is H$_2$O, hydrogen two parts, oxygen one,
But there is also a third thing, that makes it water
And nobody knows what it is.
                    —*D. H. Lawrence, "The Third Thing"*

AT T MINUS 16 SECONDS in the launch sequence of NASA's space shuttle, the launch control computers would trigger the release of water from a 290-foot-high water tank that stands next to the launchpad at Florida's Kennedy Space Center. The two pipes that delivered the water to the pad are each seven feet in diameter. Just before the shuttle's rocket motors ignited, 300,000 gallons of water would cascade across the base of the pad, eventually flowing at a rate of nearly a million gallons a minute. As the shuttle roared off the pad, the blast from its five engines poured down into the 2.5-million-pound cushion of water. The water was flowing so furiously it ran out nine seconds into liftoff.

The water actually had nothing to do with damping the heat or fire from the shuttle's motors. It was a sound suppressant. The space shuttle's rockets were so loud that without the sound-absorbing cushion of water, the roar from the engines would bounce off the metal and concrete base of the pad and ricochet back up. The sound waves would have torn the spacecraft apart before it was clear of the launch tower.[1]

We use water to baptize our children, and we use it to launch the most advanced spacecraft ever created. Water creates both the hypnotic majesty of Niagara Falls and the miniature, untouchable filigree of each snowflake. Solid water tore open the steel hull of the *Titanic*; liquid water sank her. You need great water to brew great coffee, you need great water to make great beer, you need pretty darn good water to make good concrete. Water adds the fun to water balloons, to a Slip 'N Slide, to a shower for two.

Water is both mythic and real. It manages to be at once part of the mystery of life and part of the routine of life. We can use water to wash our dishes and our dogs and our cars without giving it a second thought, but few of us can resist simply standing and watching breakers crash on the beach. Water has all kinds of associations and connections, implications and suggestiveness. It also has an indispensable practicality.

Water is the most familiar substance in our lives. It is also unquestionably the most important substance in our lives. Water vapor is the insulation in our atmosphere that makes Earth a comfortable place for us to live. Water drives our weather and shapes our geography. Water is the lubricant that allows the continents themselves to move. Water is the secret ingredient of our fuel-hungry society. The electricity you use at home each day requires 250 gallons of water per person, not just more than the actual water you use at home in the kitchen and the bathroom but two-and-a-half times more. That new flat-screen TV, it turns out, needs not just a wall outlet and a cable connection but also its own water supply to get going. Who would have guessed?[2]

Water is also the secret ingredient in the computer chips that make possible everything from MRI machines to Twitter accounts. Indeed, from blue jeans to iPhones, from Kleenex to basmati rice to the steel in your Toyota Prius, every product of modern life is awash in water. The two-liter bottle of Coke in your refrigerator required five liters of water to produce.[3]

Water is how we amuse ourselves—in the pool, at the beach, in a sauna, or on a sailboat. Water is a source of excitement on a white-water rafting trip, and an instinctive source of comfort in a steaming shower at the end of a long day.

Water is, quite literally, everywhere. When you take a carton of milk from the refrigerator and set it on the table, within a minute or two the outside is covered in a film of condensation—water that has migrated almost instantly from the air of the kitchen to the cold surface of the milk carton. Water infuses our language the way it does the air around us, the water references so common we don't even notice them: "go with the flow," "blow off steam," "wet behind the ears." The mortgage crisis that triggered the Great Recession was caused, in part, by all those homes that ended up "underwater."

And, of course, water is the most important substance in our lives because we ourselves are made mostly of water—men are typically 60 percent water, women are typically 55 percent water. A 150-pound man is 90 pounds of water (11 gallons).[4]

Everything human beings do is, quite literally, a function of water, because every cell in our bodies is plumped full of it, and every cell is bathed in watery fluid. Blood is 83 percent water. Every heartbeat is mediated by chemicals in water; when we gaze at a starry night sky, the cells in our eyes execute all their seeing functions in water; thinking about water requires neurons filled with water.

Given that water is both the most familiar substance in our lives, and the most important substance in our lives, the really astonishing thing is that most of us don't think of ourselves as having a relationship to water. It's perfectly natural to talk about our relationship to our car or our relationship to food, our relationship to alcohol, or money, or to God.

But water has achieved an invisibility in our lives that is only more remarkable given how central it is. Water used to be part of the rhythm and motivation of daily life, and there are plenty of places, including farms and whole swaths of the developing world, where it still is.

But in the United States and the developed world, we've spent the last hundred years in a kind of aquatic paradise: our water has been abundant, safe, and cheap. The twentieth century was really the first time when all three of those things were true. It has created a kind of golden age of water, when we could use as much as we wanted, whenever we wanted, for almost no cost.

Water service is so reliable that it has become completely inconspicuous. It is possible for a typical American to go a whole lifetime and never turn on the kitchen faucet and have no water come out. Indeed, water faucets that don't work are so rare, they're a little spooky. We don't even have an expression for the water equivalent of a power failure, whereas power failures are common enough that our microwave ovens are programmed to display "PF" when the electricity has gone out.

We live very wet lives, but we have no idea just how wet. The effortless way we have come to manage water is a testament to both water's moment-to-moment utility and to our own ingenuity. But unlike the time we spend at the gas pump—where we can see the gallons as they are pumped, and the instant impact on our credit card bill—the way we handle water use insulates us not just from the wonders of water, but from any sense of how much water daily life requires, or the work and expense required to deliver that water.

The good news is that most of what we know about water isn't really wrong, because we don't know that much. The bad news is that the invisibility of water in our lives isn't good for us, and it isn't good for water. You can't appreciate what you don't understand. You don't value and protect what you don't know is there.

Back in 1999, a team of researchers recorded 289,000 toilet flushes of Americans in twelve cities, from Seattle to Tampa. In fact, the researchers used electronic water-flow sensors to record not just toilet flushes but every "water event" in each of 1,188 homes for four weeks.

Although the study cost less than $1 million, it is considered so detailed and so pioneering that it hasn't been duplicated in the decade since; the U.S. Environmental Protection Agency continues to cite it as the definitive look at how Americans use water at home.

The researchers measured everything we do with water at home—how many gallons a bath takes, how often the clothes washer runs, how much water the dishwasher uses, who has low-flow showerheads and who has regular, how many times we flush the toilet each day, and how many gallons of water each flush uses.

The study's overall conclusion can be summed up in four words: We like to flush.

For Americans, flushing the toilet is the main way we use water. We use more water flushing toilets than bathing or cooking or washing our hands, our dishes, or our clothes.

The typical American flushes the toilet five times a day at home, and uses 18.5 gallons (70 liters) of water, just for that.[5]

What that means is that every day, as a nation, just to flush our toilets, Americans use 5,700,000,000 gallons of water—5.7 billion gallons of clean drinking water down the toilet.

And that's just at home.

It's impossible to get your brain around that number, of course— 5.7 billion gallons of water a day. But here's a way of thinking about it. It's more water than all the homes in the United Kingdom and Canada use each day for all their needs—we flush more water down the toilet than 95 million Brits and Canadians use.[6]

Of course, we are a big country, and we do need to flush our toilets. Or, at least, we like to.

When we think about the big ways we use water, flushing the toilet doesn't typically leap to mind. It's one of those unnoticed parts of our daily water use—our daily water-mark—that turn out to be both startling and significant.

The largest single consumer of water in the United States, in fact, is virtually invisible. Every day, the nation's power plants use 201 billion gallons of water in the course of generating electricity. That isn't water used by hydroelectric plants—it's the water used by coal, gas, and nuclear power plants for cooling and to make steam. U.S. electric utilities require seven times more water than all U.S. homes. They use 1.5 times the amount of water used by all the farms in the country. In fact, 49 percent of all water use in the United States is for power plants.

Toilets and electric outlets may be stealthy consumers of water, but they at least serve vital functions. One of the largest daily consumers of water isn't a use at all. One of every six gallons of water pumped into water mains by U.S. utilities simply leaks away, back into the ground. Sixteen percent of the water disappears from the pipes before it makes it to a home or business or factory. Every six days, U.S. water utilities lose an entire day's water.[7] And that 16 percent U.S. loss rate isn't too bad— British utilities lose 19 percent of the water they pump; the French lose 26 percent.[8]

There is perhaps no better symbol of the golden age of water, of the carefree, almost cavalier, attitude that our abundance has fostered. We go to the trouble and expense to find city-size quantities of water, build dams, reservoirs, and tanks to store it and plants to treat it, then we pump it out to customers, only to let it dribble away before anyone can use it.

One of the hallmarks of the twentieth century, at least in the developed world, is that we have gradually been able to stop thinking about water. We use more of it than ever, we rely on it for purposes we not only never see but can hardly imagine, and we think about it not at all.

It is a striking achievement. We used to build monuments—even temples—to water. The aqueducts of the Roman Empire are marvels of engineering and soaringly elegant design. They were plumbing presented as civic achievement and as a tribute to the water itself. Today, water has drifted so far from civic celebration that many people visit the Roman aqueducts without any sense at all that they moved water, or how.

In Poland Spring, Maine, there is an actual marble and granite temple enclosing the burbling spring that gives one of the nation's most popular bottled waters its name. The temple was built in 1906 to celebrate a source of water that was then so pure and so highly regarded that people as far away as Boston, New York, and Chicago could arrange to have it delivered for household use.

Many cities in the world are located where they are because of their proximity to water. For most of human history, in most settings, getting water was part of the daily routine; it was a constant part of our mental landscape. At the same time, humanity's relationship to its water supply was wary, because water often made people sick. That's why Poland Spring water was so popular in Boston and New York a century ago—it was safe.

One hundred years ago, with the dawn of bacteriology, two things happened. Cities started aggressively separating their freshwater supplies from their sewage disposal, something they had been surprisingly slow to do. (Philadelphia is just one of many cities whose sewage system, a hundred years ago, emptied into a river *upstream* of the city water supply intakes from the same river.) And water utilities discovered that basic sand filters and chlorination could clean and disinfect water supplies, all but assuring their safety.

In the decade from 1905 to 1915, as dozens of water systems around the country installed filters and chlorination systems, we went through a water revolution that profoundly improved human life forever. Between 1900 and 1940, mortality rates in the United States fell 40 percent.

How much did clean water matter? Harvard economist David Cutler and Stanford professor of medicine Grant Miller conducted a remarkable analysis, published in 2005, teasing out the impact of the new water treatment methods on the most dramatic reduction in death rates in U.S. history.[9]

By 1936, they conclude, simple filtration and chlorination of city water supplies reduced overall mortality in U.S. cities by 13 percent. Clean water cut child mortality in half.[10]

From 1900 to 1940, U.S. life expectancy at birth went from forty-seven years to sixty-three years.[11] In just forty years, the life span of the average American was extended sixteen years. A child born in 1900 could expect to live only until 1947, not even to mid-century. A child born in 1940 quite likely lived to hear the news in 2003 that the human genome had been completely mapped.

Clean municipal water encouraged cities to grow, and it also encouraged the expansion of "mains water" during the twentieth century as the way most Americans got their water. (By 2005, only 14 percent of Americans still relied on wells or some other "self-supplied" water.)[12]

That first water revolution ushered in an era—the one we think we still live in—in which water was unlimited, free, and safe. And once it was unlimited, free, and safe, we could stop thinking about it. The fact that it was unfailingly available "on demand" meant that we would use it more, even as we thought about it less.

The figures are dramatic. In 1955, the U.S. Geological Survey's water-use analysis said that rural Americans without running water in their homes used ten gallons a day per person. (That same year, each cow used twenty gallons per day.) For newly "electrified" farm families, with pumps, and for city families that number was already sixty gallons per person.[13]

Today, it's a hundred gallons per person at home, and to say that we take water for granted is, in some ways, to give us too much credit. We don't take water for granted, because we don't notice it enough to take it for granted. It's like gravity except that, in some ways, water is actually more important than gravity. We do, in fact, have people living round-the-clock without gravity, and have for a decade, in the International Space Station.

Most people have no idea where the water they brush their teeth with comes from, or how it gets to them. The pipes are hidden; the sources of the water put into those pipes are remote.

Our home water bills, which are less than half our monthly cable TV or cell phone bills, provide almost no insight into how much water we use, or how we use it—even if we study them.[14]

The pricing of water has a kind of invisibility—or opaqueness, at least—all its own. Ten gallons of tap water, at home, costs on average 3 pennies.[15] That's the equivalent of getting seventy-four of those $1.29 half-liter bottles of water we love so much for less than a nickel. We happily pay three thousand times that price at the convenience store—one bottle for $1.29. But when the water bill goes from $30 to $34 a month, customers react as if they'll have to choose between their prescription drugs and their water service. We gladly pay perfectly silly prices for productized water because of the aura that's been created around it. But we have so little appreciation of the effort required to get water into our homes that any price increase inspires outrage.

The way water infuses the products we use every day, and the volumes those products require, is invisible to us.

The ownership and control of water is invisible—in some places, for instance, you don't actually own the water that falls on your roof or in your yard. (If someone has rights to the rivershed runoff, he owns it.)

The problems of water are, for the moment, mostly invisible. When we're done with the water we use every day, it simply disappears. What becomes of it is invisible to us.

The new class of micropollutants we are beginning to hear about—infinitesimal, almost molecular, traces of plastics, birth control pills, antidepressants—have literally been invisible even to chemists until very recently; you certainly can't tell if they're in your water by looking at it or drinking it. The impact of those micropollutants on our health, if any, may remain invisible for years—and may be almost impossible to predict or trace.

Even our emotional connections to water have become submerged and camouflaged—the ease with which water enters and leaves our lives allows us an indifference to our water supply. We are utterly ignorant of our own water-mark, of the amount of water required to float us through the day, and we are utterly indifferent to the mark our daily life leaves on the water supply.

Our very success with water has allowed us to become water illiterate.

But the golden age of water is rapidly coming to an end. The last century has conditioned us to think that water is naturally abundant, safe, and cheap—that it should be, that it will be. We're in for a rude shock.

We are in the middle of a water crisis already, in the United States and around the world. The experts realize it (the Weather Channel already has a dedicated burning-orange logo for its drought reports), but even in areas with serious water problems, most people don't seem to understand.

We are entering a new era of water scarcity—not just in traditionally dry or hard-pressed places like the U.S. Southwest and the Middle East, but in places we think of as water-wealthy, like Atlanta and Melbourne. The three things that we have taken to be the natural state of our water supply—abundant, cheap, and safe—will not be present together in the decades ahead. We may have water that is abundant and cheap, but it will be "reuse water," for things like lawn watering or car washing, not for drinking; we will certainly have drinking water that is safe, and it may be abundant, but it will not be thoughtlessly inexpensive.

We are on the verge of a second modern water revolution—and it is likely to change our attitudes at least as much as the one a hundred years ago.

The new water scarcity will reshape how we live, how we work, how we relax. It will reshape how we value water, and how we understand it.

We have ignored water—neglected our water supplies and our water systems, taken for granted the economic value of abundant water, and become blasé about the day-to-day convenience of easy water. We may well go directly from the golden age of water to the revenge of water.

IN THE SPRING OF 2008, the water situation in Barcelona, Spain, had become quite desperate. A drought across Catalonia had extended to eighteen months and become the worst in sixty years. Barcelona, Catalonia's capital, is a metro area with a population of 5 million, famous for its beaches, its wide boulevards, its Seussian architecture, its fountains, and its food. The city was looking at reservoirs that were 80 percent empty, and it ordered the fountains and the beachside showers turned off. The government negotiated the rights to water from a nearby river, and had begun work on a thirty-seven-mile emergency water pipeline, down the center of a main highway, that would deliver the water at a cost of $280 million.

A new desalination plant was also under construction, big enough to supply a quarter of the city's needs, but it was a year from being finished.[16]

And so the nervous leaders of Barcelona's water company did something unprecedented. They decided to bring in water, by ship, from the Spanish city of Tarragona, down the coast from Barcelona, and from the French city of Marseille, up the coast.

It was an effort that was simultaneously heroic and silly. The first ship to tie up at Barcelona's docks was the nondescript red freighter *Sichem Defender,* which arrived from Tarragona. Barcelona's port had been specially adapted to accommodate the water-bearing ships, at a cost of $32 million, and the water flowed straight from the freighter's tanks into Barcelona's water mains. A couple days later a second ship arrived—the *Contester Defender,* from Marseille.

Barcelona was scheduled to receive several shiploads of water a week throughout the summer, at a cost to the city of $30 million.

The first two ships hauled in what seemed to be miniature reservoirs: It was a water version of the Berlin airlift, just keep the tankers coming.

The *Sichem Defender* put 5 million gallons of water into Barcelona's municipal system, and its sister ship brought in 9.5 million gallons. But given Barcelona's scale, the first ship's water lasted thirty-two minutes. The larger ship provided sixty-two minutes of water. In fact, even supertankers don't carry city-size quantities of water.[17]

And not everyone was supportive. Miguel Angel Fraile, secretary general of the Catalan Federation of Commerce, said, "The arrival of a boat full of water is the image of the absolute failure of government administration which neither Barcelona nor Catalonia deserves. You can understand a boat bringing water to an island, but not to a continent."[18]

The water-bearing ships were a symbolic gesture—an expensive one—but it's not quite clear what the symbolism was. Perhaps, "We're doing everything we can to make sure Barcelona has water." Or, "Wake up! Even if we had two of these ships coming a day, every day, Barcelona could still end up thirsty."

As Barcelona's drought problems were becoming acute, a small city in Tennessee completely succumbed to the drought that was then in its seventh year across the southeastern United States.

The city of Orme, Tennessee, is located forty miles into the hills west of Chattanooga, a hamlet just two miles from Alabama, right where Tennessee, Alabama, and Georgia come together. Once a flourishing mining town, Orme has for a hundred years gotten its municipal water from Spout Springs, a mountain stream that ends in a dramatic two-hundred-foot cascade at the base of Orme Mountain.

The drought slowly dried out the whole area—60 percent of the streams in Tennessee were at historic low flows. And Orme was just 150 miles northwest of Atlanta, which was itself the epicenter of a water crisis much like Barcelona's. On August 1, 2007, Orme's spring dried up, and Orme's 17,500-gallon water tank quickly dried up as well. Orme was out of water.

The then-mayor, Tony Reames, organized his own version of Barcelona's virtual water main. Volunteer firefighters drove the town's 1961 fire truck down the road two and a half miles to Bridgeport, Alabama, tanked up the fire truck's 1,500-gallon capacity from a fire hydrant there, then drove back to pump the water into Orme's water tank. Sometimes the firefighters got help from a second truck, loaned by another nearby city. Bridgeport, which gets its water from the robust Tennessee River, running along its eastern side, didn't charge Orme for filling its fire truck.

Every few days, the fire truck would make ten round-trips to Bridgeport's fire hydrant to fill up the town's water tank. And every night at 6 p.m., for months, Mayor Reames would go to the tank and open a valve that turned on water service to the town. Orme residents got three hours of water each night—three hours to shower, do the laundry, cook, and fill whatever containers of water they needed for getting ready for work and school the next morning. Mayor Reames turned the supply valve off every night at 9 p.m. and the town's taps and toilets went dry for another 21 hours.

Orme is tiny—forty families, 142 people—and remote; cell phones have no reception in the valley where the town sits. Still, on a typical day before the drought, the town would empty its water tank and refill it from Spout Springs. Oddly, although "the town that ran dry" in the drought briefly became one of those novelty stories that the national news networks like CNN and Fox News love, no one came to the rescue of Orme.

It is almost unheard-of for a U.S. town of any size or scale to simply run out of water. It's equally exotic for the residents of a U.S. town to be limited to three hours of water a day—twenty-four-hour-a-day water is the unquestioned standard. On Thanksgiving—three months after the spring ran dry—Orme was still on the three-hour-a-day water ration.[19]

Barcelona was rescued by several weeks of drenching rain. Within a month of the arrival of the first water ship, the city's reservoirs were back to 54 percent full. The emergency water pipeline that would have cost $7 million per mile was canceled. Ships quietly continued to deliver water through the summer, because they had been contracted to.

Orme was rescued by replacing the virtual fire-truck pipeline to Bridgeport, Alabama—the fire truck ultimately hauled in 1 million gallons of drinking water, one 1,500-gallon trip at a time—with a real pipeline. The U.S. Department of Agriculture had, years earlier, provided a $378,000 grant to lay a pipe from Bridgeport to Orme, which was never built because bids came in too high. During the fall, utility crews finally laid that water main, and on Friday, December 7, 2007, 128 days after the creek stopped flowing, twenty-four-hour-a-day water service was restored.

"If it can happen in Orme, where we have a waterfall," said Mayor Reames, "it can happen anywhere."[20]

The availability of water is the symbol of a civilized society—can you give your child a glass of clean, safe water when she's thirsty?

By that standard—as the 5 million folks in cosmopolitan Barcelona can tell you, and the 142 folks in rural Orme can tell you—water is in trouble, and we're in trouble.

The big numbers are so big, and so scary, as to have become cliché. They've lost their ability to shock us, or to move us. The world has 6.9 billion people. At least 1.1 billion of us don't have access to clean, safe drinking water—that's one out of six people in the world. Of those, 700 million (twice the population of the United States) live on less than $2 a day. So it's not as if they could afford a bottle of Evian if we could just get a case to their neighborhood.[21]

Another 1.8 billion people don't have access to water in their home or yard, but do have access within a kilometer.[22]

So at least 40 percent of the world either doesn't have good access to water, or has to walk to get it. Forty percent—look to your left and your right, that's four out of ten people. In the developed world, we don't know anybody who has to walk to get her water. Walking to get your water, by the way, is no fun. The basic minimum requirement for a family of five is 100 liters (26 gallons)—that's 220 pounds of water that needs to be toted every day.

Of course, even the 1.1 billion people without access to clean, safe water do drink water every day—everyone on Earth finds water every day. They have no choice. Those 1.1 billion people use about 5 liters a day each.[23] In the United States, on average, each one of us uses 70 liters (18.5 gallons) just to flush our toilets. And to just goose the stunning contrast between water wealth and water poverty one more notch, the 1.1 billion people who subsist on 5 liters of water a day are drinking water we wouldn't wash our dishes in; whereas we are peeing into pristine drinking water, and flushing it away. (In fact, just a single flush of a low-flush toilet in the United States—1.6 gallons—uses 6 liters of drinking water.)

Every year, according to the World Health Organization, 1.8 million children die either from lack of water or from diseases they got from tainted drinking water.[24] That number, too, is hard to absorb. It's 5,000 children a day. A typical U.S. elementary school has about 500 kids—so every day, the equivalent of ten entire elementary schools of children are dying, just because they lack clean water. Put another way—a somewhat unsettling way, but vivid nonetheless—the number of children who die every year just for lack of a daily glass of clean water is equal to the number of elementary school children in Florida. It's like losing every kid in Florida between the ages of five and twelve—every year, year after year.[25]

Hardly the symbol of a civilized society.

Even if those were our only problems with water, and if those problems were static, we'd have a water crisis.

But that's where we find ourselves at the end of what we will come to regard as the golden age of water. We don't have things quite in hand now, and in the next fifteen years, by 2025, the world will add 1.2 billion people. By 2050, we will add 2.4 billion people. So between now and forty years from now, more new people will join the total population than were alive worldwide in 1900. They will be thirsty.[26]

In fact, during the golden age of water, during the last hundred years, the population of the world has gone up by a factor of four; our total water consumption has gone up by a factor of seven.[27]

And then there is the unpredictability of climate change. Water availability is intensely weather- and climate-dependent, in both the developed world and the developing world. At one point in 2008, during the years-long drought across the southeastern United States, 80 percent of the residents of North Carolina were living under water-use restrictions. Lake Mead is the largest man-made reservoir in North America, created by Hoover Dam, sprawling an apparently endless 110 miles through the desert of Nevada and Arizona. Lake Mead is the source of water for 20 million people, and it is half empty. It is also the source for almost all of Las Vegas's water—for homes, golf courses, swimming pools, and those spectacular Vegas Strip fountains. The Las Vegas area has 2 million residents and 36 million visitors a year, and its water source in January 2010 was lower than it had been in any January going back to 1965. At that time, Las Vegas had about 200,000 residents; today, on a typical day, there are twice that many tourists in town.

The climate-change models show that India as a whole may well get more rain if global warming proceeds as predicted, but that new rain will come in a band across the north, the parts of India already inundated during the monsoon season. The rainfall-change map shows that two-thirds of India's land area will actually receive less rain, an area where half the country's people live, tens of millions of whom are already trapped in water poverty.

Australia has had its climate completely transformed in the last twenty-five years. A nation that looks and feels much like the United

States, including its water consumption and its per capita GDP, Australia is struggling to quickly adapt its economy and lifestyle, one that assumed a certain amount of water, to a completely new, and much reduced, water budget.

Water problems now literally circle the globe.

Chinese soldiers were dispatched in early 2010 to help deliver water in southwest China, where drought had left 11 million people without adequate water for themselves or for 8 million head of livestock.

In Venezuela, President Hugo Chávez declared an electricity emergency in early 2010 and imposed daily electricity blackouts across the country, because lack of rainfall had so drained the country's hydropower reservoirs.

Drought across the Caribbean Sea forced several island nations to impose routine water rationing. Antigua was providing water only a few hours a day to homes, on a rolling schedule, after its main reservoir fell to just a two-week supply in mid-February 2010. In St. Lucia, the government declared a water emergency in late February 2010, temporarily outlawing not only such activities as filling swimming pools and outdoor watering but also water-intensive commercial activities like making concrete. In Jamaica, the national water utility said the island's two main reservoirs stood at only half full, and some areas of the country would receive water only every other day for eight hours; some would receive water just one day a week.

And the Syrian government reported that during 2008 and 2009, a quarter-million Syrians abandoned farming because drought conditions made it impractical.

There has never been a moment when drought did not plague the world. Mayan civilization may have been undone by climate change and water shortages; a thousand years later, the Dust Bowl in America drove 2.5 million people to leave the Great Plains states.

But three things have come together to make this moment different from the last hundred years, and perhaps different from any previous moment.

The number of people we've got, and the pace at which we're adding people, is unprecedented. Las Vegas's source of water may be disappearing, but sixty thousand new residents have been arriving there every year, requiring 100 million gallons of additional water a week from Lake Mead. India is staring down dramatic shifts in water availability, even as it grows by nearly 20 million people a year.

Beyond population and climate change, the other huge and growing pressure on water supplies is economic development. China and India are modernizing at a whirling pace, and together those two countries account for one out of three people in the world. Economic development requires rivers full of water, not just because people want more secure and more abundant water as their incomes improve but because modern factories and businesses use such huge volumes of water. In an ironic twist, that "modern" economic development doesn't just consume huge new volumes of water, it damages the very sources of water the development is depending on with new sources of unregulated pollution.

India's economy has grown between 5 and 9 percent per year in the last five years, a stunning rate of modernization. And yet, not one of India's major cities provides twenty-four-hour-a-day water. In fact, the cities most associated with India's modernization—including Bangalore, Mumbai, and the capital, Delhi—provide just one or two hours of water a day to their tens of millions of residents. The global economy allows Americans to receive tech support from customer service representatives in Bangalore, but the person on the other side of the world helping us with our computer headaches goes home each day to less routine water access than the people of Orme, Tennessee, had after that town officially ran out of water.

WATER HAS A ROBUSTNESS, a durability, even a strength of character that we rarely appreciate, even as we rely on it.

Although we think of the Earth as getting crowded, and it is, the 7 billion people alive now represent a tiny slice of the history of human beings. Demographers estimate that over the course of the last fifty thousand years, about 100 billion people have lived on Earth.[28] A typical person needs a minimum of 3 liters of drinking water a day. If we imagine that, stretching way back into prehistory, the average life span of those 100 billion people was a conservative thirty years, that means that all the people who have ever lived on Earth have drunk 3,300 trillion liters of water.[29]

And that's just the people.

The animals outnumber the people 1,000:1.[30] And they stretch back in time hundreds of millions of years. An elephant drinks 150 liters of water a day. How much water did a *Tyrannosaurus rex* drink each day? It may not be known for sure, but scientists have found a spot where a dinosaur paused one day in the Mesozoic era to pee on a sandy patch of ground. The resulting trench, from just a single squat, is at least the size of a modern bathtub, 40 to 50 gallons.[31]

More important, people have been around for just 50,000 years. Tens of millions of dinosaurs lived on Earth for 165 million years. Drinking water every day, and peeing. The total water consumption of all the animals who've ever lived is hard to even conceive, but at the low end, it is certainly 10 million times the total human water consumption (and that doesn't include the plants).

Together, the creatures that have lived on Earth have easily required a thousand times the amount of liquid fresh water available on the planet.[32]

And we only have that one allotment of water—it was delivered here 4.4 billion years ago. No water is being created or destroyed on Earth. So every drop of water that's here has seen the inside of a cloud, and the inside of a volcano, the inside of a maple leaf, and the inside of a dinosaur kidney, probably many times.[33]

Every glass of water you pour—whether it's coming from an Evian bottle, a filtered refrigerator spigot, or the kitchen tap—has a rich history. Americans like to debate the palatability of what's called "toilet to tap"— taking a city's wastewater, purifying it to drinking water cleanliness, and putting it back into the water mains. Almost no municipalities have the fortitude to do that. But in the larger context, whatever place you find least appealing to imagine your water, well, your water has been there. More than once.

That's not gross. In fact, it points up two central facts about water, and our relationship to water.

The first is, water can be cleaned, always. The spinning weather machine that is Earth's climate, in which water is a full partner, does a great job of turning swamps and oceans into rushing, crystalline mountain streams. And almost no matter how dirty we humans make water—and we've gotten much more sophisticated in making water dirty in the last hundred years—we can clean it back to the point that it's drinkable again, history notwithstanding.

The second point is, you can't use up water. It has become fashionable to talk about the "water footprint" of this or that product—the amount of water required to raise the beef in a Burger King Whopper, the amount of water required to produce the five pounds of paper in the Sunday *New York Times*, the amount of water required to raise the cotton and manufacture the denim that goes into a pair of Levi's stonewashed 501 blue jeans. This is often called "virtual water"—the water embodied in the products, if not actually contained in them.

But it's a fundamentally misleading concept—completely different from the idea, for instance, of a "carbon footprint." It requires both water and diesel fuel to grow rice—but the consequences are completely different. The diesel fuel is, in fact, consumed in the process. The tractor and the harvester burn the diesel, and it no longer exists as fuel.

The water used to raise the rice, on the other hand, isn't lost at all, except to the person downriver from the rice farm, or the competing irrigator across the road, or the underground aquifer from which the rice farmer's wells draw. The water goes into the rice-growing process—and it is completely recovered, back into the ground, or the atmosphere, or back into the river into which the farm runoff flows. With a bit of water into the rice itself.

That's not to minimize or trivialize water shortages, which are urgent, often catastrophic. Nor is it to minimize the importance of smart, careful water use and water management. How much water the rice farmer uses is critically important, as are how he uses it and what happens to the runoff. We get all the water back, but where it reappears, and on what timescale, is often not well controlled. Water scarcity is often the direct result of bad water management by people.

But neither is it trivial that the rice farmer does not "consume" the water—he doesn't use it up—any more than the steel mill (1 ton of steel, 300 tons of water) or the nuclear power plant (30 million gallons of cooling water an hour).[34] None of the 1 million ways we use water each day actually consumes the water, including, of course, drinking it ourselves.

Water is tirelessly resilient. Water participates in a mind-bending array of physical, chemical, biochemical, geological, and human-created processes every minute of the day—water is essential to creating soup and computer servers, it drives both hurricanes and erosion, it is the essential element in human beings maintaining our body temperatures at 98.6 degrees—and yet water emerges from every one of those processes intact, undamaged, unchanged, ready to make a fresh cloud or a fresh drop of sweat, an iceberg or a jellyfish, as the occasion requires.

Water's indestructibility, its reusability, will be vital as we confront an era where water scarcity becomes more common. Water itself isn't becoming more scarce, it's simply disappearing from places where people have become accustomed to finding it—where they have built communities assuming a certain availability of water—and reappearing somewhere else.

That points up a fundamental problem with water. If all Americans were to swear off bottled water, for instance—if we were to give up the 1 million gallons an hour of Poland Spring and Dasani that we drink—not one person in the world who desperately needed water would get it.[35] Likewise, if we were all to switch to low-flow toilets, and save 3 billion gallons a day of the drinking water we flush down the toilet now, that doesn't help get water to a single village in India or China or Haiti.

All water problems are local.

That allows us a certain shrugging resignation about water problems somewhere else, anywhere else. One billion people don't have access to water as good as what's in my toilet tank, and that's terrible, but there's not much I can do about it.

In this, water is different than many other things. It does help a broad swath of the world to drive a more fuel-efficient car, for instance, because the impact of using less gasoline isn't just local. It stretches around the world, across the whole supply and consumption chain of oil, drilling, transportation, and refining, right through the point at which the exhaust comes out the tailpipe of your Ford Escape Hybrid. It does help the world to switch your lightbulbs to efficient compact fluorescent bulbs, because power plants require huge resources to run and their operations have wide environmental impacts.

The supply chain of water is global too, of course—in the sense that the rain falling on Chicago may have gone airborne as water vapor in the South China Sea. But as the people of Barcelona discovered, you can't move community-size quantities of water around in ships. Within river basins and watersheds and aquifers, water supplies are local. Water we don't take from the Ogallala Aquifer or Lake Okeechobee or the Colorado River can't help Atlanta in a drought, let alone the water-impoverished people of Ethiopia.

Our own water problems—as the folks in both Barcelona and Orme know—are insistent, urgent, frantic. But fifty or a hundred or a thousand miles away, those people's water problems are simply unfortunate—for them.

But the idea that all water problems are local isn't quite so simple. The problems are local, but the consequences, the damage, and the costs are anything but local. The distance we imagine between ourselves and other people's water problems is just another case of not seeing water, and our relationship to water, clearly. In Las Vegas, they are building a backup water-supply pipe that is so expensive, it is costing $2.25 for every man, woman, and child—in America. You will never drink from that pipe. Across India, millions of girls are literally trapped by having to walk and fetch water each day; they don't go to school as a result. India, of course, gives up a huge pool of labor, energy, creativity, and talent by allowing girls to go uneducated just so they can walk to fetch water. That is the true meaning of the phrase "water poverty." But the consequences of that water problem stretch far beyond India—it's in no one's interest to leave tens of millions of girls and women uneducated. Poor farming practices around the world squander huge quantities of water. Agriculture uses two-thirds of all the water people use—and especially in developing countries, half that water is wasted. That hurts water management right where it happens, of course, but it also dramatically undermines global food production.

And any community or country that has experienced serious water scarcity knows that when water problems become water conflict, the consequences rarely stay local.

Water is one of those resources, one of those issues, that can only be managed for the long term. But it is a mistake to imagine that small things don't matter, or that even big water issues are not manageable.

One of the most startling, inspiring, and least well-known examples involves the United States. The United States uses less water today than it did in 1980. Not in per capita terms, in absolute terms. Water use in the United States peaked in 1980, at 440 billion gallons a day for all purposes. Twenty-five years later (the latest USGS survey is from 2005), the country was using just 410 billion gallons a day.[36]

That performance is amazing in many ways. Since 1980, the U.S. population has grown by 70 million people. It's as if we've annexed a fifty-first state the size of France, for instance (population 63 million)—while actually decreasing the amount of water we use.

Since 1980, the U.S. GDP in real (constant dollar) terms has more than doubled. We use less water to create a $13 trillion economy today than we needed to create a $6 trillion economy then.[37]

And despite living lives that are literally awash in water, Americans in 2005 used less water per person than they did in 1955.

It has been nothing less than a revolution in water use in the biggest economy in the world, a completely silent revolution. Most of the change has come in water use by power plants and farms. U.S. farmers today use 15 percent less water than they did in 1980, and produce a 70 percent larger harvest.[38]

And the water-use revolution is just getting under way in most of the economy. Companies from Coca-Cola to MGM Resorts casinos, from Intel to Royal Caribbean Cruises, are starting to track their water use, report it publicly, and reimagine it. GE and IBM are not only learning how to reduce their own water use, both companies have created new water divisions to teach other companies, and communities, how to better manage their water use.

Meanwhile, we haven't yet really tried to get Americans to install water-efficient fixtures at home, or to turn off the water when they brush their teeth, or to use their sprinklers more thoughtfully. Fifty percent of the water delivered to homes in Florida is used for lawn watering.[39]

There's another, perhaps more powerful way of looking at water use in the United States. If we were using water at the rate we did in 1980, we would be using 578 billion gallons a day instead of 410 billion gallons a day. That's 168 billion gallons a day of water we're not using today because of dramatic improvements in water productivity and management. It's hard to understand what that 168 billion gallons a day would mean, but think of it like this: It would mean every place in the United States would need 40 percent more water than we're using now.

There are plenty of water problems in the United States, plenty of ways that water is wasted, plenty of places where people are fighting about water, or will be soon. But the real lesson of the transformation in the United States in the last twenty-five years is that it is possible to grow dramatically and use less water. Water will stretch in remarkable ways, if the people handling it are smart enough and demanding enough to insist on it.

WE HAVE A COMPLICATED, conflicted, and mostly unacknowledged relationship to water. Because water has so often made people sick, because natural disasters—hurricanes, flooding, blizzards—are often caused by water, we have an ambivalent attitude about water that has only been softened by the last hundred years. It is an ambivalence with deep cultural roots.

The Quran credits God with creating humanity, and all of life, directly from water. "We made from water every living thing,"[40] the Quran says, and, "It is He Who created man from water."[41] The Old Testament considers water so primal a substance, so fundamental a tool of creation, that the Bible does not mention God creating water. Water was already present—"the spirit of God hovered over the face of the waters"—just before God says, "Let there be light."[42]

But God also uses water to destroy the world, to drown his entire creation except those riding out the forty days and nights of rain with Noah. (The Quran also relates the story of Noah, and the Hindu tradition contains the story of Manu, who saved humanity from a similar devastating flood.)

Having reached his limit of patience with humanity's inhumanity, God could have destroyed the world in any of a range of imaginative ways, as he would later demonstrate when he unleashed the ten plagues on Egypt. Instead, God used the source of life to destroy it.

The most obvious place to see our slightly sour, subtly resentful, attitude about water's fearsome power is in the way water shows up in everyday expressions, almost always with a negative tone.

Water expressions infuse our language—drained, watered down, in deep, high and dry, mainstream.[43] But we don't really have much of a language or a framework for talking about water itself. In fact, when we talk about water, it's a little like when we talk about love. We aren't really talking about water—we're talking about our anxieties, our hopes, our sense of our selves, refracted through water. We're talking about our vision of our community, about our livelihoods. When we talk about water, we're often talking about power, or about security, or both.

Just as we don't have a good language for talking about water, we don't have a politics of water, or an economics of water. In fact, the lack of all three is a function of the golden age of water—you don't need politics, or economics, or even language to manage something that is unlimited, safe, and free.

Politics, economics, and language are the tools we use to manage conflict and scarcity—and in the new era of scarcity we're entering we'll need all three to handle water. When conflict over water arises, typically, it's not about the water itself, but about the role the water is playing, the use it's being put to, who gets it and who doesn't, and what condition the water is in when all is said and done. Water is one of those unusual substances that cause people to tell each other how to behave. It is typically my way of using water that is both right and essential, and your way of using it that is inefficient and probably unnecessary.

Our everyday attitude about water is filled with contradictions.

Water is absolutely indispensable—in most of the ways we use it, from growing rice to washing our clothes to making microchips, there is simply no substitute. But water is also one of the few resources that we typically don't manage or allocate with price. It is indispensable, but so cheap as to seem free.

Technology is making it easier to solve almost any water problem. But while the water itself might be cheap, the technology to clean it, and the energy to run that technology, are not.

And while technology can solve almost any water problem, technology is also the source of a whole new wave of water problems. The kinds of pollution getting into water supplies now—from the exotic chemicals used to extract natural gas in the drilling process known as fracking, to the Prozac detected in water-supply streams and lakes across the United States—those kinds of pollutants are a consequence of mixing modern technology with water.

Perhaps the most unsettling attitude we've begun to develop about water is a kind of disdain for the era we've just lived through. The very universal access that has been the core of our water philosophy for the last hundred years—the provision of clean, dependable tap water that created the golden age of water—that very principle has turned on its head. The brilliant invisibility of our water system has become its most significant vulnerability. That invisibility makes it difficult for people to understand the effort and money required to sustain a system that has been in place for decades, but has in fact been quietly corroding from decades of neglect. Why should I pay higher taxes just to replace some old water pipes? I'll just drink bottled water if I don't like what comes out of the tap. It is almost as if tap water is regarded not with respect and appreciation but with a hint of condescension, even contempt.

Of course, you can't call Dasani if your house catches on fire. We are in danger of allowing ourselves to imagine that since we've got FedEx, we don't also need the postal service. When universal, twenty-four-hour-a-day access to water starts to slip away, it becomes very hard to bring back. But sustaining it requires more than paying the monthly water bill.

If we're going to be ready for a new era of water, we need to reclaim water from our superficial sense of it, we need to reclaim it from the clichés. We need to rediscover its true value, and also the serious commitment required to provide it. It is one of the ironies of our relationship to water that the moment it becomes unavailable, the moment it really disappears—that's when water becomes most urgently visible.

2

The Secret Life of Water

To say that water is essential hardly covers it.
—*Dr. Richard Wolfenden,*
professor of biochemistry,
University of North Carolina at Chapel Hill

EVERYONE IS AN EXPERT ON WATER.

We all know how it feels to be so grungy that nothing but a good shower will make us feel better. We know how it feels to be so thirsty that only water will really satisfy us. And we know exactly how the water will taste—really, how the water will feel—going down, in that first swallow.

We know how light shines through water in a drinking glass or a swimming pool. We know how to anticipate the arc of water from a drinking fountain, and the force of a wave at the beach, although we can be surprised by both.

You know exactly how hot you want the shower spray to be, and how strong, before you step under it. When you dip a foot or a hand into the swimming pool, you know what temperature will make diving in seem irresistible, and what temperature will give you a little shiver.

You know how far to turn on the outside spigot to get exactly the right spray from the lawn sprinkler. You remember how it feels to race through that sprinkler, and what it's like to have your big sister turn the hose on you in the backyard. We know what it's like to brace briefly before the dash through the rain from the office door to the car.

We know the sharp smell of rain misted through the air after a summer downpour. Snowy air has its own smell, chilly early morning fog has its own smell, the humidity in Key West has its own smell.

Water speaks a whole range of languages, specialized and universal, utilitarian and poetic and romantic. Sometimes they are all talking at once. The wind across the surface of a lake or bay writes clearly on the waves; the changing hues of aqua off Cape Hatteras or Cape Town are their own kind of depth sounding. Lake Superior is a dazzling turquoise that is memorable in part because it seems to belong not to a chilly northern lake but to a tropical beach.

The feel of water is as familiar as the feel of our own skin. The playful, sparkling flow of a creek through our fingers, the thrill of planing a hand across the top of the water over the side of a speedboat. The trickle of a single bead of sweat down your nose during a run, the feel of snow that gets inside your parka melting down your back, the surprisingly hard slap of swimming pool water when the dive becomes a belly flop, the faint rocking motion of a bunk on a sailboat at anchor.

Water is sprinkled through our memories, from junior high biology class to our honeymoons. It's easy to remember what it's like to use an eyedropper to squeeze a single, fat drop of water onto a microscope slide, and how big that drop looks, like an overstuffed couch cushion. It's hard to learn to read the meniscus of water measured in a graduated cylinder. It's hard to forget the first sight of Rome's Trevi Fountain, or Las Vegas's Bellagio fountain, or the fountain from your infant son, having his diaper changed. We know the shapes of shifting light that water makes on the bottom of a swimming pool (although we may not know they are called "caustics"). And we know the captivating power of a waterfall, how in the end you have to simply pull yourself away from watching the cascade, which manages to be always the same and always different, never boring.

We are all experts on water.

And yet, for all our intimacy with water, we actually know almost nothing about it—about water itself. Water is as potent in our daily lives as gravity, but also as mysterious.

For most of us, even the most basic questions about water turn out to be stumpers.

Where did the water on Earth come from?

Is water still being created or added somehow?

How old is the water coming out of the kitchen faucet?

For that matter, how did the water get to the kitchen faucet?

And when we flush, where does the water in the toilet actually go?

The things we think we know about water—things we might have learned in school—often turn out to be myths.

We think of Earth as a watery planet, indeed, we call it the Blue Planet; but for all of water's power in shaping our world, Earth turns out to be surprisingly dry. A little water goes a long way.

We think of space as not just cold and dark and empty, but as barren of water. In fact, space is pretty wet. Cosmic water is quite common.

At the most personal level, there is a bit of bad news. Not only don't you need to drink eight glasses of water every day, you cannot in any way make your complexion more youthful by drinking water. Your body's water-balance mechanisms are tuned with the precision of a digital chemistry lab, and you cannot possibly "hydrate" your skin from the inside by drinking an extra bottle or two of Perrier. You just end up with pee sourced in France.

In short, we know nothing of the life of water—nothing of the life of the water inside us, around us, or beyond us. But it's a great story—captivating and urgent, surprising and funny and haunting. And if we're going to master our relationship to water in the next few decades—really, if we're going to remaster our relationship to water—we need to understand the life of water itself.

One of the great scientific students of water is Dr. Richard Wolfenden, alumni distinguished professor of biochemistry at the University of North Carolina at Chapel Hill. Wolfenden, seventy-five, is a researcher who has spent much of his career studying how water shapes our body chemistry, inside our cells. Proteins in our cells manage every element of human life, and the key to the effectiveness of proteins is the way their long chains are folded into intricate and precise shapes, like tiny molecular origami. Proteins only work—which is to say, everything about our bodies only works—if the folding is exactly correct. That folding, it turns out, is engineered by water. "Proteins know how to fold up because they have the rules written into their atoms," says Wolfenden. "And the rules are entirely a reflection of how eager that part of the molecule is to get away from water or to cling to water. To say that water is essential hardly covers it."

Wolfenden has the e-mail address water@med.unc.edu because, he says, "of this never-ending interest in how many things water influences." His North Carolina license plate used to read "DROPLET"; he's switched to "HUMID." "That matches North Carolina," he says.

Asked if he is optimistic about our future relationship to water, Wolfenden is silent for a moment. Then he says, "I think our relationship to water is going to be one of the deciding things of the next century. I don't think water's in any trouble. But we might be."

No EXPRESSION IN COMMON USAGE is as thoroughly wrongheaded as "dull as tap water." One thing water is not is dull.

Given our intimacy with water, our dependence on it, and water's apparent simplicity—"H_2O" is surely the best-known molecular formula in human consciousness—the surprising thing is how many surprises the story of water contains, and how many flat-out mysteries. Often the two go hand in hand.

The oldest rock discovered so far on Earth—in northern Quebec—is 4.28 billion years old.[1] That's an old rock—it's getting close to the age of the solar system itself, estimated at about 4.6 billion years.

But turn on the faucet in the bathroom to brush your teeth, and the water pouring out is probably just a bit older than Canada's old rock. Scientists don't agree on the precise age of the water on earth, but it's certainly 4.3 or 4.4 or 4.5 billion years old. It's one of the more astonishing things about water—all the water on Earth was delivered here when Earth was formed, or shortly thereafter. The water around us is original equipment—it was included with the planet itself, in the first 100 million years or so. There is, in fact, no mechanism on Earth for creating or destroying large quantities of water. What we've got is what's been here, literally, forever.

Of course, everything we've got on Earth was, fundamentally, delivered as original equipment, from our cars to our granite countertops. It's all made out of the raw material around us—the atoms and molecules of which Earth, and everything on it, are composed. But here's the difference: Even the granite kitchen countertops come from rock that has been heated, pressed, and completely remade deep in the crust of the Earth.

All the water on Earth came from space in exactly the form it's in now: H_2O. Water not only came from space, it was created out in space. It is, in fact, cosmic juice, formed hundreds of millions, or even billions, of years before the solar system itself.

Once you understand the lineage of water, you realize that the ads touting Evian ("born in the French Alps") and FIJI Water ("untouched by man") dramatically understate the case. It is remarkable the space and time that routine glass of ice water has covered, in order to soften your thirst.

Here's a brief exchange with William Latter, a Caltech research astronomer specializing in studying the interstellar medium, the dust clouds between stars that will eventually condense and form new stars.

Q: So all the water on Earth came from an interstellar cloud somewhere in the Milky Way?

LATTER: Exactly right.

Q: And it was formed one molecule at a time?

LATTER: Exactly.

Q: There's a lot of molecules of water here.[2] It seems like that would take some time.

LATTER: The amount of time involved is millions of years, or depending on what we're talking about, it's billions of years. The universe is very old. It has time to do things, and space.

The universe has time and space to do things, like a master vintner, allowing the oxygens and hydrogens to find each other, crafting each H_2O molecule, patiently creating a planetful of water, always with exactly the right balance between freshness and maturity.

Gary Melnick is a senior astrophysicist at Harvard University's Center for Astrophysics who has spent years using orbiting telescopes to study star formation and water in space. He was part of the team that in 1997, using the European Space Agency's Infrared Space Observatory (ISO), discovered an interstellar spring of water of astonishing scale.

They pointed the ISO telescope at Orion, a constellation that is quite easy to spot with the naked eye on a clear night. The stars of Orion are part of our own Milky Way, and one shining spot in Orion—the middle dot in Orion's sword—is in fact not a star but a massive, glowing cloud of gas and dust, the Orion Molecular Cloud. That cloud is the kind where new stars form, condensing out of the hydrogen gas.

"Orion is the closest region to Earth where massive stars are being formed," says Melnick. The full cloud of hydrogen, which looks to the eye like just that one dot in Orion, is of such vast scale that it is giving birth to thousands of stars at once.

When Melnick and his colleagues used the telescope to look at part of the Orion Molecular Cloud, he says, "What we found was that there is enough water being formed sufficient to fill all of Earth's oceans every twenty-four minutes."

As the stars coalescence and collapse in on themselves, they send shock waves out through the clouds of gas, which contain lots of loose hydrogen and oxygen. When the shock waves slam the hydrogens and oxygens into each other, they often form water. Hydrogen, for the record, is the most common element in the universe; oxygen is the third most common.

The result, right there in the sword hanging from Orion's belt, is a water factory that is making the equivalent of all the water on Earth, sixty times a day. Have a look at the Pacific Ocean, or the Atlantic Ocean, or something small like Lake Michigan, and think about creating just that much water three times an hour—the scale is really almost hard to credit. All the water on Earth, sixty times a day? That's one swampy patch of the Milky Way.

Not quite, says Melnick, with the bemusement of a guy used to thinking in cosmic distances that the rest of us find literally unimaginable. "The cloud is creating enough water *molecules* every twenty-four minutes to fill all of Earth's oceans," says Melnick. "The density of the water, well, it's not like if you were floating out there, you'd get hit with a bucket of water in the face or anything."

While the cloud is making sixty Earth waters every twenty-four hours, it is doing it across a span of space 420 times the size of our solar system.[3] As a place in space, the Orion Molecular Cloud is pretty gassy, pretty dusty, and very wet. For us, not so much. Even the dustiest parts of the cloud—

the places with the most particles—are emptier than any vacuum that people can create on Earth.

"It is the same as cool mist," says Melnick. "It's just a lot less dense." A mist of water so fine that it resembles the emptiest space you can possibly imagine. But a lovely cool mist of space water, nonetheless. And a productive mist. In the fourteen years since its discovery by Melnick and his colleagues, the cloud has created enough water for about 300,000 planets as wet as Earth.

Water forms in interstellar clouds in another way, equally familiar and equally alien. The hydrogens and oxygens literally mate on the surface of tiny grains of dust that are part of the interstellar clouds. In the cloud, says Melnick, "hydrogen hits the dust grains quite frequently. And every once in a while oxygen will hit the dust grains too, and linger."

What happens next isn't quite clear, says Melnick, "but somehow the hydrogen finds the oxygen and forms OH. Then another hydrogen finds it and forms H_2O." The new water molecules form a thin coating of ice on the dust grain, like what you find on ice cream that's been in the freezer too long.

A water molecule is about one-thousandth the width of one of these dust grains, so the dust grain can carry millions of water molecules if the hydrogens, the oxygens, and the dust grains can find each other. "They could form multiple layers of ice over time," says Melnick, "if they are left undisturbed."

And sometimes the water from the cool mist lands on a dust grain, and clings. "If the dust grain is cold, in almost all of those cases, the water will stick to it and not come off," says Melnick, "just the way your tongue sticks to a cold pole in the middle of winter."

Your tongue, of course, is covered with saliva that is 99.5 percent water, water which was itself once floating around in interstellar space, perhaps frozen onto a dust grain.[4]

And that's it, as far as the astrophysicists and the astrochemists understand it at this point: All the water on Earth—the thunderheads, the snow-covered ski slopes, Old Faithful, and the current of the Mississippi River—started out as the finest mist, the smallest ice cubes, drifting around inside an interstellar dust cloud.

In general, there is no dispute about that.

Here's the mystery. Scientists don't actually know how that water gets from the interstellar cloud to Niagara Falls. And perhaps most startling of all, they don't know how much water there is on Earth.

ONE OF THE GREAT MYTHS about water is that it is the most common substance on Earth. Indeed, you can Google search "the most common substance on Earth," and water pops up repeatedly. The Earth's surface is 71 percent covered in water, and water is the primary force shaping every element of the character of the planet—the geology, the weather, the range and variety of life, the planet's gleaming profile in space.

But unless you're playing a children's game in which you mean that water is, quite literally, the most common substance sitting on the surface of the Earth (as opposed to the most common substance making up Earth's composition), then the amount of water on the planet's surface is trivial in every way except its impact. The total water on the surface of Earth (the oceans, the ice caps, the atmospheric water) makes up 0.025 percent of the mass of the planet—25/100,000ths of the stuff of Earth. If Earth were the size of a Honda Odyssey minivan, the amount of water on the planet would be in a single, half-liter bottle of Poland Spring in one of the van's thirteen cup holders.[5]

Put another way, if the oceans on Earth were as deep, in relative terms, as the skin on a typical apple is thick, they would average ten kilometers deep instead of four kilometers, and all the land on Earth would be inundated except the planet's tallest mountains.[6]

Once you actually pause and appreciate how fine the film of water enveloping Earth is, water's impact is even more dramatic.

What's more, most of the water on Earth is not the water we're familiar with. Most of Earth's water is not the water on the surface—in the clouds, in the great lakes and rivers and oceans, the ice caps and aquifers. And most of the water on Earth does not exist in any of the three familiar states—ice, liquid, or vapor.

Water exists in a fourth form, one that is so exotic that despite its abundance and its importance, it almost never merits a mention outside of scientific circles. This vast reservoir of water—at least as much as in all Earth's rivers and oceans and glaciers, perhaps four or six or ten Earth oceans' worth—is locked in the rock deep in Earth's mantle, in a layer about 410 kilometers (255 miles) below your feet. (All the water on Earth's surface is referred to by scientists, for convenience, as "one Earth ocean.")

Some of this exotic fourth state of water may be hiding in plain sight in the middle of your kitchen, if you have the green stone countertops made from the mineral serpentine. A kitchen island countertop of serpentine (say, 4 feet by 3.5 feet) could easily weigh two hundred pounds. Not just as solid as a rock—it is a rock. But of that two hundred pounds of serpentine, twenty-two pounds is H_2O—ten liters of water fused into the stone. Not mixed in the way, say, you'd mix an egg into pancake batter. The water is baked into the very molecular structure of the stone itself, tucked in among the magnesium and silicon and oxygen that make up the lattice of the serpentine.

Much of the rock 410 kilometers deep in the Earth has some water squeezed into it in this way, but it's not the familiar H_2O. At that pressure and temperature—the weight of 255 miles of solid rock piled up, heated above 2,000°F—one of the hydrogens peels off the water, leaving the OH and the separate H to wriggle into the structure of the stone. Scientists call the resulting water-infused rocks hydrated minerals, or hydrous minerals (literally, "watery rocks").

Here's the thing that's a little hard to grasp: Once you squirt that oxygen-hydrogen pair and that lone hydrogen into the crystal lattice of a rock, buried three hundred miles down, in what way are those atoms still water?

Steven Jacobsen, a geophysicist at Northwestern University, is a gracious host to this world of heat, pressure, and darkness that is almost completely inaccessible to humans. Jacobsen has used an enormous press (the kind you can make artificial diamonds with) to mimic the pressures and temperatures three hundred miles down and to squeeze water right into rock. He is devoting a large chunk of his career to studying the importance of wet rocks deep inside the Earth.

Q: So this idea that there is water inside these rocks—inside the structure of the minerals—is this a theory or a fact?

JACOBSEN: Oh, this isn't a theory. It's reality. Absolutely.

Q: And if the water is inside the molecular structure of the rocks, why do you scientists think of it as water? In what sense is it still water?

JACOBSEN: Well, it's not water by any stretch of the imagination, of course. We use that term very colloquially.

Q: But it really is water, in fact?

JACOBSEN: Yes, it's water, unquestionably. If you release the pressure and temperature, the hydrogen and the OH come out as water. If it's not in the rock, it's water. It is where most of the planet's water might be, in fact. In the rocks.

The graphic-novel version goes something like this: In the right conditions of temperature and pressure, certain kinds of rocks literally suck water into their structure, much the way a sponge sucks up water. As the water goes into the rock, it dissociates—the H goes here, the OH goes there.

There is absolutely water in these rocks, and the scientists know it in at least three ways: These hydrated minerals are literally more pliant, more puttylike, than in their unhydrated state; the scientists can measure water's pieces inside the structure of the rocks using infrared spectroscopy; and most important of all, when the pressure and temperature on the rocks are released in the right way, the H and the OH come squirting right back out of the rock, and they come out as water.

And here's the thing: Scientists think they have figured out that these kinds of watery rocks are common in a band inside the Earth stretching from about 250 miles deep to about 400 miles deep, a layer 150 miles thick, a lot thicker than the film of water on Earth's surface.

"Even if only 1 percent of that rock is water," says Jacobsen, "that's a lot of water, several times Earth's oceans, in fact."

Hundreds of scientists around the world are studying the physics of Earth's deep water, and its impact. While water in this fourth state, this deep water, may be out of sight, while it may be harder to study and harder to understand than the water NASA discovered in 2009 on the Moon, Earth's deep water is directly connected to the water crashing ashore at the

Santa Monica Pier or the White Cliffs of Dover, the storm clouds crowding the horizon in Johannesburg or Shanghai.

EVERY SCHOOLKID IS FAMILIAR with the cheerful drawing illustrating the basics of the water cycle: Clouds drop rain or snow on the flanks of mountains; water runs off into streams and rivers and lakes, and then into the ocean, from which the beaming sun evaporates it (often in the form of lines that look like wiggling snakes rising straight into the air) to become clouds again. Precipitation, evaporation, precipitation.

The diagrams are always a little cartoony. The actual process itself is awesome, even majestic. The volumes of water that the Earth and the Sun are moving around are Olympian, so large they are measured in a unit rarely heard in the ordinary world: cubic kilometers. A single cubic kilometer, an imaginary cube one kilometer on a side, holds an incredible amount of water: 260 billion gallons, enough to cover the island of Manhattan to a depth of thirty-seven feet.

Every hour, on average, Earth's oceans are evaporating 50 cubic kilometers of water into the air (13 trillion gallons). The entire United States uses only 410 billion gallons of water a day for all purposes—so every two minutes, the oceans create more clouds of fresh water than Americans use in twenty-four hours.[7]

Just the leaves from a single acre of trees might send eight thousand gallons of water up into the air in a day, enough to fill two-thirds of a typical backyard swimming pool.[8]

A molecule of water that evaporates into the air—from a fountain, from a puddle, from your skin—spends about nine days floating in the sky before returning to Earth as rain or snow.[9] Half the Earth's surface is typically covered by clouds, with the life of a particular cloud usually being no more than an hour.[10]

And how much water is floating up there in those fat black rain clouds, literally defying gravity until the rain falls? It is lakes full of water. If just an inch of rain falls on your half-acre yard, that's 13,577 gallons of water—one inch, one storm, one small patch of ground.[11]

Perhaps the most mind-bending fact that shows up on the standard water distribution charts is something identified cryptically as "biological water." It's a small number, just 1,120 cubic kilometers—one-tenth the water cycling through the atmosphere at any moment, enough water to fill just one of the five Great Lakes (Erie).[12]

What is "biological water"?

It's the amount of water ziplocked into the bodies of everything alive on the planet—earthworms, squids, pelicans, mosquitoes, pythons, giraffes, sardines, hippos, the swine flu virus, not to mention all the Earth's trees, ferns, flowers, and grasses. And inside us too. That 1,120 cubic kilometers comes to 300 trillion gallons of water. Given that the average human—considering adults and children—contains about 5.5 gallons of water, people account for only about 38 billion of the 300 trillion gallons of "biological water."[13]

Putting aside the question of how a scientist could calculate the total amount of water inside all the creatures in Earth's biosphere, it's a humbling number. Of all the water doing life-support duty, 99.9987 percent of it is inside creatures besides us.

One funny thing about the numbers describing how much water is streaming through the world—the total water volume for the Earth's surface, the total frozen in glaciers, the total evaporating annually from land, the total inside crocodiles and poodles—is how precisely the numbers agree, no matter which source you consult. On the Web site of the U.S. Geological Survey, in grade school curriculum materials, in science textbooks, across the Internet, and even in a handmade table taped to the wall of a professor of natural resource sciences at the University of Adelaide in Australia—everywhere the numbers are exactly the same. It's not just hard to believe the precision and lack of variance, it's impossible. You can't get complete agreement on a number as fixed as the diameter of the Earth, a measurement that doesn't change as much as, say, the annual evaporation of water from the Indian Ocean. Almost none of the charts with the amazing numbers list sources—or they reference each other—but one indicates that the numbers are the work of a man named Igor Shiklomanov, from a chapter he wrote for a book edited by the American water expert Peter Gleick in 1993 called *Water in Crisis*.

And if you go to Gleick's *Water in Crisis*, there on page 13 is *exactly the same chart everyone else prints.* Except this one is the original. Igor Shik-lomanov, a highly regarded Soviet water scientist, prepared the chart based on his own analysis, and the work of Soviet colleagues, with some of the data originally published in 1974. Right in the text of his essay in *Water in Crisis,* above the chart, Shiklomanov writes with all modesty, "It should be noted that the data on the amount of water on earth (as the authors of the cited monograph themselves note) should not be considered very accurate; they are only approximations of the actual values."

Given that very clear caution, it's not just amazing how widespread the water data have become. What is so startling is that given the incred-ible leaps in computer modeling, water measurement technology from space, and computing power—not to mention the intensity and impor-tance of climate change science and its dependence on moisture in the atmosphere—no one has come up with a fresh set of calculations. Shik-lomanov's seventeen-year-old chart, based on data almost four decades old, remains the standard.[14]

The real gap in the water cycle drawings, of course, is not the uni-formity or precision of the numbers, but as Steven Jacobsen and his deep-water colleagues would point out, that most of the water is missing. It's in the mantle, and it, too, cycles.

Joseph Smyth is a geologist at the University of Colorado, one of the pioneers in trying to understand the dynamics and significance of deep water (he was also Steven Jacobsen's thesis adviser).

The water in the deep interior rocks of the Earth's mantle gets there through the oceans. "The most significant way this happens is in the ocean crust," says Smyth. Along the ocean floor is a mineral called olivine. As it happens, olivine reacts with seawater to create serpentine—the green stone that might show up as your kitchen counters. Then, at the places where the continents are grinding into each other, the ocean floor is "subducted," it's shoved downward into the Earth's interior by continental drift. The water-saturated serpentine dives into the crust, taking its load of water with it.

You could release the water from your kitchen counters by heating them (it would ruin the counters, however). That, in fact, is exactly how the deep water comes back to the surface, says Smyth. "It returns mostly through volcanoes. When there's an eruption in the Andes or at Mount St. Helens—that big eruption cloud is largely water, with ash mixed in." A volcano's eruption cloud is often 70 percent or more water. "What's making the explosion, in fact, is water coming out of the magma," says Smyth.

Exactly how much water has come to be stored in the mantle is a mystery—and one of the questions scientists are trying to answer.

"What's going on there is extremely inaccessible," says Steven Jacobsen. To say the least. The deepest hole humans have ever drilled is 12 kilometers (7.5 miles)—and that took the Soviets twenty-four years of effort and $100 million.[15] As of 2008, the world's deepest mine, that is, the deepest people can actually travel inside Earth, is just 3.9 kilometers (2.5 miles).[16] Both of those are barely finger scratches on the surface of the Earth. The interesting action in deep water is at about 410 kilometers. So all the research is done using sound waves and seismology, and also huge, powerful presses that can mimic the pressures and temperatures 410 kilometers inside the Earth, allowing scientists to create samples of the kinds of rock found there, which they can then study. Re-creating those conditions requires so much effort that a hydraulic press two stories tall creates rock samples the size of the period at the end of this sentence.

Even if there are four or five earth-oceans of water deep in the Earth's mantle, it's not like finding a huge reservoir of oil or natural gas. Deep water isn't something humans can sample even to study directly, let alone tap it to irrigate a dry patch of the Sahara.

But the water has at least three critically important roles.

First, Smyth, from Colorado, thinks that this fourth state of water, locked inside rock, may be how the Earth actually got its original supply of water.

"I think most of the water came here as hydroxyl (OH) in the primitive meteorites called chondrites," says Smyth. "There isn't universal agreement on this. And some of the water came as molecular water"—as cosmic juice, that is—"but I think most of the water came from these chondrite meteorites, with the water in them as hydroxyl." (It would still have to have been formed in space first, of course.)

The water that has come to blanket the Earth would then have been released by the planet's early volcanism.

It's an intriguing theory, and one possibility among several. But the question of how Earth's water got delivered is a messy scholarly splash fight at the moment, with several passionate camps. Distinguished research scientists will actually shout, referring to colleagues who disagree with them, "Well, has he read my latest paper on isotope distribution? Has he? You tell him to read that paper and he'll understand how it happened!"

Second, there is little question that the "wet rocks" deep in Earth's mantle are vital for plate tectonics—the water reduces the viscosity of the rocks, and their resulting "plastic" quality enables the continents to slide beneath and over each other. Those sliding plates create the geology of much of Earth.

Finally, and perhaps most important, the deep water may be the only reason Earth is a blue planet at all.

"We've had relatively constant ocean volumes over time, going back at least 500 million years," says Jacobsen. "Sea level does rise and fall, yes, but on the scale of dozens of meters." Geologists call the part of the continents that crowns above the oceans "continental freeboard." If you don't fret too much about the edges of the continents (which, unfortunately, are where most of the people live), the amount of continental freeboard is remarkably stable going back hundreds of millions of years.

Considering that there might easily be five earth-oceans of water stored in the planet's interior, that sea-level stability is intriguing. Even the release of a single earth-ocean—doubling the surface water on the planet—would swamp everything.

"Maybe the water in the mantle is why we have those oceans," says Jacobsen. "We can't extract and drink that water, but maybe the water in the mantle is buffering the amount of water in the oceans."

The mechanism isn't understood. But the importance is.

Jacobsen's mentor, Smyth, goes back to Earth's beginnings. "Early in Earth history, when it had these big violent impacts from meteors and comets, the atmosphere got blown off a few times during the first 100 million or so," Smyth says. "It may be that Earth has retained water through the last 4.3 billion years by having this reservoir of water in the interior.

"If the water were just on the surface, there might not be any water on Earth now."

~~~

PERCY SPENCER, an executive with Raytheon Manufacturing, was a self-educated orphan whose formal schooling ended at age twelve. An intuitive and brilliant engineer—Spencer ended up with 130 patents—he worked alongside MIT's physicists developing technology for World War II, and his practical sensibility helped figure out how to mass-produce magnetrons, the electronic guts of the radar units whose widespread use helped win the war.[17] As the war was wrapping up in 1945, Spencer was touring Raytheon's lab in Waltham, Massachusetts, where magnetrons were being tested. He noticed that the Mr. Peanut candy bar he routinely carried in his shirt pocket to feed the squirrels was melting.

It was the kind of moment for which Spencer was famous—he knew that high-frequency radiation from the magnetrons had melted the candy bar. But rather than pass it off as a messy inconvenience, he was intrigued at the possibilities.

As the story is told, Spencer immediately dispatched a Raytheon office boy to buy a package of unpopped corn kernels. He put the unpopped corn in range of the magnetron, and in a satisfying moment familiar to every hungry office worker, the popcorn popped all over the lab.

Percy Spencer had discovered both the microwave oven and its single most distinctive cuisine in a single moment.[18] Today, the working heart of a $39 microwave from Wal-Mart—the magnetron, which generates the microwaves—is the same as the technology that helped the Allies use the then-new radar to defeat Nazi planes and U-boats.

Percy Spencer and Raytheon went on to patent the new cooking technology, but it took them decades to figure out how to deliver the convenience of the microwave to the kitchen counter, and Raytheon had to purchase appliance maker Amana to finally make it a success. The first Amana microwaves honored the technology's roots—they were called "Radarranges," a brand Amana still uses, now with just a single "r" in the middle. Today, 96.4 percent of U.S. homes have a microwave oven—more homes than have a landline telephone or a computer—and a typical family of four goes through the equivalent of forty regular-size bags of microwave popcorn a year.[19]

When you use a microwave oven, to reheat coffee or puff a bag of popcorn, you're really cooking with water—specifically with water molecules.

The microwave oven only cooks because of microwaves' affinity for water at the molecular level. Microwave radiation—the same kind of radiation as radio waves or light waves—moves at a frequency that water molecules absorb. When you microwave a baked potato or a cardboard tray of frozen macaroni and cheese, it is the water molecules that get energized, and that do the cooking.

In fact, each individual water molecule is really a tiny magnet—the three joined atoms look a bit like Mickey Mouse's head, two hydrogens as the big ears, one oxygen as the head. The hydrogen "ears" create a positive side, the single larger oxygen is the negative side. When microwaves come zinging through, each water molecule tries to orient itself in the waves of radiation, and ends up spinning. Water molecules inside a cup of coffee, or a baking potato, can spin 1 billion times a second in response to the microwaves.

The water molecules' motion creates heat, which cooks the surrounding food. Microwave popcorn pops when the 14 percent of each kernel that is water vaporizes into steam, and expands to pop the kernel's hull loose. (Conveniently, most plastics, dishes, and things like paper plates are transparent to microwave radiation, and a paper plate doesn't contain many water molecules.)

Despite its sturdy simplicity, in fact, water is a complicated, unusual, almost enchanted substance—not in the emotional or cultural sense, but literally, physically, starting right at the molecular level, with the very magnetic quality that allows Percy Spencer's "radar range" to work.

Scientists refer to molecules that are tiny magnets—one end with a small positive charge, one with a small negative charge—as polar molecules. In the case of water, the polarity has much more significance than making microwave popcorn possible.

The polarity creates a stickiness among the water molecules, like the clinginess of socks in the clothes dryer.[20] The stickiness of liquid water molecules is called hydrogen bonding. Those hydrogen atoms—the ears on the mouse head—are bonded tightly to the oxygen atom in their own water molecule. But because of their slight positive charge, they are also attracted to the slightly negative charge of the oxygen atoms in the water molecules floating nearby. Like the socks in the dryer, the fabric in each sock is a unit knitted together, but the socks also have a clinginess for neighboring socks.

The hydrogen bonding isn't something theoretical or ephemeral. Hydrogen bonding is as basic to the character of water as its sparkle, its splash, its very wetness. That clinginess of water molecules for each other is, in fact, the glue that holds much of the natural world as we know it together.

Scientists have been able to measure that liquid water molecules are up to 15 percent closer to each other because of hydrogen bonding than they would be without it. And the stickiness of the molecules for each other gives water a whole range of surprising qualities, which we have come to depend on.

Because of the extra stickiness from hydrogen bonding, water's reaction to changes in temperature is quite unusual. Water is a liquid through an incredibly wide range of temperature—it is liquid for 180°F, from 32° to 212°—and it is liquid over a range of temperature hospitable to life. Much of the chemistry and biology that we rely on takes place in liquid water—life requires not just water, but liquid water. As it happens, water is a liquid at what we think of as room temperature.

Hydrogen sulfide, a substance much like water chemically (H_2S instead of H_2O), is a gas at room temperature. If water were a gas at room temperature, there likely wouldn't be any rooms, because there would be no creatures to make them.

The stickiness allows water to absorb and hold heat well—it is a good insulator, which allows living things to use it to regulate their own temperatures more easily. It also means the vast liquid ocean film that covers 71 percent of Earth's surface tempers the globe's climate—ocean temperatures vary just one-third as much as land temperatures.

And almost everyone knows the most familiar anomaly of water: Solid water is less dense than liquid water.

Ice floats.

Every basic lesson in physical chemistry would tell us that ice should sink in a glass of water (and that icebergs should sink to the bottom of the ocean). In liquids, the molecules are packed tighter than in a gas; in solids, the molecules are packed tighter still.

But in the case of ice, the hydrogen bonds do the opposite of what they do in liquid water. As water freezes, the tiny water-molecule magnets want to put some space between each other as they lock into place. The hydrogen bonding enforces a distance that causes ice to form a crystal lattice that is 9 percent less dense than the same amount of liquid water.

The result is that ice cubes in a tray grow 9 percent as they freeze. And ice floats. If ice didn't float, rivers, lakes, and even the oceans would freeze—the ice piling up from the bottom in winter, never melting fully in summer—and most aquatic life would die. As it is, the ice layer across the top of lakes and rivers does just the opposite—it acts as a layer of insulation, keeping the rest of the water warmer than it would otherwise be, keeping it liquid, and allowing aquatic life to survive each winter.

Water's stickiness as a liquid, its lower density as ice, and all the things those qualities make possible—those are the kinds of qualities that cause even scientists who study water to go a little mystical on you. It makes perfect molecular sense—if you're a chemist or a physicist—that water lightens a bit when it freezes. And yet nothing else in the regular world does that, and it is precisely that quality of water that enables life to go on living.

Liquid water exists at ambient temperature because of its somewhat quirky molecular structure. Liquid water just happens to be the perfect medium for the raw materials of life to get together, to find each other, and every bit of life depends on liquid water to be alive. And then, when water freezes, it does so in a surprising way that, almost miraculously, happens to provide protection for the very life that water itself makes possible.

Water is a great solvent—almost anything will dissolve in water, in part because of its polar molecules. All the chemistry of our bodies—the conversion of food to energy, the conversion of energy to a ballerina's pirouette or an opera singer's aria—all that chemistry takes place in the hospitable environment of tiny drops of water inside cells that allow all kinds of molecules to move freely back and forth doing the work keeping us alive requires. (The solvent qualities also mean water is easy to pollute, and that water is a great harbor of all kinds of things that make people sick, from germs to heavy metals.)

For the record, water molecules are conveniently tiny—the interior of a single human red blood cell, pretty small on the scale of cells inside our body—that one red blood cell is the equivalent of a vast domed football arena, with a single H_2O molecule being about the size of a children's party balloon. One blood cell can hold 3 trillion water molecules.[21]

The physics and chemistry of water—which amount to the alchemy of life—generally get short shrift, in high school, in college, in daily life. Few of us know enough to appreciate that the ice cubes floating in the glass of iced tea are like some kind of cosmic magic trick, let alone know enough to marvel that the whole sentient universe hangs on that trick.

Water really is the genesis ingredient for life at all levels—water is so fundamental to everything involved in creating, reproducing, and sustaining life that it's possible to imagine that God created water, and let water do the work to create life.

WATER HAS A SECRET LIFE that goes beyond its birth in deep space one molecule at a time, its quiet accumulation by the oceanful in the deep rock of Earth, and its role in bringing each bag of microwave popcorn to life. Water is so adaptable, so nimble, that as we humans have gotten more inventive and more demanding, water has come right along with us, becoming as crucial a tool in the digital era as it is to a farmer.

Every modern electronic device—from the simplest desktop calculator, to our iPhones and the computers that control our car engines, our medical diagnostic machines, and the servers that run the Internet—relies for its creation on water, but a kind of water so exotic that it exists nowhere on Earth, except inside microchip factories.

In Burlington, Vermont, at one of only two IBM semiconductor factories in the world, IBM turns ordinary water into a liquid so alien that it's not safe to drink, and it accomplishes that by doing nothing more to the water than cleaning it.

IBM creates huge quantities of this purified water—2 million gallons a day, 80,000 gallons an hour, without stopping, because you can't make microchips without water, and for the microchip water to do its job, it has to be water of absolute purity.

The IBM chip plant in Vermont—in Essex Junction, just northeast of Burlington—has eighty acres of floor space, and with a staff of five thousand, is the largest employer in Vermont. The buildings are two and three stories of brown brick, with narrow smoked-glass windows that make the place look like a down-at-the-heels community college from the 1970s.

Inside, however, is one of those space-age facilities that make the modern world possible. IBM Burlington produces semiconductors that give intelligence to printers and cell phones, TVs and GPS handhelds and cameras. Everyone has an impressionistic sense of what computer chips look like—sometimes you get a glimpse of them when you change the battery on your TV remote control or when you drop your cell phone and it cracks open. Silvery lines and dots on a green plastic board.

But the actual features on the chips themselves—the pathways that the electrons follow to deliver our text messages or flip the flat-screen TV between HBO and ESPN—those pathways aren't something you've ever seen. They are just 90 nanometers wide. They are so thin, not only can't you see them with the naked eye, you can't see them with a microscope. The pathways can't be seen with visible light, because the narrowest visible light waves are 400 nanometers—light skates right past something 90 nanometers wide without noticing. You have to use a scanning electron microscope to see the pathways.

And yet the pathways that can't be seen are designed with the utmost architectural precision, and they must be laid down with the same precision, or 2 + 2 will not, in fact, routinely equal 4. The round semiconductor wafers that IBM Burlington's technicians are making often require as many as seven hundred steps to manufacture, over days or weeks. Between the steps, the chemical solutions used to etch patterns and lay down layers of circuits have to be washed away.

With water.

But the water itself is a problem, because water coming from the tap—clean enough to drink, and quite refreshing in Vermont—is just filthy from the perspective of a computer chip. Precisely because of water's invaluable solvent properties, the water coming into IBM Burlington is awash in chunky debris of every kind—minerals, ions, bacteria, viruses, and just plain old bits of dirt way too small to bother a person but boulder-size to a computer chip. You'd no more wash your computer chips in tap water than you'd ladle water from your toilet to make lemonade.

Water is, in fact, the thing those chips need to be washed with. It is the only thing they can be washed with, but it literally has to be nothing but water. Nothing but water molecules.

Here's how sensitive a silicon wafer with 90-nanometer pathways is. Imagine that the tiny pathway is, in fact, the width of a sidewalk. In that case, if a human hair were lying across it, you'd be in trouble, because the hair would be three-quarters of a mile in diameter. You'd have to climb it like a three-thousand-foot mountain to get to the sidewalk on the other side. If you were walking along your 90-nanometer sidewalk and happened to encounter a single red blood cell—just one—you wouldn't miss it. That single cell would be as long as a football field—you'd have to detour fifty yards to the right and then back to resume walking on the sidewalk. Even if, while walking on your 90-nanometer sidewalk, you encountered something really small, a single particle of influenza virus, even that would force you off the path. A particle of flu virus would be a spherical blob four feet across.

Indeed, if you were to stumble on a single water molecule on your walk, you might well notice it: A water molecule would be about the size of a single, green M&M's candy.[22]

Making microchips is a demanding, complicated, unforgiving business, a profession all its own, and the clean water is essential, but hardly the hard part. Making the water for the microchips is also a demanding, complicated, unforgiving business—it requires a factory all its own, in fact.

Deep inside IBM Burlington is a big industrial area known as the Central Utilities Plant (CUP). In a sprawling space dense with pipes, pumps, and equipment, regular tap water that starts in Lake Champlain is turned into what the semiconductor industry calls ultra-pure water (UPW).

"Ultra-pure water is 10 million times cleaner than regular tap water," says Lindsey Stahl, who is an ultra-pure water engineer at IBM Burlington. It's hard to know what "10 million times cleaner" means, exactly. Stahl's boss is a veteran IBMer named Janette Bombardier, who is director of site operations at IBM Burlington, everything from cleaning the bathrooms to cleaning the microchip water.

"UPW means taking every trace element and every ion out of that water," says Bombardier, "so there is literally nothing in it but the water."

This isn't like taking your tap water and pouring it into the Brita pitcher.

The basic process to make UPW involves eighteen steps. The sixth of those steps is reverse osmosis (RO)—the high-intensity, high-energy process that the everyday world views as "taking everything out" of water. Huge reverse-osmosis plants in Israel and Florida and Australia turn seawater into city-size volumes of drinking water. Coke and Pepsi are routinely mocked for taking perfectly clean municipal water and putting it through RO plants in order to "make" their branded bottled waters—Coke's Dasani and Pepsi's Aquafina. RO is so effective at cleaning water that Coke actually adds back a frisson of minerals to give Dasani what the company calls "a pure, crisp, fresh taste." Unpolished RO water—that is, RO water with nothing added back—is so clean that it feels flat on the tongue.

For ultra-pure water, though, a river of RO water is the basic raw material. RO cleans the water to the point that IBM Burlington's UPW factory can really get down to business.

There are a dozen cleaning steps after the reverse-osmosis process—huge, specialized filter beds to take out ions; tubes filled with UV light to blast apart any organics. The final step—the eighteenth—is a filter with pores that are 20 nanometers. Those holes are smaller than anything but individual molecules and parts of cells—although still seventy times bigger than a water molecule.

Water is a good cleaner precisely because "cleaning" is another way of saying "dissolving"—water dissolves stuff and carries it away. Water's supersolvent qualities are what make it so hard to clean, in fact—while you're taking stuff out, water is looking around for anything it can enfold and carry away.

That's one of the things that make ultra-pure water in particular great for microchip manufacturing. "UPW is hungry," says IBM's Steve Blair. "It's a really good cleaning agent. It will take anything it can get."

Cleaning the water makes it a much better cleaner.

The electron-carrying pathways on microchips have gotten smaller and smaller over the last twenty years—that's why our electronic devices have been able to get smaller while doing more. As the pathways have gotten smaller, the water used to make them has had to get cleaner.

"Every year, the purity requirements are increased," says Lindsey Stahl, the IBM water engineer. "What we called UPW ten years ago would be laughed at today. But we're at our limit of the cleanest we can do. It's almost as clean as you can make it."

Ultra-pure water is found nowhere on Earth because water out in the world is constantly flowing past, over, and through things, from which it is absorbing particles. While every microchip plant uses UPW, water so clean is a purely human idea, and a purely human creation.

Which raises an irresistible question: What does ultra-pure water taste like?

Janette Bombardier, the site operations manager, is an engineer who has worked for IBM in Vermont for thirty years, including six years in charge of water operations. She's friendly, with more the demeanor of a middle-school music teacher than a high-powered high-tech executive.

Her staff makes 83,000 gallons of ultra-pure water an hour, the cleanest water on Earth. So what's it taste like? Has she ever opened a spigot and drawn herself a glass?

There is the longest pause in the conversation.

"I have never had a glass," she says. "I have no idea what it tastes like." She pauses again. "It has never occurred to me to taste it."

In fact, in many factories that use UPW, it's not just regarded as an industrial solvent, but it is considered akin to a poison. A swallow or a glass of it won't hurt you, but as it does with the microchips, UPW is "hungry"—it will leach minerals right out of your body tissues.

Eric Berliner, an environmental manager for Bombardier, pipes right up. "I've tasted it," he says. "It was horrid. I stuck my tongue in it. It's very bitter. Horrible."

Even when it comes to water, clean isn't everything.

WATER HAS A REMARKABLE RANGE OF QUALITIES—it can be playful or comforting, it can be annoying, it can be powerful or erotic. But in terms of water's fundamental character, its personality, water is elegant, it's smart, it's a little sly in terms of what it can pull off when you're not looking (the Grand Canyon, for instance). But water is dependable. Water doesn't let you down; it rarely disappoints.

In 2008, the folks at PUR water filters went looking for a voice to bring water to life, and after sifting through hundreds of candidates, they chose Zach Braff, the actor who plays the recovering-nerd doctor J.D. on the TV series *Scrubs*.

Okay. He does a reasonable job. But Zach Braff's a little . . . goofy for water. Water doesn't often descend to the Old Testament baritone of James Earl Jones. But water isn't really ever goofy—even coming out of a pink squirt gun, or splattering from a water balloon, water isn't goofy.

Water is more Paul Newman or George Clooney, Faye Dunaway or Annette Bening—not urgently in need of drawing attention to itself, but perfectly capable of holding center stage. Water is a grown-up, quietly forceful, not self-important, with both a playful twinkle and a graceful wink, as the occasion requires.

Water is charming.

Part of water's charm is the way it combines simplicity and complexity, in both its physical nature and its qualities. Water is just three atoms—one oxygen, a pair of hydrogens—but with those three atoms, it is the elixir of life in all its variety and beauty. A single molecule of DNA, by comparison, has 204 billion atoms.[23]

Water's stickiness—its socks-in-the-dryer quality—is a result of its slight polarity, one side a little positive, one side a little negative. But that simple stickiness unfurls a set of complicated properties—the ability to absorb heat, the willingness to dissolve almost anything, the fact that ice floats—that make water the force that shapes the world.

Water's personality, in fact, is layered with polarity, both inherently and in the ways we approach and manage water.

Water is transparent, and also reflects light.

Water is soft and soothing, and also hard as concrete.

Water is comforting, and also threatening; gentle, and fierce.

Water is the source of life, and also often a source of death.

Water is all-important, indispensable, but almost always free, or essentially free.

Water is the most basic necessity to human life, and also a symbol of luxury and indulgence.

Water is sexy and alluring, and also often appalling and repugnant.

Water is as natural and wild as anything in the world—from white-water rapids and waterfalls to the power of hurricanes—and yet water is thoroughly domesticated in everyday life.

Water is a team player—a partner in a cosmic gallery of natural events—and yet water has an independence of both body and spirit. It participates in all kinds of processes but emerges again as simply water.

Water is as familiar as anything in ordinary life, and yet largely ignored, misunderstood, overlooked.

We often keep the two kinds of water separate in our brains and in our day-to-day stewardship—utility water and beautiful water, water's immediate functional role, and water's larger context. That, in part, is why we're drifting into trouble, failing to get water to people who don't yet have it, and doing an indifferent job of managing water in places that take abundant water for granted.

Our relationship to water goes way beyond what we know about it. The facts about water, the science, the chemistry, the geology—those are both fascinating and important. There would be no advanced civilization today without that understanding—we would have long since poisoned ourselves.

But our relationship to water is at least as much emotional as it is analytical. That's why a bottle of Evian tastes so good that we pay a thousand times more for it than for the same amount of water from the kitchen faucet. It's the reason that water pipes hidden beneath our streets are poorly maintained, it's why people around the world get so angry when their water bills go up.

We need to understand that the science of water goes only so far in explaining how we deal with water every day, both as individuals and as a society. And our feelings about water are often so powerful, so visceral, that we need to be sure they don't prevent us from seeing water clearly.

3

Dolphins in the Desert

When I came to work here, our attitude was, "Just shut up and turn on the faucet." That was a huge mistake. We were all going to run out of water in 1995.

—*Patricia Mulroy,*
water czar of Las Vegas since 1989

IF YOU START AT THE SOUTHERN END of the Strip in Las Vegas, the first resort you come to is Mandalay Bay, on the left as you walk north. Mandalay Bay's main entrance is flanked by waterfalls cascading over tall, faux boulders and cliffs into a tropical lagoon. Inside, the hotel is home to a 1.6-million-gallon slice of the Pacific Ocean, the Shark Reef Aquarium, which includes clear tunnels that allow you to walk among the hundred resident sharks. Mandalay Bay is also known for a pool area designed like a beach, with five pools, set among eleven acres of what the hotel describes as "real sand"—5 million pounds of it.

As you stroll north, just a block up from Mandalay Bay, you come upon a half-size version of the Statue of Liberty, standing on her pedestal in the water of New York harbor, torch raised high over the corner of Tropicana Avenue and Las Vegas Boulevard (as the Strip is officially named). Right next to Lady Liberty floats a New York City fireboat, with five jets of water arcing festively into the Hudson River twenty-four hours a day. Just beyond the fireboat, you are alongside a miniature version of the Brooklyn Bridge, which naturally spans the East River.

One casino farther along is the brand-new CityCenter complex, which includes the Aria hotel. Although you can't quite see it from the sidewalk, the Aria's main entrance is anchored by not one but two fountains, both from the world-renowned water design firm WET. One of the fountains is a curving wall of water 281 feet long, almost as long as a football field, that flanks the hotel's entrance. Waves of water pour over the top ledge of the twenty-four-foot-tall wall, and are slowed and shaped by tiny squares of stone along the wall's surface. Sometimes torrents cascade over the top, sometimes just modest pulses. Visitors stand transfixed by the waves coursing along the face of the wall, as if watching waves break onto a beach. In a Mojave Desert town where there are seventy-two days a year when it's 100 degrees or hotter, but just nineteen days a year when it rains, the Aria's sweeping horizontal fountain creates the mood of a Zen water garden at the entrance to a four-thousand-room casino hotel.[1]

In the same block as CityCenter is the Bellagio, a five-star luxury hotel that is as famous for its fountain as for its rooms, its service, and its cuisine. The Fountains of Bellagio are a Vegas attraction all their own, drawing thousands of spectators each night, who surround a man-made lake of 8.5 acres in which the fountain sits. The lake holds 22 million gallons of water, and the fountain is composed of 1,214 jets, shooters, and fan sprayers, all computer-controlled to perform in tight coordination with music, from show tunes to Handel's "Hallelujah Chorus." The fountain, whose arcs of jets stretch across the length of three football fields, creates rippling curtains of water that soar up 200, 300, even 400 feet high. Each night, starting at 8 p.m., the fountain gives four absolutely captivating performances an hour—you can stand on the sidewalk, in a town where not a single month averages even one inch of rain, and watch a fountain that, at any given moment, lofts 17,000 gallons of water into the air. The fountain is matched inside the Bellagio by an 1,800-seat theater whose stage is a tank of 2 million gallons of water, where each night acrobats from Cirque du Soleil perform a water opera called *O* (for the French word for water, *eau*).

And that's just the beginning, just the first third of a "water walk" up the Vegas Strip. In the next block up from the Bellagio is the Mirage, which has a signature volcano out front that erupts on the hour every night and sits in the middle of a lagoon. Deep inside the Mirage is a dolphin habitat—yes, another aquarium inside a casino—home to what the hotel calls "a family of Atlantic bottlenose dolphins," eight in all, including a couple born in captivity right on the Strip, all now living in a 2.5-million-gallon seawater habitat presided over by celebrity animal trainers Siegfried and Roy.

Across the street from the Mirage is the Venetian—a hotel built around the theme of Venice's canals. Right in front is the Venetian's lagoon, where you can ride around in a gondola poled by a singing gondolier in Italian costume.

Let's pause just a moment and take stock. In the space of a two-mile walk through the desert, and only crossing the street once, we've encountered three lagoons, a set of tropical waterfalls, the shark-infested Pacific, the dolphin-dappled Atlantic, an aquatic theater with a 2-million-gallon stage, a water-spouting fireboat, a ninety-yard-long wall of pulsing waves, the canals of Venice, and what was until 2009 the largest fountain in the world. Without noticing, actually, we walked right by what may be the only successful artificial rainstorm in the United States (inside the shops at the Planet Hollywood hotel). And we can't forget the sexy water—this stretch of vacation fantasyland also includes five adults-only topless swimming pools, each with a suggestive name, including Venus at Caesars Palace, Bare at the Mirage, and Liquid at the Aria.

Some people think that the obsession in Las Vegas is money—or, more precisely, easy money.

Las Vegas itself wants the obsession in Las Vegas to be regret-free sin ("What happens in Vegas stays in Vegas").

But the real obsession in Vegas is water—displaying it, unfurling it, playing with it, flaunting it.

Say what you will about Vegas, about the shows and the showgirls, about the slot machines at the airport gates and the craps tables that never close and the images of Donny and Marie Osmond plastered twenty stories high across the face of the Flamingo hotel—the most amazing thing may be that a hundred sharks and eight bottlenose dolphins live right on the Strip, and some of them are Nevada natives.

There is no two-mile stretch of ground anywhere in the United States that has such a density of water features, water attractions, and sheer water exuberance. Las Vegas, which can invest something as routine as break-

fast with outlandish extravagance, has taken our most unassuming substance and unleashed it as the embodiment of glamour, mystery, power,
and allure. In the way that only Las Vegas can, it has created a whole new
category—ostentatious water.

The Strip is a demonstration of water imagination, of water mastery,
and also of absolute water confidence.

It's all the more remarkable because Las Vegas is the driest city in
the United States—indeed, it's not even a contest. Of the 280 cities in the
United States with at least 100,000 people, Las Vegas is No. 280 in precipitation and No. 280 in number of days each year that it rains. Las Vegas gets
4.49 inches of precipitation a year. And it rains or snows, on average, just
nineteen days a year.[2]

Even places of legendary dryness are two or three times as wet as
Las Vegas—Phoenix, Arizona (8.29 inches, 28 days), El Paso, Texas
(9.43 inches, 42 days), Los Angeles, California (13.15 inches, 33 days).

Las Vegas is no El Paso, though—it is a metropolis with 2 million residents and 36 million visitors a year. Ninety percent of its water comes from
a single source, Lake Mead, the spectacular, man-made reservoir created
on the Colorado River by Hoover Dam. Lake Mead is the largest reservoir
in the United States, winding for 110 miles through the desert canyons of
Nevada and Arizona. When Lake Mead is full, it holds a sixty-year supply
of water for Las Vegas. Indeed, Lake Mead is so big, if every municipal water system in the United States could somehow tap it, it could supply all the
water delivered by water utilities in the United States for 210 days.[3]

But Las Vegas is legally allowed to take only a tiny sliver of Lake Mead
water—300,000 acre-feet a year, 98 billion gallons. All the water Las Vegas
is allowed lowers the lake between two and three feet.[4] Las Vegas's allocation
is about 4 percent of what everybody else gets to take from Lake Mead—
96 percent of the water people use from Lake Mead goes to either California
or Arizona. And Las Vegas's allocation is fixed in law, just as the allocations
of California and Arizona are fixed—so the amount of water Las Vegas has
access to hasn't changed even as Las Vegas's population has doubled, and
doubled again, even as the city has added 100,000 new hotel rooms, along
with fountains and waterfalls, swimming pools and shark tanks.[5]

If you're running the Las Vegas water system, it has been a harrowing twenty years, watching with a combination of fascination and queasiness as your desert city has grown so fast that, from 1990 to 2007, it added 60,000 new residents every single year, without adding any new sources of water for them. Even according to Las Vegas's most conservative water-use figures, 60,000 new people require 5.3 billion gallons of new water a year. And 60,000 new residents was only typical—in 2004, 105,000 new residents settled in town, as if everyone in Berkeley, California, had picked up and moved to Las Vegas, along with their need for 9 billion gallons of water a year.[6]

In big-picture terms, between 1990 and 2009, Las Vegas nearly tripled in population.

What's even scarier is that for the last ten years, the rainfall and snowfall that everyone along the Colorado River had become accustomed to for the last century fell off dramatically. By 2010, Lake Mead was down to 41 percent of capacity—which is to say, the largest reservoir in America was more than half empty. As recently as January 2000, Lake Mead was within seven feet of being full. Lake Mead in mid-2010 is 125 feet lower than it was ten years earlier—a lake that is 110 miles long has lost a stunning 125 feet of water depth in just a decade. Forget how much drinking and irrigation water that is—it's enough water to put the Strip, and every one of its hotels, at the bottom of a lake a half-mile deep.[7]

It is just as easy to make fun of Las Vegas as it is to have fun there—who really needs to market "real sand" in the middle of a desert? Who builds the world's largest fountain in the middle of the driest city in the country? Why visit a fake Eiffel Tower and a fake Brooklyn Bridge when it's just as easy to visit the actual Eiffel Tower and the actual Brooklyn Bridge?

But that easy snickering ignores the most important fact. We love Las Vegas. The town gets 36 million visitors a year—86 percent of whom are Americans; 10 percent of the country visits Vegas every year.[8]

And so although you'd never know it on the Strip or in the new subdivisions of the bedroom community of Henderson, Las Vegas doesn't have water challenges or water troubles. Las Vegas has a water emergency. People are working on the emergency twenty-four hours a day, in fact, desperately trying to stave off what could become a catastrophic water crisis.

The first of Las Vegas's two big water intakes—the huge pipes through which the city literally sucks its water out of Lake Mead—is just 38 feet from breaking the surface. A couple more dry years and Intake 1 could literally be sucking air. It's normally safe under 125 feet or more of water.[9]

With the fountains jetting and the waterfalls cascading and the gondoliers serenading, it would be easy to add an absolute water heedlessness to the sins of Vegas. As the water is literally disappearing into the sand, the main Las Vegas tourism Web site still opens to a picture of a gorgeous come-hither brunette, in front of a shimmering turquoise swimming pool, with the line "Welcome to Lake Dowhatyawanna."

But as is often the case in Las Vegas, everything is not quite as it looks. Las Vegas may seem to be gambling away the last of its water currency without either concern or a plan. But in fact, Las Vegas is far more advanced in both water consciousness and water management than almost anywhere else in the country.

PATRICIA MULROY HAD ALREADY been head of all water in metropolitan Las Vegas for thirteen years, she had faced down the casino moguls and the real estate developers, she had outmaneuvered, and lassoed, the West's water bureaucrats with the speed and agility of a cowgirl, and she had applied both imagination and blunt force to getting Las Vegans to use less water.

But Mulroy is the first to admit she's never quite done learning about people's relationship to their water. It was 2003, and the drought that is now a decade old was just beginning to settle in, the surface of Lake Mead had fallen sixty feet in just three years, and Mulroy decided it was time to tackle the small but symbolically wasteful fountains spraying across the Las Vegas Valley—the ones in office building lobbies and bank parking lots and shopping center plazas.

"This is funny," says Mulroy. "We got into an aggressive campaign to limit fountain use. What we really tried to do was separate out the economic value of fountains from the purely decorative use of fountains. The Bellagio fountain versus a little fountain at a shopping center."

This is one of those topics that Mulroy warms to. She is as fierce, intelligent, and independent as if she were a descendant of a pioneering Nevada ranching family; in fact, she is the daughter of an American father stationed with the U.S. Air Force in Germany and a German mother. Mulroy is both the conjurer of Las Vegas's water and its protector; she's as aware of the fountains, lagoons, and canals watering the Strip as she is of her own breath.

"You start turning those water features off, and it will have an effect on visitors. You might be able to turn off that pathetic little fire hose in front of [the casino] New York New York, but you turn around and you dry up the canals in the Venetian, and you watch occupancy drop. How do you discern what's a good water feature from a bad water feature?"[10]

For Mulroy, the difference isn't aesthetic, it's economic: The water on the Strip is part of the allure of the place—an allure that is magnified precisely by the astonishment of having so much free-flowing water in the middle of the desert—whereas the fountains installed across the civilian part of Las Vegas are just the water equivalent of Muzak.

"It's perfectly natural to say less people will come visit the city if that fountain at the Bellagio is not here," says Mulroy. "But I will still go to the grocery store without a fountain. I will still go to the dentist without a babbling brook in the lobby."

Las Vegans did not agree.

"The banks went absolutely crazy about us telling them to take out their fountains," she says. "I got the poop beat out of me. A psychologist called me up one morning. He said his patients needed the sound of the babbling brook to do their therapy.

"Really? *Really?*" Mulroy, a woman with a penetrating gaze, rolls her eyes. "I sent him a CD with the sounds of running water."

Then Mulroy backed down, kind of."

"We said, okay, you can keep your fountains—if you take out enough grass to save fifty times the amount of water the fountain uses." When it comes to Mulroy and Las Vegas's water, that counts as a compromise. Which is to say, the bankers and therapists and building superintendents might have walked away with their fountains, but Mulroy walked away not just with victory, but a victory fifty times greater than before she started to compromise.

Still, if the silly babbling fountain at the supermarket and the doctor's office sends the wrong message, don't the waterfalls of Mandalay Bay and the fountain at the Bellagio send exactly the same wrong message—on a grand scale?

Mulroy narrows her eyes. "People don't understand water," she says. Mulroy is working on that.

Las Vegas is a very different place than it was when she reluctantly took over as the general manager of the Las Vegas Valley Water Authority (LVVWA) in September 1989. That year, Las Vegas residents used 348 gallons of water per person, each day. Twenty years later, in 2009, Las Vegas residents used just 240 gallons per person, per day. Under Mulroy, per capita water use in the desert metropolis has dropped 31 percent.[11]

Even more amazing, total water use in 2009 for the whole metropolitan area was almost exactly what it was in 1999. Las Vegas and its suburbs grew by 685,000 people—Las Vegas grew by 50 percent—without having to use any more water. Mind you, there were some big years in between 1999 and 2009—all-time water use peaked in 2006—but that's the kind of water performance that literally makes Las Vegas's continued growth possible. It's the kind of water performance that has made Mulroy the best-known water manager in the country, and perhaps in the world. Water managers in Australia and India mention her admiringly without prompting.

Lots of people get credit for creating Las Vegas—Bugsy Siegel, Meyer Lansky, Howard Hughes, and in the modern era of moguls, Steve Wynn, who has built five signature Strip megaresorts—the Mirage, Treasure Island, the Bellagio, Wynn, and Encore. But without water, there would be nothing like the Las Vegas we know, and so in very real terms, Patricia Mulroy invented modern Las Vegas, or as she would put it, she allowed modern Las Vegas to invent itself.

The woman who has had such an impact on Las Vegas is in both the city and the job she has by accident. Mulroy's dad was a civilian personnel manager for the air force stationed mostly in Germany, and although the family lived briefly in the States, Mulroy did all of high school and three years of college in Munich. She wanted to finish school in the United States and got a full scholarship to the University of Nevada, Las Vegas. "I had to go to a map to see where Las Vegas was," she says. She spent her first night in Vegas in the old Desert Rose hotel, where New York New York is now.

When she finished, she was admitted to a graduate program in German literature at Stanford, but family finances dictated that she get a job. She stayed in Vegas, and started in 1978 as an administrative assistant working for Clark County—the county that surrounds Las Vegas and includes the Strip. Before long she was on the county's lobbying team, working lawmakers during the legislative session in Carson City. She eventually moved on to a job in the county courthouse. But when the deputy general manager job opened at the Las Vegas Valley Water District (LVVWD)—second in command, in charge of engineering, operations, and administration—Mulroy grabbed it.

"It was just an escape," Mulroy says now. "I had decided that I would rather slit my wrists than work for judges." The water department she joined in 1985, she says, "was a standard, traditional water utility: Our No. 1 job was to remain invisible. We had a very parental attitude. We knew better than anybody."

In those days, the Las Vegas metro area had seven separate water utilities, and a "use it or lose it" division of the available water, in which every municipality needed to use up its allocation of water every year or run the risk of having the allocation reduced. "There was every disincentive to conservation," says Mulroy. There was also rivalry and competition.

When her boss—the general manager of Las Vegas water—became a victim of local water wars, Mulroy wasn't inclined to try to replace him. "I didn't want the job," she says. "I lacked self-confidence. I was the mother of two small children, a three-year-old and a two-year-old. It took some talking." The accidental Las Vegas resident became the reluctant water boss nevertheless, in September 1989, at age thirty-six.

Her hesitation was justified. "When I took over in 1989," says Mulroy, "water was organized chaos. The house of cards was caving in."

In 1990, any resort installing a waterfall or fountain as often as not used municipal water to feed it, straight from Lake Mead. Golf courses used city water too. Real estate developers were building subdivisions with artificial lakes and filling their fake lakes with water from Lake Mead. Residents and businesses watered their lawns with a casual profligacy that was described in a *New York Times* story from a year after Mulroy took over. "Sprinklers send small rivers of water into the gutters daily all over Las Vegas," wrote Robert Reinhold. "Acres of grass surround the new Citicorp Nevada building, a credit-card processing operation, on West Sahara Avenue. Huge sprinklers spray water during hot daylight hours into the air and, on a recent windy day, thousands of gallons drained into the streets."[12]

If Mulroy was originally reluctant to take on the general manager's job, there was nothing tentative about her command of it from the very start. In her first year she did three things that revealed her style and her agenda. Pat Mulroy would not talk or act like previous water managers— she wasn't going to ask permission to do what was necessary to keep the water flowing, she wasn't going to let a little whining cause her to second-guess her judgment. And Las Vegans were going to start changing how they used water, because Pat Mulroy was going to show them they had no choice.

In her first few months in the job, she moved to raise water rates— reducing the monthly fixed charge but increasing the rates based on volume, including the highest-volume residential customers, those using 30,000 or 40,000 gallons of water a month watering lawns. It was the opening salvo in an effort to change the attitude about the availability of water, the use of water, and the cost of water. One home builder told Mulroy that big water users in the desert should pay *less*—that they should get a "volume discount."[13]

Mulroy had done the math on Las Vegas's water use, on the Lake Mead allocation, on the galloping population growth, and she knew that the predictions on which the water department was relying had become scarily, ridiculously optimistic—and out-of-date. In 1990, Las Vegas's population was 750,000, and it was predicted to reach 1 million ten years later. In fact, the population reached 1 million in 1995. By 2000 it was 1.4 million.

"When I came to work here, our attitude was, 'Just shut up and turn on the faucet.' That was a huge mistake," she says. "We were all going to run out of water in 1995."

Nevada has an unusual system of water rights. Groundwater is not exclusively connected to the land under which it is found. Groundwater that isn't being used is controlled by the state, and can be claimed and used by anyone who can make "beneficial use" of it. With just months in charge, Mulroy had the staff of the LVVWD scour nearby counties for unclaimed groundwater rights, and file 146 applications in the state capital for unused water rights in four counties.

At the time, the water Vegas was laying claim to was estimated to total a stunning 865,000 acre-feet—almost three times what Las Vegas was entitled to from Lake Mead. The plan was to secure permission to use the water, and then, over twenty years, dig wells and build a big underground pipeline to bring the water three hundred miles south. When rural ranch owners, county officials, and conservationists learned of the water applications, there was an explosion of anger and opposition.

Rural residents feared that as Las Vegas's new wells sucked away the groundwater, natural springs would dry up, and their communities would be turned into miniature dust bowls. What was for Mulroy a savvy—and totally legal—strategy for securing new water for her thirsty city looked to the rural counties like water theft. "The development of rural Nevada is dead now because of Las Vegas," said one county commissioner at the time.

Retorted Mulroy: "If [Las Vegas] dies, Nevada dies." [14]

That effort to find new water, in the first year of Mulroy's tenure, is still being fought. In June 2010, the Nevada Supreme Court allowed Las Vegas's twenty-one-year effort to gain access to the groundwater to move forward, although the court ruling hardly guaranteed success.

Just by filing the applications for the water, Mulroy helped do something that has turned out to be both pioneering and essential—it gave her the credibility and the muscle to unify the seven wrangling Las Vegas–area water utilities. In the second year of her tenure, they created a new entity, the Southern Nevada Water Authority (SNWA), which is now the regional water power. The SNWA is the area's water wholesaler, in charge of securing enough water for the region, as well as planning for and managing the region's water supply. SNWA's customers are the water utilities themselves, who are now retailers, supplying water to homes, apartments, businesses, and hotels.

And the head of the new regional water power? Patricia Mulroy. She runs both the SNWA and the Las Vegas Valley Water District. That early, lightning-strike filing for the unused groundwater north of Las Vegas gave Mulroy's colleagues a taste of her ability to look ahead—in terms of demand for water, in terms of finding supply, and in terms of fearless political strategy.

The final thing Mulroy did that first year was sit down with Steve Wynn, the man who created the model for the modern Vegas casino—big, luxurious, themed like a cruise ship or a resort, and drenched in water. Wynn's Mirage hotel, with its lagoon, waterfalls, and volcano, opened just as Mulroy got the water chief's job. In her first few months, he was working on his second casino, Treasure Island, where he wanted to feature a nightly "battle" between pirate ships in a lagoon.

In early 1990, Wynn summoned Mulroy. The water district was getting ready to regulate—or even disallow—huge water features that relied on drinking water for their supply.

"I was asked to come to the Mirage," says Mulroy, "and we talked for two hours." It was a blunt conversation, from both sides.

"Steve Wynn was the first to say to me, 'People don't come to the desert. They come to the oasis in the desert. Who escapes to rocks?'

"And he's right about that. Las Vegas sells virtual reality. We're creating virtual realities that are no different than Cinderella's Castle at Disney World. And when it's 118 degrees outside, water becomes an integral part of that."

For her part, Mulroy knew already that Las Vegas had to change the way it approached water—that you couldn't use millions of gallons of purified drinking water, from your limited stock of Lake Mead water, to stage mock pirate-ship battles in an artificial lake every night.

"What Wynn said to me was, 'Don't tell me I can't have fountains and water features. Tell me what I can do to do it.'"

So she told him: Use recycled water in Treasure Island—use purified wastewater for the fountains and lagoons. "He double-plumbed Treasure Island," says Mulroy, "and he put a water treatment plant in the basement."

It was another moment when a potential Mulroy adversary walked away with apparent victory. Steve Wynn went from wondering who this woman was who was going to ban his lagoons and fountains, to opening a hotel that did, for its first decade, stage nightly pirate battles. Mulroy, though, walked away with her own victory. First, she had won an ally. And second, she set the precedent that has become the water law of the Strip, the secret of its splashy, extravagant water features. Developers must either use wells from their own property, the rights and permits for which date back before the modern era, or they must use treated wastewater, or they must offset their water features with bigger water savings elsewhere, like the banks and shopping centers.

That, at least in part, is how you get both water ostentation and water conservation.

As YOU STEP ONTO THE INDUSTRIAL FLOOR of Mission Industries Plant 50 in North Las Vegas, the thing that strikes you first is the smell: the aroma of freshly laundered hotel towels, thick, soft, and inviting.

For a lot of the hotels on the Strip, this is where that smell, and those crisply creased pillowcases and fluffy bath towels, come from. Las Vegas is a place of fantasy, but someone has to wash the dirty sheets. Mission Industries handles the laundry for more than half the Strip resorts—sending out fleets of trucks each day to collect bins of dirty linen, and then re-delivering carts piled with perfectly pressed and stacked linens to every level of hotel—from those charging $49 a night to four- and five-star resorts like Mandalay Bay and the Bellagio.

Mission Industries Plant 50 is really a single laundry room the size of a Wal-Mart Supercenter, with a twenty-five-foot-high ceiling, lines of industrial washers, dryers, and high-speed ironing and folding machines, and at the loading dock, carts of dirty linen lined up twenty deep waiting to be emptied. It's clear why the big Strip resorts outsource this task to specialty houses. Laundry at this level is an art in itself. Before the economic downturn, Plant 50 was running twenty-two hours a day, seven days a week. The process starts with a sort—to pluck out the teddy bears, underwear, and miniature shampoo bottles that end up in the piles of towels and sheets. Anything that's not getting laundered goes into the garbage.

The four washing machines, each longer than a school bus, look like comically elongated Laundromat washers, and are called tunnel washers.

Loads of sheets and towels move from front to back in a series of chambers, washed in 150-degree water.

The machines that do the finish-work on queen- and king-size sheets look like huge printing presses—wet sheets get fed in at one end of a fifty-foot-long line of machinery; dry, wrinkle-free, razor-edge piles of folded sheets emerge at the other end, all moving almost too quickly for the eye to follow. Sensors on the folding line scan the sheets for stains and pluck out any that need rewashing.

Plant 50 is one of the largest laundry facilities in Las Vegas. When the Strip hotels are well booked, it does 3.5 million pounds of sheets and towels a week, linens from twenty thousand rooms a day. All that washing requires a lot of water. Plant 50 is a "light soil" facility—cleaning sheets and towels involves nothing like cleaning the stains and grease in restaurant linens, which go to the "heavy soil" plant. Even so, Plant 50 uses nearly a million gallons of Lake Mead water a day.

Or it did, until Ralph Barbosa convinced his bosses to install a water reuse system. It was a test to see if Mission could clean its own dirty wash water on-site and send it back into the washing machines for a second run, saving both water and money. Barbosa, who is director of engineering for Mission, has worked for the company since May 1973. "It was my idea," he says. "It took me a year and a half to talk the owner into it.

"You know, the lake is about 120 feet below its usual level—there's a lot of white rock showing. It's really a sight that is shocking. I think we can have an impact."

Barbosa spent months talking to equipment companies, looking for the right water recycling system at the right price. "We used anywhere from 2 to 2.5 gallons of water per pound of laundry. We were shooting for 1.5 to 2 gallons instead."

According to Barbosa, the typical room in a Strip resort produces twenty-two pounds of dirty sheets and towels a day—requiring 44 gallons of wash water. Mission's water and sewer bill is $9 for every 1,000 gallons—to save $1,000, Barbosa had to reduce water use by 110,000 gallons, about what a family uses in a year.

The system Mission ended up installing, in May 2007, is an elaborate array of filtration and disinfection steps that amounts to a small water treatment plant. Sitting outside next to Plant 50's water storage tanks, the system looks like a miniature oil refinery: Water drained from the tunnel washers—which isn't very dirty, whatever is washed off the sheets and towels, mostly hair and lint—goes through a series of five increasingly fine filters. The water is disinfected twice, with ozone and with UV light. Then the water is ready to do its washing again.

The system cost about $800,000 to buy and install. In its first year of use, it saved Mission 240,000 gallons of water a day—$2,000 off the water bill each day. And, Barbosa says, there was a surprise or two.

The recycled water helped cut the amount of time thick towels needed to dry in the big, natural-gas-fired dryers. How can reusing wash water cut drying time?

Mission's recycled water stays very hot—it comes out of the treatment process almost as hot as when it comes out of the washers, about 130 degrees. Reusing such hot water reduces the cost of heating the wash water up to 150 degrees in the first place, of course. But Mission also uses that 130-degree recycled water in the rinse cycle of the tunnel washers. The rinse water used to be only 95 degrees. The towels emerge from the washers hotter now.

"The hotter rinse water opens the pores of the terry cloth," says Barbosa, "so the towels dry faster." Hotter towels literally release their moisture faster. "It saves us three or four minutes on the drying cycle for the towels, out of a typical drying time of twenty-two minutes."

So Mission saves money on its water bill, on its water-heating bill, and also on its natural gas bill. It's exactly the kind of unexpected, cascading benefit people often discover when they start managing their water use more closely. Much of the cost of using water in the first place isn't in the water itself but in what you do to the water in order to use it, and when you reuse the water, you often retain things, like heat or cold, that you paid to put into the water.

Mission's recycling system has paid for itself in a little more than a year, and Mission was impressed enough that before the recession hit, the company installed a second system at another of its four Las Vegas laundry plants. The recycled water, in fact, has helped Mission weather the downturn by reducing routine operating costs: The cost of the water to wash one room's linens has gone from 40 cents to 30 cents.

When the Strip hotels are generating a typical load of linens, the two systems save Mission 360,000 gallons of drinking water a day—a little more than 1 acre-foot. Which is to say, Mission's water reuse systems will save not just Mission, but Las Vegas, 400 acre-feet of Lake Mead water a year, out of the 300,000 the city is allowed—permanently. One company's savings is everyone's savings, because when a Las Vegas water customer reduces water use, it frees the water for someone else.

Barbosa says that the recession, which for a while cut laundry loads in half, put on hold plans to install the water recycling systems at Mission's other plants, but that when business picks back up, there's no reason not to install two more.

"It's going to be huge, this kind of thing," says Barbosa. "It's going to take off."

In the occasionally counterintuitive world of cutting-edge water management, no one was happier to see Mission Industries reduce its water use than the woman who gets Mission its water, Pat Mulroy. As part of its conservation efforts, the SNWA gave Mission a check for $150,000 to help pay for the recycling system—about 20 percent of the cost—after Mission showed how much water it wasn't using.

"At the moment," says Barbosa, "the bean counters are ecstatic."[15]

THE WATER USED TO CLEAN HOTEL LINENS in Las Vegas is trivial compared with the much more visible water required to keep Las Vegas's golf courses green. Laying down a carpet of turf, and keeping it alive, on the red rock of the Mojave Desert doesn't require watering, it requires irrigation, and obsessive irrigation at that. The average high temperature in Las Vegas in June is 99°F, in July it's 104, in August it's 102. For a quarter of the year, the temperature is 99 or more every single day.

Angel Park is a popular course in Las Vegas, open to the public, with thirty-six holes designed by golf legend Arnold Palmer, and a smaller twelve-hole course lit for nighttime play. By the time a golfer steps up to the first tee, Angel Park has used 2,507 gallons of water to make that round of golf possible, 139 gallons of water for each hole of an eighteen-hole round, for each golfer, each day. A foursome of Angel Park golfers playing eighteen holes requires enough water to supply the needs of a typical U.S. family for a month.[16]

That 2,507 gallons of water isn't "virtual" water—on midsummer nights, Angel Park's sprinkler system can spray 2 million gallons of water onto the course.

And that's all good news: Angel Park, under superintendent Bill Rohret, has been among Las Vegas's most aggressive courses at finding ways to use water more carefully. "We've taken ourselves off the potable water grid," says Rohret, on a walking tour of Angel Park on a morning in May. "Everything you see in terms of water use is reclaimed water. It's tertiary treatment, clean enough that you could bathe in it."

In its early years, Angel Park used much more water than it uses today—in 1996, the course used 644 million gallons of water, enough for a city of ten thousand people—and all the water it used was drinking water, straight from the Lake Mead allocation. Since 2001, Angel Park has irrigated exclusively with recycled wastewater. Rohret is using four times the recycled water that Mission Industries uses, and the golf course is buying it from a Las Vegas city sewage treatment plant. The 1 million gallons of water the sprinklers have put on the course this May day, says Rohret, "was in someone's house twenty-four hours ago."

More dramatically, Rohret and Angel Park have changed the nature and texture of the course. Since 2007, the club has taken out 76 acres from the course's original 260 acres of turf, removing almost the equivalent of a whole golf course's worth of grass, and replaced it with sophisticated desert xeriscaping. In some places, like the islands in the parking lot, removing grass had no impact on golfers. In other places, like the fourth hole of Angel Park's Mountain course, the patches of stark, arid landscape in the midst of golf holes take some getting used to. The shot from the tee to the green now crosses a shallow desert canyon, an arroyo, where there had previously been a "water feature," a lake holding millions of gallons of water. The patches of desert look as out of place in the midst of a lush golf course as the golf course itself looks out of place in the desert.

"It's a new kind of golf course," says Rohret, who has lived in Las Vegas since 1987. "It's funny, twenty years ago when we opened, we took out the desert to put the grass in. Now we're taking grass out to put the desert back in."

Angel Park has a computer-controlled irrigation system linked to weather sensors, and like most golf courses, it has the equipment, the knowledge, and the staff to keep the course immaculate using 36 percent less water per acre than a typical Vegas homeowner.[17] Every shrub, plant, and tree has its own individual sprinkler head—the turf emitters, spraying wide areas of grass, run at 30 gallons per minute; the drip irrigation heads run at about 1 gallon an hour. Replacing seventy-six acres of grass with rocks, gravel, and desert plants, says Rohret, he hoped to save 80 million gallons of water a year. "Last year, we were down 120 million gallons, way more than I dreamed." That takes $280,000 off Angel Park's water bill (Angel Park pays $2.33 per thousand gallons of recycled water).

And taking out 30 percent of the grass to save water has had other benefits—just as Mission Industries discovered. "That's eighty acres of turf we don't have to mow," says Rohret, "at $3 a gallon for fuel. It's eighty acres we don't have to seed. It's eighty acres we don't have to fertilize." He's buying 120,000 pounds of grass seed a year, down from 160,000 pounds, and the fertilizer budget is down 20 percent. Rohret took out a thousand of the course's seven thousand sprinkler heads, and has saved on the replacement costs for them. "The coyotes and the rabbits eat the emitters," says Rohret.

Relandscaping an Arnold Palmer–designed golf course is not like replacing a patch of lawn at home—it requires planning, design, materials, work, and heavy equipment. The SNWA has an incentive program to encourage homeowners and businesses to take out turf. Nicknamed "cash for grass," it pays between $1 and $1.50 for every square foot of grass removed and replaced with desert landscaping—whether the grass is in the front yard of a blackjack dealer's subdivision house or on a golf course. That comes to about $45,000 an acre. For a course like Angel Park, where the desertification program cost almost $3 million, the SNWA's turf-removal program essentially covered the cost of reconstruction. And Angel Park plans to xeriscape another twelve acres of turf.

Rohret loves Angel Park's new personality. "It looks like a desert," he says. "There's no need to have all this turf. The mind-set has changed—the old course is like a dinosaur."

The water features of the casinos on the Strip are splashy and ostentatious, but the real indulgence in Las Vegas are the sixty-one golf courses. Orlando, Florida, with a third more visitors than Las Vegas and the same number of residents, and which gets a foot of rain for every inch Las Vegas gets, has just twenty-one golf courses.[18]

Angel Park, which is home to two full-size golf courses, uses 376 million gallons of water a year. The Fountains of Bellagio, which need 22 million gallons to fill and operate, only require 12 million gallons of water each year to replace water lost to leaks, wind, and evaporation. So one golf club, of Las Vegas's forty-five, uses more water in twelve days than the Fountains of Bellagio use in a year.

The SNWA now rigidly regulates the water golf courses can use—every golf course has a mandated water budget. The water budgets have not only reduced the amount of water courses use but also reduced the creation of new courses. Still, to grow grass on Las Vegas's hardpack, golf courses are using the same amount of water that farmers in the Imperial Valley of California use to grow lettuce and wheat, carrots, broccoli, and onions, in one of the most productive agricultural regions in the country—and the Imperial Valley farmers, in fact, are using the very same water, Lake Mead water delivered by canal to California.[19]

Growing grass in Las Vegas isn't just an indulgence, it's the biggest problem. According to the SNWA, 70 percent of the water Las Vegans use at home is used outdoors, mostly for lawn and garden watering, but also to wash cars and fill swimming pools. In most of America, of course, we don't use 70 percent of our home water for outside consumption. It is one of the ironies that in popular suburban desert communities like Las Vegas and Henderson, and Phoenix, and Scottsdale, and Tucson, homeowners feel the need to use so much more water on their landscaping in precisely the places where that water is less available. Because homeowners account for almost half the total water use in Las Vegas, the outdoor use means literally 31 percent of the water the city takes from Lake Mead ends up lost to the arid landscape and the desert air.[20]

That's why Mulroy has been so focused on paying people to take out grass. "You know, every time it snows in Buffalo, people pack and move to Phoenix and Las Vegas," she says. "They come here and they think they can bring their Kentucky bluegrass and their magnolias with them. They want the benefits of no precipitation, and the things they were used to, too."

Or as Steve Wynn put it so crisply, people don't come to the desert, they come to the oasis in the desert.

Mulroy's "cash for grass" program has paid out $155 million to Las Vegas residents—almost $80 per person—and Mulroy likes to put the turf removed in terms of square feet: 140 million square feet of grass removed. While that sounds dramatic, it's not huge: 3,214 acres. That's the equivalent of just 13,000 homeowners each taking out a quarter-acre of turf in a town with 400,000 homeowners.

Of course, it's not so much the grass removed as the water saved that matters. The SNWA concluded that every square foot of grass removed saves 55 gallons of lawn watering, every year, forever. So the SNWA says the turf Mulroy has paid to take out is saving Las Vegas 7.7 billion gallons of water a year—8 percent of the Lake Mead allocation, hardly trivial.

The real cultural change that Mulroy and her colleagues have pulled off, though, has been much more dramatic than paying people to give up their grass. Mulroy, the transplant from Germany, is trying to teach Las Vegans to live in the desert, and to act like they live in a desert.

Driven by the drought, the SNWA has imposed a whole series of water-saving rules—the kinds of things that make sense in a desert city, but that would have been unthinkable when Pat Mulroy took her job, and that would astonish Americans in any other community. And though inspired by the drought, the rules aren't temporary, they are permanent.

Some seem relatively small. In order to wash your car in your driveway, you have to use a nozzle with one of those spring-loaded automated shutoffs. No running hoses allowed.

Every residence in Las Vegas is assigned mandatory watering days. In the spring and fall, you can only water three days a week. In the winter, outdoor watering is limited to a single day a week. In the brutally hot summer months, watering is allowed every day, but never between the hours of 11 a.m. and 7 p.m.

In a nod to Mulroy's skepticism about little fountains in shopping centers and office parks, new outdoor water features at commercial establishments are forbidden (except for resorts following offset rules, or businesses with their own well water and permits to use it).

Most dramatically, in all of metropolitan Las Vegas, newly built homes are forbidden to have front lawns. Only half the backyard of a new home can be planted in grass. New commercial developments are forbidden to have turf at all.

The scenes from the 1990s, of sprinkler systems sending wide arcs of water across the green lawns of office parks, spilling the excess into the scorching gutters—those days are simply over. In fact, it's not only illegal to water your lawn when it's not your day, or at 1 p.m., or to have a front lawn at a new house, it's illegal to let water from a misaimed (or broken) sprinkler run down your driveway into the street or spray onto the sidewalk. And the local water utilities have water cops who patrol the city's streets, looking for water waste and writing violations.

Perhaps the most surprising, and least well-known, element of Las Vegas's water system is that at this point, it's only the outdoor water that really matters at all. Las Vegas reuses or recycles almost every gallon of water that is used indoors in any context. Any water anywhere in the metropolitan region that goes down a drain—water from car washes, water from dishwashers, from showers and mop buckets, from elementary school water fountains and hot tubs and toilets—ends up treated and reused at a golf course or a park, or it ends up cleaned and sent directly back to Lake Mead.

More than 90 percent of the water used indoors is captured and recycled. That, in fact, is the only way that Las Vegas survives.

The city first broke through the federal ceiling of the 300,000 acre-feet of water a year it is allowed back in 1992—that year, Las Vegas took 306,000 acre-feet. But that same year, the city returned 128,000 acre-feet of treated wastewater back to Lake Mead, so in fact it had only "used" 178,000 acre-feet.

For each of the last ten years, Las Vegas has taken more than 450,000 acre-feet of water, but these days it returns about 210,000, keeping it under the cap. Overall, Las Vegas is sending directly back to its source of water about 40 percent of the water it takes.[21] (Lake Mead itself helps keep Las Vegas's water use in perspective. Every day, twice as much water evaporates from the surface of the lake as Las Vegas uses total.)

For Las Vegas, the absolute inviolability of the return-flow credits is crucial. For every gallon of water Las Vegas is able to collect, clean, and return to the lake, it gets to take a new gallon of water to use.

As a result, getting water back into the sewers is a bit of an obsession in Las Vegas, because water that goes into the sewers is water that can, ultimately, be taken back out of Lake Mead. The SNWA's Web site, for instance, gives detailed instructions for how to empty your backyard swimming pool, not into the gutter but into the sanitary sewer. It is illegal in Las Vegas to drain your pool or hot tub into the street or the storm drains.[22]

And that's why the SNWA is so focused on managing outdoor water use, and slowly laying the groundwork for a community where the habits of outdoor use are permanently reduced by laws, by practice, by the very landscape that people get used to seeing every day.

It's just one more oddity of Las Vegas that, in fact, it doesn't really matter how long your shower is, whether you are a resident or luxuriating in your suite at the Bellagio. The shower water is all going right back to Lake Mead.

But the irony cuts both ways. Although it's great that Angel Park and other golf courses now use treated sewage to stay green, it's not nearly as great as it first seems. That water has been used once, to be sure, for showers or laundry or flushing the toilet. It is cleaned, and pumped to the golf course for reuse as irrigation water. But in a back-channel way, the golf course water still comes right out of Las Vegas's drinking water.

If the wastewater treatment plant weren't pumping the water to a golf course, it would be sending it back into Lake Mead, and Las Vegas would be getting water credits for it. Once it's sprinkled onto a green, at least in terms of Pat Mulroy's water ledgers, it's water that's gone for good. The golf course reuse water only gets used twice—once in someone's home or business, and once to water the golf course. That's why it still urgently matters not just where golf courses get their water but how much they use. A typical toilet flush in Las Vegas, on the other hand, can be repeated hundreds of times—really, infinitely—and the very act of flushing the toilet regenerates the water necessary to refill it.

~~~~~

As Las Vegas was watching its reservoir evaporate, all the way across the country, another energetic, sprawling American city was having exactly the same experience. In February 2008, in the midst of the worst drought in Atlanta's history, Lake Lanier, the city's drinking water reservoir, was just two feet from the lowest it had been since being filled in 1958.[23] With the Atlanta metro region scrambling to figure out how to survive another dry spring and summer, the Georgia state legislature took bold action.

Citing a flawed land survey from 1818, both the Georgia House and the Georgia Senate voted to move their state's border with Tennessee 1.1 miles north. That would put Georgia's northwest corner in the middle of the Tennessee River, and entitle Georgia to a slice of that big river's water, magically solving Atlanta's water shortage. The Senate vote, to establish a commission to redraw the boundary—states can't, of course, unilaterally change their boundaries—was unanimous; the bill was signed by Governor Sonny Perdue. As Republican senator Dick Shafer, the bill's sponsor, rose to speak before the Senate vote, his fellow senators sang, "This Land Is Your Land."

The mayor of Chattanooga, the big Tennessee city that sits on the border with Georgia, and through which the Tennessee River runs, responded by dispatching an aide to the Georgia Capitol in Atlanta with a truckload of two thousand bottles of water to ease Georgia's thirst. "Today they come for our river," the Chattanooga mayor said. "Tomorrow they might come for our Jack Daniel's."

Tennessee state representative Gerald McCormick was less amused. Of his Georgia legislative colleagues, he said, "They're idiots."[24]

As easy as it is to shake your head at Las Vegas's attitudes about water, Las Vegans themselves have spent the last twenty years taking their relationship to their water supply seriously in a way that almost no one else in the United States has. They may be in trouble, but it's not because they don't realize how precarious their supply of water is.

What's really remarkable is that a community like Atlanta, which sits in an area of water wealth and prides itself on its sense of civic-mindedness, has water troubles at least as urgent as those in Las Vegas. And where Las Vegas's challenges come mostly from the struggle to sustain a metropolis in a desert, Atlanta's troubles spring from a much more universal problem: water complacency.

Atlanta is now a metropolitan area of 5.5 million people. In each of the last two decades, it has added more people than any other metro area in the United States—in sheer numbers, it has for twenty years been the fastest-growing place in America. Just from 2000 to 2009, the city added 136,000 people each year. That's nearly 400 new residents every single day.

Las Vegas has grown mind-bogglingly quickly any way you measure it. But here's how dramatically Atlanta has grown: Since 1990, the equivalent of everyone in Las Vegas has picked up and moved to Atlanta. The city has added 2.4 million residents since 1990.[25]

And how much new water has Atlanta added for those people?

Not a drop.

Odd as it may seem to compare the positively lush geography of Atlanta with the desert landscape of Las Vegas, in some important ways involving water, Atlanta is very much like Las Vegas.

Atlanta, too, gets most of its water—75 percent—from a single source, Lake Lanier, a reservoir created by a dam, just like Lake Mead.

Lake Lanier, like Lake Mead, is managed by the federal government, and its water is committed to a wide variety of uses, from keeping a nuclear power plant in Alabama running to keeping alive a threatened species of mussel called the purple bankclimber, which lives down in the Florida Panhandle.[26]

So Atlanta, like Las Vegas, doesn't control its main source of water, or even control how much water it can have from its main source of water.

The differences between Atlanta and Las Vegas, in water terms, are equally important, but they have nothing to do with climate, or even with water. The real difference between Atlanta and Las Vegas is attitude.

There is no water czar—no Patricia Mulroy—in the Atlanta metro region. Indeed, there is no real water leadership. The Atlanta metro region includes 28 counties, 140 municipalities, and 52 separate water utilities.[27] Going back fifty years, water planning and water management in the area have consisted mostly of wishful thinking, rain dances, and litigation.

Atlanta sits on the Chattahoochee River, which begins in the Blue Ridge foothills a hundred miles northeast of the city, winds through Atlanta's western suburbs, and then flows south, forming the border between Georgia and Alabama and slicing through the Florida Panhandle, where it becomes the Apalachicola River and flushes into Apalachicola Bay.

The Chattahoochee's reassuring presence—along with an average of 50 inches of rain a year, ten times what Las Vegas gets—has always made Atlanta rather blithe about its water supply. Atlanta takes 500 million gallons a day from Lanier (more than Las Vegas uses total), and 3 million Atlanta area residents rely on it.[28]

Back in 1948, when Congress was getting ready to build Lake Lanier, it asked Atlanta to contribute to the construction cost, which would have secured the city permanent drinking water rights to the reservoir. Mayor William B. Hartsfield, a legendary figure after whom Atlanta's airport was named, wrote back rejecting the idea of helping to pay for the dam and reservoir. "In our case the benefit so far as water supply is only incidental and in case of a prolonged drought. The City of Atlanta has many sources of potential water supply in north Georgia. Certainly a city which is only one hundred miles below one of the greatest rainfall areas in the nation will never find itself in the position of a city like Los Angeles."[29]

Fifty years later, when that prolonged drought drained Lake Lanier almost to the point of disaster, Georgia officials were simultaneously imperious and baffled at their lack of leverage. With the lake's level dropping a foot a week in November 2007, Atlanta area officials couldn't understand why Lanier's federal managers were releasing 1.5 billion gallons a day of water through the dam, into the downstream channel of the Chattahoochee, and on to the Apalachicola. Each day, the feds were literally flushing away three times the water Atlanta needed.

The Chattahoochee and Apalachicola wind for 542 miles, and Atlanta often seems to behave as if the water sent downstream is a gift. In fact, the Army Corps of Engineers is legally required to maintain certain flows of water down the river in part to support the habitat in Florida for three endangered species—the purple bankclimber mussel, the fat three-ridge mussel, and the gulf sturgeon.

As Lanier's condition became perilous, Georgia's congressional delegation introduced legislation to suspend the federal Endangered Species Act. "This legislation will help to ensure that the Endangered Species Act doesn't turn the people of Georgia . . . into an endangered species themselves," said Representative Phil Gingrey.[30]

In fact, over the river system's four hundred miles below Atlanta, seven cities and counties—with 130,000 residents—along with nine power plants and five factories, all take their water from the flow of the Chattahoochee and the Apalachicola. And the river ultimately empties into a Florida bay that produces 10 percent of the U.S. mussel harvest each year.[31]

"Atlanta has based its growth on the idea that it could take whatever water it wanted whenever it wanted it, and that the downstream states would simply have to do with less," said Alabama governor Bob Riley. "They should not be able to shut off the water coming into a state."[32]

Remarkably, during the congressional hearings in 1952 about approving construction of Lanier, with Atlanta having opted out of helping foot the bill, a young congressman from Michigan named Gerald Ford precisely foresaw what would eventually happen, even though Atlanta was then smaller than Buffalo or Kansas City. Questioning an official from the Army Corps of Engineers, Ford asked, "Is it not conceivable in the future, though, when this particular project is completed, that the city of Atlanta will make demands on the Corps because of the needs of the community, when at the same time it will be for the best interests of the overall picture . . . to retain water in the reservoir?"[33]

Atlanta ultimately skated through the 2007–2008 drought with the help of stringent outdoor watering restrictions, and rain that came in 2008. It was the area's fourth drought since 1980, and although there was much talk of both conservation and fresh sources of water, once Lake Lanier refilled and the recession took hold, the urgency about water faded.

The backdrop to not just the drought, but to Atlanta's water future, was a series of federal lawsuits in which Alabama and Florida had challenged the amount of water Atlanta was taking from Lake Lanier. Although Mayor Hartsfield had dismissed the idea that the Atlanta area would come to depend on Lanier, by 1990, Atlanta was using hundreds of millions of gallons of water a day, and the states downstream sued, claiming that Atlanta's galloping use of Chattahoochee water was impinging on everyone else's use. Alabama and Florida made a much more dramatic claim, as well. They said Atlanta wasn't legally entitled to any Lanier water, because when the reservoir was originally constructed, supplying Atlanta with drinking water wasn't one of its purposes. Atlanta had been offered the chance to drink from the reservoir, and pay a few million dollars for the right, and had specifically declined. Only Congress could give Atlanta water from Lanier, and it would have to pass legislation to do so.

Those lawsuits ground through the federal court system for decades, accompanied by years of side negotiations among the governors of the three states. The lawsuits took so long, in fact, that while they were being litigated, Atlanta's population grew from 3 million to 5.5 million people. And then in July 2009, just four months after Atlanta's drought was formally declared over, came a blow that would make the drought seem minor.

U.S. District Court judge Paul Magnuson, in a ruling consolidating and resolving six separate cases involving eighteen litigants, agreed completely with Alabama and Florida. He declared that most of Atlanta's withdrawals from Lake Lanier, going back thirty years, had in fact been unauthorized and illegal.[34]

In a meticulous ninety-seven-page decision, Magnuson ordered the Corps to stop providing water to Atlanta, and ordered Atlanta's use of Chattahoochee water cut back to what had been permitted in 1976. That figure instantly reduced Atlanta's communities to surviving on sixty-six gallons of water per person per day for all purposes, less than half what they used during the driest period of the drought (and just a quarter what Las Vegans use).[35]

Magnuson offered a single-paragraph lecture on the water-planning failure of Atlanta and Georgia leaders. Zeroing in on Atlanta's growth and lack of effort to match it to the available water, Magnuson wrote:

> Too often, state, local and even national government actors do not consider the long-term consequences of their decisions. Local governments allow unchecked growth because it increases tax revenue, but these same governments do not sufficiently plan for the resources such unchecked growth will require. Nor do individual citizens consider frequently enough their consumption of

our scarce resources, absent a crisis such as that experienced in the [Apalachicola-Chattahoochee] basin in the last few years. The problems faced in the [Apalachicola-Chattahoochee] basin will continue to be repeated throughout this country, as the population grows and more undeveloped land is developed. Only by cooperating, planning, and conserving can we avoid the situations that gave rise to this litigation.[36]

And Magnuson took a subtle swipe at a certain confidence on the part of Georgia officials that, after twenty years, Atlanta couldn't possibly lose the court cases—because where would all those innocent people drink from?

"The Court recognizes that this is a draconian result," wrote Magnuson, but the law doesn't say, "Changes shall be made only upon the approval of Congress unless it is inconvenient to do so."[37]

He did offer Atlanta and Georgia one break. Acknowledging that Atlanta's communities "cannot suddenly end their reliance on the water," Magnuson gave Atlanta three years to either get Congress to approve Atlanta's use of Lanier water or negotiate an agreement with Florida and Alabama to share the water. The three-year clock started ticking the day he issued his decision, July 17, 2009.[38]

The ruling went off like a bomb in Georgia. Sam Williams, president of the Atlanta Metro Area Chamber of Commerce, said, "It would perhaps have a Katrina-size effect on the metro economy. We've got to make sure this gets solved before the ultimate deadline hits us."[39]

Georgia governor Perdue, while ordering an appeal, formed a task force to quickly study what Atlanta's options would be if the appeal failed, and the CEO of Coca-Cola's largest bottler cochaired the study.

Perdue—aware that Georgia's congressional delegation was badly outnumbered by the combined delegations of Florida and Alabama—vowed to negotiate a sharing agreement with his fellow governors that would end the dispute once and for all.

And Perdue signed a bill that moved rapidly through the Georgia legislature called the Water Stewardship Act of 2010, which he said would create "a statewide culture of conservation" and was also intended to impress Magnuson with Georgia's newfound seriousness.

Then something strange happened.

Perdue's emergency Water Contingency Planning Task Force produced a forty-two-page report. But the task force's main recommendation is the second sentence of those forty-two pages: "The Task Force evaluation reaffirms that Lake Lanier is by far the best water supply source for the metro region." One sentence later, the report says, "The Task Force does not foresee the ability of the metro region to meet the potential water shortfall in 2012, when Judge Magnuson's ruling could take effect, even with extremely aggressive mandated conservation."

In accepting the report, Governor Perdue said, "Lake Lanier is absolutely our best option." [40]

Facing a cutoff of water from Lake Lanier, which would produce "Katrina-like" devastation in Atlanta, the governor of Georgia and his eighty-member task force concluded that, really, the water from Lake Lanier is the best solution to their water problems after all.

As for Perdue's vow to end the problem by finally getting an agreement with the governors of Alabama and Florida, by the first-year anniversary of Judge Magnuson's order, there was no agreement and not much prospect for one. Georgia found itself without much negotiating leverage, since both Florida and Alabama had little to lose by using Magnuson's decision as a starting point for negotiations. [41]

The governors of all three states were scheduled to leave office in January 2011. There seemed almost no chance that the three would resolve one of the knottiest political problems facing their states in their last year in office. When three new governors take office in January 2011, Magnuson's thirty-six-month cushion will have been cut to eighteen months.

As for the Water Stewardship Act of 2010, Perdue signed it a year after Magnuson's ruling, on the shores of Lake Lanier. But by the standards of places that are taking water conservation seriously, Georgia's law is relatively modest. It does limit outdoor lawn watering across Georgia to the hours of 4 p.m. to 10 a.m., but exempts home vegetable gardens, athletic fields, even golf courses. The bill requires new apartment buildings to have a "submeter" for each apartment, not just a single meter for the complex, and further tightens low-flow shower and toilet standards dating back to 1992. The oddest part of the bill, given the effort to impress everyone with Georgia's new conservation attitude, is that many of its provisions won't be in force until July 2012, the month Judge Magnuson's order would take effect.[42]

Almost incredibly, a year after being delivered a devastating court ruling that could dramatically reduce Atlanta's access to Lake Lanier, the metro area's leaders were down to two strategies: to insist that they need water from Lake Lanier, and to appeal the judge's ruling. They had returned to the water strategy established by Mayor William Hartsfield: wishful thinking.

Governor Perdue's spokesman, Bert Brantley, summed it up: "I don't think anybody . . . believes the ruling will go into effect in 2012."[43]

As it happens, the folks in Atlanta and Georgia did at least have the foresight to consult an expert on water scarcity and water sharing: Patricia Mulroy. As the drought in Atlanta was ending, and before Magnuson's decision, several members of Governor Perdue's water negotiating team flew out to Las Vegas to consult with her. They went in December 2008, during the annual meeting of the Colorado River Water Users Association (CRWUA), so they could hear from officials from Colorado and California too. They wanted to know how Nevada, California, and Colorado successfully shared the Colorado River.

"We met in a suite in Caesars Palace," says Mulroy. "The people from Georgia were sitting on the couch, it was 'hear no evil,' 'see no evil,' 'speak no evil.' One of the women from Georgia said, 'We don't have a water problem. We have an endangered species problem.'

"This is the state's lead negotiator!

"I lost it," says Mulroy. "I said, at the end of the day, the Atlanta metropolitan water district has to be able to serve its population with less water. What caused that is irrelevant!"

Mulroy shakes her head exactly the same way she did when talking about the psychotherapist who thought his patients needed the sound of a babbling brook to solve their psychological problems.

"It was hilarious," she says. But not really very funny.

ALMOST EVERY COMMUNITY in the United States has water problems. The good news is, water problems can be solved, and the sooner we start thinking about them, the less expensive those solutions are. The bad news is, water problems can't be solved quickly, and when there's a water crisis, the quick solutions are expensive. Water requires thinking about the future not in sunny, optimistic terms but in frankly realistic terms.

Patricia Mulroy has been unafraid to look at the water future, and she's been unafraid to try to master it. She is not counting on rain in the Rockies to save Lake Mead. With Las Vegas's two water intake pipes in danger of being left dry by Lake Mead's continuing fall, Mulroy pushed to install a third intake, the so-called third straw.

The third straw is an extraordinary effort any way you look at it—in engineering terms, political terms, water terms. The third straw is designed to come up into the bottom of the lake, like a bathtub drain, as opposed to coming in from the side, as Las Vegas's first two intakes do. Coming in from the bottom means you have access to water almost no matter how low the lake falls; it also means you have a huge engineering problem, which is that you have to dig deep enough beneath the lake bottom, and build your pipe strong enough, to withstand the crushing weight of the water in the largest reservoir in the United States. The Italian company awarded the contract to build the third straw, Impregilo S.A., built the flood-control gates for Venice, and is helping to build a third set of locks for the Panama Canal. Just to go deep enough to start tunneling under Lake Mead, workers had to dig a staging hole six hundred feet down. The single pipe they ultimately install will be twenty feet in diameter, wider across than the fuselage of a Boeing 747, and be capable of supplying three times the amount of water Las Vegas currently uses. The machine that will dig the tunnel for that pipe will be lowered in pieces, six hundred feet into the working pit, and assembled in place.[44]

Mulroy is worried enough about the rate at which Lake Mead has fallen that work on the third straw goes on twenty-four hours a day, seven days a week. Lake Mead would only have to fall thirty-eight feet—that's two or three really dry years—for Intake 1 to become useless, leaving Las Vegas utterly dependent on Intake 2, which is another fifty feet down.

In an illustration of the speed with which water problems seem to be accelerating, the water staff in Las Vegas first started talking about the third straw in 2002, as Lake Mead's level started to drop. That year, Las

Vegas's second straw first came online—literally as the city was finishing Intake 2, it was embarking on its third.

Intake 3 will cost $700 million—$350 for every man, woman, and child in Las Vegas. Intake 2, and the related pumping stations, cost $109 million a decade ago. In fact, as part of getting approval for Intake 2, the residents of Las Vegas voted themselves a sales tax increase to pay for Intake 2 and subsequent major water projects. The tax is small—one-quarter of one percent. To pay a penny in water sales tax, you have to spend $4. But that dedicated water tax (which is also paid by all the visitors to Las Vegas) brings in between $35 million and $50 million a year. Not many of us pay a water tax every time we shop, even a little one.

It is remarkable that in the space of fifteen years, one city will have spent $809 million on just two pipes—that's $1,200 for every family in Las Vegas—and those two pipes, despite their cost and complexity, accomplish nothing in water-future terms. They don't help Las Vegas get the water it will need for its next million residents or the next 25,000 hotel rooms—they are just a two-pipe emergency backup system. It's like fire insurance on your house: It costs money—a lot over time—and if you have to use it, you're in trouble. But without it, you're in much worse trouble. Says Mulroy, "You can't afford not to build it."

Back in 1990, just after taking over as the head of Las Vegas's water system, Mulroy instinctively realized two things: that Las Vegas couldn't continue to use water the way it was if it wanted to grow, and that securing new water supplies would be both difficult and essential. The interesting thing about both those apparently banal insights is that doing anything about them requires not just insight but a vision for what the future might look like, and a kind of long-haul planning and long-haul determination that is mostly absent from public policy today, in water and everywhere else.

Las Vegas has filed for the unused groundwater supplies in the counties north of it, water that could prove essential to Las Vegas's future growth, and a water supply that would for the first time be independent of Lake Mead. Whether or not the city could take that water without hurting the land under which the water sits, the political process to decide whether Las Vegas even gets to try has moved at a glacial pace.

In the twenty-one years Mulroy has been water chief, use of water in Las Vegas has dropped by 108 gallons per person per day—Las Vegans aren't using 216 million gallons of water a day they would be without the dramatic conservation regime that has taken hold since 1989. As it happens, that's almost the total amount of water the city was using in 1989—Las Vegans today save almost as much water as they used when Mulroy got her water job. But those kinds of water-use habits change very, very slowly. In any given year, per capita water use in Las Vegas only fell four or five gallons.[45] You can't save a hundred gallons of water per person per day with posters and TV commercials urging people to take shorter showers and skip a car wash. You save that much water by thinking twenty years ahead and imagining what it would take to change how a whole community operates, how it thinks about itself—and by giving people time to think differently about themselves, their community, and their water.

That's what's missing from the conversation in Atlanta—where the most serious water conservation law limits the hours you can water the grass, unless of course you're watering important grass, like on a golf course. And it's missing from the water conversation in most places that have serious water problems—not how to get through the dry spell or the drought, but how to think differently about the available water and change how we use it.

What the officials in Las Vegas realize, and what even the residents have come to appreciate by imposing a daily water tax on themselves, is that the cost of water isn't what it takes to clean it and pump it to your house. The cost of water is, first, the cost of running out of water. What happens to an economy—in Las Vegas, in Atlanta, in Australia—built assuming a certain level of available water, when that water drops off dramatically?

And the cost of water is the cost to add the next gallon of supply—the effort and money, the time and political will required to find new water, secure it, and deliver it.

Neither of those costs is in your monthly water bill.

Patricia Mulroy, for all of her success both in helping Las Vegas change and helping Las Vegas grow, doesn't sound optimistic. After she has spent twenty-one years trying to give Las Vegas a sense of both water appreciation and water security, the precipitous drop in Lake Mead makes the city seem less water secure than ever. "Are things kind of a mess now, again?" says Mulroy wryly. "Yes. Maybe I should have retired in 2000"—when Lake Mead was full.

Mulroy is worried about the impact of climate change on water supplies and water infrastructure. She's helped form a climate-change alliance composed of some of the largest water utilities in the country.[46] She's worried about micropollutants—the residue from birth control pills, pharmaceuticals, pesticides—that seem to be slipping into drinking water supplies, including her own. "Does it need to be dealt with?" she says. "Absolutely."

She's worried about attitude. "Future generations are not going to be living the same way we've lived," she says. "They're not going to have the luxury of the abundance of supply per person that we enjoy." In testimony to the U.S. Senate in 2009, Mulroy said, "We know that the way we've been managing water resources for the last hundred years is obsolete."

By that, Mulroy means that, as individuals, as companies, as communities, we use water in ways that are wasteful and shortsighted, that don't actually get real value out of the water available. That's something we can only change by paying more attention to how we use water. No one person in Las Vegas, no one company in Atlanta, is causing the water problems in those places—everyone together is.

But there's a much larger sense in which Mulroy thinks we've been mismanaging our water and will need to change. For Mulroy, water is just another resource—like electricity, or coal, or gold—and romanticizing water, or investing it with special status, doesn't help you get it, manage it, and use it.

Mulroy briefly considered, for instance, building her own desalination plant on the Pacific coast in either California or Mexico, and piping the water three hundred miles inland to Las Vegas. What dissuaded her? The electricity just to pump the water from the coast to Las Vegas would cost $400 million a year, more than the entire operating budget of the SNWA. Not that she's given up what seems like a wild idea, though. Mulroy has started talking to officials in both Mexico and California about building and operating a desalination plant, and trading the desal water to her hosts for a piece of their Lake Mead water.

Mulroy told President Obama he should consider a major public works water project, on the scale of Hoover Dam itself. She thinks the federal government should create a system of canals to capture, then divert Mississippi River floodwaters straight to the Midwest. When the Mississippi River floods, whole communities are devastated; the water they are devastated by is largely wasted. Mulroy isn't talking about diverting the Mississippi itself—she knows something about the politics of river water—just the floodwaters.

The result, in theory, could save the Mississippi River basin from periodic natural disasters, it could allow states east of the Rocky Mountains to access a new source of water, it could even allow the United States to replenish the desperately falling Ogallala Aquifer in the high plains. "One man's flood control is another man's water supply," says Mulroy. "You could capture 9 million to 14 million acre-feet of water a year." That's more than all the water Arizona, California, and Las Vegas take from the Colorado River.

"It's an engineer's idea," Mulroy says. "People are still stunned by it. But the interstate highway network, what did that cost?" And where would we be without it?

One of the things that makes Mulroy positively angry is a territorial protectiveness people have about water. Perhaps that's understandable, given that she and Las Vegas will never have any water that other people try to take. But she clearly thinks we have an immature attitude about how we manage water as a resource. The Great Lakes states, in the last five years, have passed a new agreement (that also includes Canada) forbidding any water from within the Great Lakes basin to be diverted in any way outside the basin. The compact was motivated in part by fear that dry western states would try to purchase water from Great Lakes communities and pipe the water west.

"We take gold, we take oil, we take uranium, we take natural gas from Texas to the rest of the country," says Mulroy. "We move oil from Alaska to Mexico. But they say, 'I will not give you one drop of water!'

"They've got 14 percent of the population of the United States, and 20 percent of the fresh water in the world—and no one can use it but them? 'I might not need it. But I'm not sharing it!'

"When did it become *their* water anyway? It's nuts!"

Imagine Texas passing a law that no oil drilled in Texas could leave the state.

Not, mind you, that Mulroy really thinks there would ever be a pipeline from Chicago to Las Vegas. She's girding for the process of getting permits to build a pipeline to bring groundwater from only two counties away, if she ever gets permission to take it. But she finds irrational the attitude that regards transcontinental oil and natural gas pipelines as routine, even essential, and transcontinental water pipelines as unthinkable, even sinful. "Nothing makes better cheap politics than water," she says.

Emerging from hours of conversation with Patricia Mulroy, you certainly end up with a heightened alertness to how water is being used and misused. And yet, right there on the Strip, in the midst of all the waterfalls and lagoons and fountains that Mulroy can both explain and justify, there appears to be a glaring irony. In the town that is paying golf courses to rip out acres of turf and specifying the hose nozzle you use to wash your car, the wide median along the city's main street is covered in vivid green grass. How in the world does Mulroy justify cultivating turf down the center of Las Vegas Boulevard?

It's actually a bit difficult to get out to the middle of the Strip—in most places, it's flanked by four lanes of traffic in each direction. But if you dodge the tour buses and rental cars to dash out to the grassy median, and reach down and stroke the turf, you'll find it soft, cool, and appealing to the touch. And then, if you grab it and pull, you'll find that the blades are stubbornly planted and a whole patch tugs up when you grab just a bit.

The grass in the median of Las Vegas Boulevard turns out to be like many of the other water features on the Strip: like the canals of Venice and the shark reefs of Mandalay Bay and the topless allure of the adults-only swimming pools. The grass is soft, beautiful, perfectly cultivated. And it's fake. Of course.

Las Vegas is a potent lesson in not taking water for granted, and also in not taking for granted that you understand the water you think you see.

The water that Las Vegas returns to Lake Mead—the purified wastewater that gets Las Vegas its critical return-flow credits—runs back to the reservoir in an old creek bed called the Las Vegas Wash. Right near the shore of Lake Mead, you can stand on the bank of the Las Vegas Wash and watch the water flowing back to the lake. The Las Vegas Wash has no natural flow. All its routine flow comes from the outfalls of the city's wastewater treatment plants (and from storm water when it rains). The Wash looks like any other creek at the edge of an urban area. It meanders through the landscape, its banks creating a greenway through the desert. About four feet deep, the Wash is often not more than fifteen or twenty feet across.

When you stand on its banks, what you're seeing is something re-markable: It's the real-time indoor water use of all of Las Vegas, the water that 2 million Las Vegans are using right now to shower, to cook, to wash clothes, to clean the bathtubs in the suites of the Bellagio. The water moves at a good clip—it turns out to be about a million gallons every eight minutes—but part of what's extraordinary about the current in the Wash is how utterly unremarkable it is.

In a city that prides itself on water wonders of all kinds, this quiet stream is in some ways the most wonderful of all. It's a display not of the new tricks we can think of for water to do, but of a new way of thinking about water. We can use it, we can clean it, we can return it, and we can reuse it. That's what the world itself has been doing since long before we arrived, and many water problems would get much easier to handle, or disappear altogether, if we could finally accept that lesson.

When you stand and watch the Fountains of Bellagio, it's hard not to be amazed at what people have managed to get water to do. When you stand and watch the current of the Las Vegas Wash flowing back to Lake Mead, it's a quiet reminder of what water can, eventually, get people to do.

4

Water Under Water

I'M STANDING IN THE DARK, at the bathroom sink in my Comfort Inn hotel room in Galveston, Texas, washing my hands. Or I hope I'm washing my hands. It's a Monday night, September 22, about thirty minutes past sunset, and the whole room is dark. The power has been out for ten days, since Hurricane Ike shredded Galveston, and flooded it, the previous weekend.

The beachfront Comfort Inn, mostly undamaged, is renting rooms because it has a generator, but the generator routinely cuts off, sometimes for a couple minutes, sometimes for hours.

It's my fourth day in Galveston, and I've spent it watching plumbers and electricians, city workers and navy sailors struggle to bring water service back to the city before its furious and impatient residents are allowed back to their homes at 6 a.m. on Wednesday. The city manager doesn't want to delay the reopening of the city to its own inhabitants even one more hour, and he also has no intention of allowing sixty thousand residents back if the water isn't turned on, if the toilets can't be flushed whenever necessary.

The race to get the water restored started before Ike had even finished lashing the island—it's been ten days of methodical but frantic effort, with just thirty-six hours to go. From the moment water service to Galveston mysteriously failed four hours before the eye passed over the island, nothing has gone as expected, and nothing has gone smoothly.

I'm in the bathroom in the dark, washing away the day's grime. I'm holding a flashlight in my mouth, and thinking about what it means to wash your hands with water that isn't safe to drink. Hurricane Ike has obliterated Galveston's ability to provide water. Two of the three main water-pumping stations were crippled, as was the city's main sewage treatment plant. A small building containing the electrical controls for the treatment plant has a sludge line on the walls four feet up, right across the fuse boxes and electrical circuit boxes, indicating that for hours the building and its vital motor and pump controls were awash inside with a mix of seawater and sewage.

Galveston's water pipes, the ones in the ground, are mostly undamaged—except for outlying beachfront areas, beyond the protective seawall, where the water pipes have in many cases been scoured away.

Bringing all this back to life—the pumps, the sewage treatment plant, the pipes—is turning out to require more than just money or manpower or a sense of urgency. It's requiring an exhausting level of persistence and ingenuity.

One of Galveston's inland pumping stations is undamaged, and it is running at partial capacity using patched-in electrical generators so big they come on 18-wheelers. That means that even though the city's water system—inbound and outbound, faucets and drains, clean and dirty—is basically a dead fish, a narrow strip along the beachfront is receiving emergency water pressure. The Comfort Inn is inside that emergency zone.

The limited water available is officially non-potable. No one knows how safe it is, or what's in it, no one has time to treat it or test it, and the official word to the hundreds of government staff and recovery workers double-bunking in a few beachfront hotels is not to drink the water, or even brush their teeth with it. There is a "boil water" order out for those places receiving city water. The notice advises that the water should be heated to a "rapid boil" for four minutes. Straight from the tap, it is considered unsafe even for pets to drink.

This is the water I'm using to wash with, ice cold, by the light of a flashlight. One place I've been this Monday is a spot few humans ever get to stand: the bottom of a wastewater treatment tank normally filled with about half a million gallons of sewage, a cylindrical concrete pool ten feet deep and ninety feet across. The bottom was still clogged with the kinds of debris you'd expect, from both the sewage treatment plant and the hurricane. Most vividly, the debris pile was dotted with dozens of condoms, bright red and green.

What does it mean to soap up with dirty water? My dog routinely slurps from thoroughly ugly puddles. What does it mean to wash your hands with water that is officially unfit for a dog?

Once you start wondering, the water seems to hum quietly with peril, with invisible infectious possibility. A few hours earlier, I walked the bottom of a sewage treatment tank—but now I'm thinking, Is washing with this tap water making me cleaner or dirtier? I have been showering every morning, forcing myself beneath the icy spray, and wondering whether the water that splashed on my face, into my eyes, in my mouth was dangerous.

It isn't often that a good-size American city has both its water system and its sewage treatment facilities completely obliterated in a single swipe—it isn't often, that is, that you get to see what a modern American city with no routine water service feels like.

People routinely make do without electricity; we improvise around having no working refrigerator, or microwave, or traffic lights.

But without routine water service, it is hard to imagine civilization proceeding. Water is basic. It may be the most fundamental need beyond air, the one thing without which we cannot make it through a single day. And in the modern developed world, most people have no independent source of water, no simple, safe alternative that is the equivalent of a flashlight or a camp stove, a home generator or a cooler packed with ice. What's more, in a devastated city—particularly a city devastated by water—the first thing people want to do is clean up, and even basic cleaning up requires clean water. The average American uses 99 gallons of water a day at home; that's 750 half-liter bottles of water, the most common size in which we buy our indispensable Poland Spring and Evian. Seven hundred and fifty half-liter bottles per day, per person. In a crisis, even in a pinch, bottled water will not save us.

Like so much of modern life, safe, reliable water and sewer service is both essential and a complete mystery. We have no idea where our water comes from, we have no idea what happens to it when the dishwasher is done with it. We have no idea what effort is required to get the water to us, and no idea what's required to get rid of it.

That ignorance doesn't matter, until things start to go wrong.

In Galveston, Hurricane Ike literally washed away the Hooters restaurant built on pilings over the Gulf of Mexico—when dawn broke on Saturday morning, it was gone, as if it had never been there. Ike's storm surge filled the first floor of the University of Texas Medical Branch hospital, rendering the ER and the operating rooms of one of the nation's premier level 1 trauma centers useless for months. The city of Galveston decided civilization would not, in fact, proceed without water service.

As Hurricane Ike whirled toward the coast of Texas in early September 2008, it had the dimensions of a monster storm with a monster's personality.

But Hurricane Ike didn't get much attention. September 2008 was one of the most momentous months in modern U.S. history, and unless Ike was staring you in the face, you were paying attention to something else.

September 2008 was the month that the U.S. government let Lehman Brothers fail, and it was the month in which the sub-prime mortgage crisis nearly unraveled the world's economy. The historic, unprecedented events came day after day, one more astonishing than the previous. The arrival of a hurricane on the Gulf Coast couldn't compete with the financial tornadoes that seemed to be flattening one institution after another. Insurance giant AIG was taken over by the federal government. Mortgage giants Fannie Mae and Freddie Mac were taken over by the federal government. Wachovia was rescued by Citibank, and was then re-rescued by Wells Fargo. The Bush administration proposed the first $700 billion bank bailout, and the U.S. House rejected it. Washington Mutual failed—the largest bank failure in U.S. history—and was seized by the FDIC.

And, of course, as Hurricane Ike was gathering itself into what would become a category 4 storm, Sarah Palin made her national political debut with a speech at the Republican convention in St. Paul, Minnesota (and it was revealed that her daughter Bristol was pregnant).[1]

In a quieter moment, Ike's swath of destruction would have lodged itself in American consciousness the way Hurricane Andrew did, if not Katrina. Ike is one of the largest Atlantic Ocean hurricanes, in terms of diameter, ever. Hurricane-force winds stretched across 240 miles, the distance from Orlando to Miami. Sustained tropical storm winds—above 39 mph—spread 550 miles, making Ike a brutal giant whose winds would easily have filled the entire space from Atlanta to Detroit.[2]

Approaching the Texas coast the second week in September, Ike completely filled the Gulf of Mexico, and it inspired one of the most sharply worded warnings ever issued by the National Weather Service.

On Thursday, September 11, at eleven-thirty in the morning, twenty-four hours before Ike would make Galveston a difficult place to find shelter, the NWS issued an unvarnished "leave or die" hurricane warning for Galveston and the nearby coastal communities.

"Persons not heeding evacuation orders in single-family one- or two-story homes will face certain death."

That's as blunt as a government weather warning gets.[3]

Approaching hurricanes do make you think about water—about driving rain and flooding, about what might happen to the inside of your home or business if a window is shattered or a door gives way, if the roof peels off or the storm surge starts pouring in over the windowsills. And people know hurricanes will bring down the electricity—they expect it. But as with so many other things involving water, ordinary people don't even think about their water supply during a hurricane. We just expect those invisible water mains to keep working.

Eric Wilson wasn't underestimating Ike. Wilson was then Galveston's director of utilities—water and sewer were his main responsibilities—and like almost all of Galveston's officials, he rode out the height of Hurricane Ike in what seems like an unlikely spot, the posh San Luis Resort, right on the beachfront, with its rooms broadside to the beach and to the storm's wind. The San Luis, ten stories tall, is one of the safest places on Galveston Island in a storm. It is built in part on the remains of a retired U.S. military installation, Fort Crockett, and the belowground floors are elegantly appointed conference space that also happen to be a windowless concrete bunker, protected behind old gun emplacements.[4]

Wilson and the Galveston water staff prepared for the hurricane's assault in ways both routine and unexpected. All utility vehicles were filled with fuel. Water service was valved off to the entire western end of Galveston Island, where homes and businesses sit right on the beach, and water mains are buried in three feet of barrier-island sand, unprotected by the city's famous seventeen-foot-tall seawall. Beachside breaks could suck the pressure from the city's mains during the storm. "I didn't want to bleed to death," Wilson says. He had a big electrical generator staged on the mainland, to provide power to run water pumps if they needed it.

And although all the city's critical water valves had been pinned down with GPS coordinates, city workers went out before the storm and added a nondigital backup. They pounded metal street sign poles into the ground next to thirty-nine critical water valves. "We drove the poles four feet into the ground," says Wilson. "So we wouldn't lose the poles, or the valves."

As Ike's northern edge was coming ashore on Friday afternoon, Wilson drove the beachfront road for one last check before heading to the San Luis hotel. "I could barely get out there," he says. "Water was already crossing the road. Waves were breaking on the truck."

Hundreds of staff and recovery workers spent Friday night at the San Luis. The power failed at about 9:30 p.m., as did the water service. The eye of Ike passed directly over the resort sometime after midnight. Wilson was sharing a room with his boss, deputy city manager Brandon Wade. "We got used to the noise, and I fell asleep. I woke up to silence sometime after midnight. I remember thinking, Am I dead, or am I in the eye?"

As dawn broke Saturday morning, the winds across Galveston were still hurricane force or better, and the scenes of waterborne destruction and chaos stretched from the Interstate 45 causeway connecting the island to the mainland—with both north- and southbound lanes completely blocked by boats lying on their sides—to the far western end of the island, where whole blocks of houses were simply gone.

Wilson was impatient to get out of the hotel Saturday morning and figure out what had happened to his water system. He was surprised that the water service to the San Luis had failed. Two of the three city water-pumping stations had backup engines powered by natural gas. Those natural gas engines—they look like ungainly motors pulled from 1965 farm tractors, and they roar like the sound of a hurricane itself—should have kept the water running right through the storm.

So at 9 a.m., with Hurricane Ike's backside still slashing across Galveston at 70 to 80 mph (just stick your hand out the car window on the interstate sometime to see what a 70 mph wind feels like), Wilson stepped out into the storm. With one of his deputies, Dennis Stark, he walked out the rear entrance of the San Luis and climbed into a City of Galveston dump truck to visit whatever water facilities they could get to.

Wilson, who has a pilot's license, took the wheel. His first goal was to get to the city's hundred-year-old 30th Street Pump Station. It's a handsome red-brick building with the proportions and styling of a nineteenth-century courthouse. It also has the distinction of being the nation's second-oldest working water pump station. It would probably be the oldest, but it had to be rebuilt after Galveston's catastrophic 1900 hurricane. The white granite cornerstone at 30th Street is inscribed "DESTROYED Sept 8 1900. REBUILT 1904."

Thirtieth Street Station is about three miles from the San Luis, and Wilson drove slowly. "When we were coming down 30th Street, the water was four, four and a half feet deep," he says. Imagine the condition your own home would be in if the street out front were cresting with four and a half feet of water. "A Cadillac was floating along, and as we drove past, the wake from the dump truck sent it floating into the side of an apartment building."

What Wilson and Stark found when they climbed the stairs to 30th Street was discouraging. Except for the sound of the wind howling outside, the cathedral-like main room was silent. No motors running. The room is dominated by two large motor pits, each the size of a backyard swimming pool, with tall putty-colored walls to protect the four motors and pumps from flooding. Each pit had three or four feet of water in the bottom. Two small motors were submerged; two large motors crowned above the surface, as if floating. The natural gas backup motors were mounted higher, and were mostly dry, but the heavy-duty truck batteries used to start them were covered with seawater.

Wilson wasn't immediately concerned with 30th Street's regular

motors—four high-speed electric motors sitting in pools of stagnant seawater aren't going to be turning anytime soon. Nor did he stop to wonder
how thousands of gallons of storm water had gotten inside the high-walled
pits, in the otherwise sturdy, dry 30th Street Station.[5]

Wilson scrambled staff by radio to pull batteries out of city equipment
to replace the waterlogged batteries, and to change out the contaminated
oil in the natural gas motor. "I just thought, We gotta get these things running again."

That Saturday, as he and city staffers worked to get the gas engines
in 30th Street Station online, Wilson didn't know how big the city's water
problems were, what it would take to get them solved, how long even temporary patches would take. But Saturday provided a flavor.

Eric Wilson is unflappable. He was forty-three when Ike arrived, and
he knows water systems from the ground up—he got his start walking
from house to house, reading water meters. He has tightly cropped hair,
and his expression defaults to an open-faced smile, making him seem
cheerful almost all the time. Up close, you can see some crow's-feet that
speak of a low-key seriousness and a steamroller determination. Wilson
has a sense of humor about the bureaucracy, a wryness that includes an
appreciation for the fact that he *is* the bureaucracy. He's a large man, his
weight ranging between 250 and 300 pounds.

Eric Wilson's core utility operating philosophy is, in his words, "You
gotta have a Plan B." Which is his way of saying, to himself and everyone
else, We're in a no-excuses business. Things will go wrong, and I have to
have a plan for how to keep the water flowing even when they go wrong.
His Plan B typically has a Plan B of its own—Wilson thinks three or four
failures out beyond Murphy's Law. On a particularly trying day, asked how
it's going, he will chuckle and say, "I'm on Plan Z."

On Saturday, the Plan B—two backup motors, already installed, with
a power supply separate from the electrical grid—was all wet before Wilson could even turn the switch. He got the new batteries installed, he got
the electronic controls fixed, he got the dirty oil swapped out, and in early
afternoon, they started trying to fire up the gas motors. Galveston had been
completely without water for fifteen hours.

"They just wouldn't fire," says Wilson. "Finally I said, Let's break a gas connection and see if we have gas. That's when we came to the stark realization that somebody had turned off the gas supply."

So much for Plan C.

AS MUCH DRAMA AS WATER CREATES—crashing ocean waves, towering waterfalls, great rivers, monumental dams, floods, blizzards, hurricanes— there is nothing quite so lackluster, even disappointing, as a modern water pump station. The Roman aqueduct system inspires awe, and tourism, two thousand years after its creation; no one will be visiting twentieth-century municipal pump houses in a hundred years.

The Airport Pump Station of the city of Galveston is a perfect example of the modern state of the art. It's an austere, well-lit room, about the size of a suburban home's garage. On one long wall, four fat pipes emerge at about head height. The pipes take an elbow-turn down, and exit into the concrete floor. The pipes are aqua. Each pipe, in its brief course through the room, has a motor and a pump built into its length.

There are flow and pressure gauges with needles that are rock steady. There are electronic controls. Because the Airport Pump Station is relatively new, the room is quiet except for the low hum of modern motors.

There is no sign of water. There is no sound of water. There is not only no drama, there doesn't appear to be anything going on; if you wait around for fifteen minutes, you can imagine that nothing will happen in the room. Ever.

But the appearance is deceiving in at least two important ways. First, of course, this is exactly the kind of place where you do not want drama. You can stand looking at an electrical substation for hours and never see any electricity. If you do, it's bad news.

The utilitarian, even pedestrian, nature of most water facilities isn't great for creating public pride and appreciation, in the way the drama of the arched Roman aqueducts did.

But trying to understand a modern water system by looking at a key facility like a pump house is like trying to understand the Internet by looking at a rack of servers. You can gaze as long as you want at the servers, from any angle, but you will never get any inkling of either the workings of the Internet or its usefulness and charisma. But try tapping the Internet's usefulness and fun without the racks of servers.

Today's water facilities aren't dramatic; they aren't particularly pretty. But in the developed world, they are part of a system that is so seamlessly integrated into daily life that they have come to seem part of the landscape, almost organic. And, in fact, city water systems do substitute for natural systems—the pipes tame the river or the lake or the well, domesticate it, bring it right inside the house. The product is the same as it was a hundred years ago, or a thousand years ago. But we are completely removed from the source of the water and the work required to deliver it.

In some ways, the systems' very reliability undermines both public awareness and public support. The pipes are in the ground. The water comes from the pipes. Just like in the pump room, nothing ever happens. What's the problem?

In Galveston, the problem turned out to be a slow-motion catastrophe—the natural water system overwhelmed the man-made water system in a way that was both offhand and humbling. Quite simply, the tide came in, and it came in high, crested by waves. One railroad bridge connects Galveston to mainland Texas, and Galveston's water supply comes across the same bridge, in a thirty-six-inch pipe alongside the rails. The Burlington Northern Santa Fe railroad tracks—rails and ties—were lifted and shoved over five or six inches toward the water main by Ike's storm surge and the waves crashing atop it, leaving the rails wriggling across the bridge like so much overcooked pasta. The railroad bed is seventeen feet above the bay below.

If you don't count the water main itself—which was undamaged—Galveston has four major water facilities, and water completely over-whelmed three of them; it came within just six inches of overwhelming the fourth. What was left when Ike passed was mud and silence.

Galveston's 59th Street Pump Station is a two-story stucco building, long and narrow. Built in 1950, it has a sturdy, tropical air. The building's most distinctive feature takes a moment to dawn on you: There is no ac-cess on the first floor. The building's doors are all on the second floor—the door for people is accessible by a flight of metal stairs bolted to the outside wall, and there are service doors on one end, the size of barn doors, only useful if you have a tall ladder or a forklift. This design was made with water in mind. Galveston Bay is four-tenths of a mile from the stucco building, in two different directions.

Through the second-story door, you step into a long, narrow room that looks and feels like an engine room. You are on a balcony, and over the railing is a deep floor, several feet below ground level. Four big motors, with pumps attached, are lined up on the floor, surrounded by water pipes. The wall opposite the balcony is covered with electrical conduits and boxes; a set of steps leads down from the balcony into the motor pit.

What happened at 59th Street is as simple as it is astonishing: The building filled up with Ike's storm water. Faintly visible encircling the outside is the scum line Ike left behind. A scum line is, simply, a bathtub ring. It's the level to which the water rose, and where it lingered, leaving behind a line of dirt and debris when it started to drain away.

The scum line on the outside of Galveston's 59th Street Pump Station is nine feet off the ground—108 inches up. A typical U.S. doorway, for comparison, is 80 inches high. Ike's storm surge completely enveloped the little stucco pump house, and cascaded into the building from the second story, pouring in around both the people door and the service doors.

When you stand on the road and look at the pump station, it's hard to take in. Fifty-ninth Street became a new inlet of Galveston Bay that was, however briefly, at least ten feet deep. Inside the pump house, standing on the gallery overlooking the pumps, the water would have been pouring in across the balcony, around your feet, and over the edge into the pump bay below. It would have come in from the side, around the service bay doors. No one was inside 59th Street during Ike, but the experience would have been both terrifying and surreal. During the height of Ike, almost 100,000 gallons of water may have poured into 59th Street.[6]

Hurricane Ike quite simply killed 59th Street Pump Station. All four motors were dead; every foot of electrical wire was fried. And the repairs would begin in the hazy window-light that didn't reach too far into the motor pit, because the building's own electricity was also gone.

Eric Wilson didn't try to reach 59th Street that Saturday morning as Ike headed north. "I knew I was looking at needing a boat to get there," he says.

Galveston's wastewater treatment plant is completely different from the pump stations—an open expanse of grassy land, sitting right on Galveston Bay, with a dozen open-topped tanks of all shapes and sizes buried in the ground. In the tanks, a combination of aeration, gunk-eating bacteria, gravity, and a smattering of chemicals routinely turns raw incoming sewage into water clean enough to be released directly into Galveston Bay, at about 4,500 gallons a minute. Wastewater, for the record, is almost all water—at least to look at it. If you think about what leaves your own home from the drain and sewer pipes, well, most of it is soapy or dirty water. Just 1 percent of the waste arriving at Galveston's plant is solids.

But the wastewater treatment plant—while a completely different kind of place than a pump station—was in exactly the same condition as 59th Street and 30th Street. Seven or eight feet of water had washed across the place—the waves easily topped the plant's six-foot fence—leaving behind everything from uprooted trees to navigation lights from a boat. A mobile home used for office space was torn from its anchors and pitched on its side against the back fence. Tons of sand and mud lined the bottom of the treatment tanks. The motors, the pumps, the spidery skimmer arms, the electrical controls—all drowned. Even the happy bacteria that eat the waste had been washed away.

The best symbol of the wastewater treatment plant's condition was the simplest: If you needed to go to the bathroom, you had to use the Porta-Potty that had been delivered. Not even at the sewage treatment plant could you use the toilet.

The missing bacteria were a reminder that a wastewater treatment plant, any wastewater treatment plant, is both uniquely human and, like the water supply system, an imperfect effort to duplicate a natural system. Squirrels and seagulls, hippos and salmon all produce copious waste, but none have resorted to creating waste treatment facilities. And even people, despite the density of our civilization, have had a lot of faith in nature's recycling system until recently. Memphis, Tennessee, was the last major city in the United States without a wastewater treatment plant. Until 1975, three years after passage of the federal Clean Water Act, it simply piped its untreated sewage into the Mississippi River.[7] But even in a sophisticated wastewater treatment plant, most of the treatment is coming from the cili-ates and metazoans—from the bugs, that is, from nature.[8]

In fact, before the place could get going again—in addition to the pumping, the vacuuming, the stringing of fresh electrical circuits, the sheer physical labor of removing whole trees—Galveston would have to coax the bacteria back into action.

IN THE FALL OF 2007, metropolitan Atlanta came within eighty-one days of running out of water. The potential for catastrophe was staggering—Atlantans use about 500 million gallons of water a day, about 21 million gallons of water an hour. Even supplying one-tenth that amount of water on an emergency basis—just 15 gallons of water per person—would have required five thousand tanker trucks a day.[9] Leave aside where you could find a fleet like that—it's hard to imagine where you could fill them with water, since the city itself would in fact be dry.

In response to the crisis, the state banned almost all outdoor water use for Atlanta and its surrounding counties, and public officials implored residents to cut back use voluntarily. Still, cities and counties routinely overshot their conservation targets, using between 10 and 20 percent more water than they pledged.

In fact, city, state, and federal officials had no plan for what to do if Atlanta actually did run dry. They were simply betting that it wouldn't. In the midst of the drought, Major Daren Payne, a senior official for the Army Corps of Engineers, which manages Lake Lanier, offered this reas-surance: "We're so far away from [running out], nobody's doing a contin-gency plan. Quite frankly, there's enough water left to last for months."[10]

On April 29, 2008, in the comparatively tiny town of Emlenton, Penn-sylvania, the nine hundred residents, all customers of the Emlenton Water Company, were issued a boil-water order by the state—they weren't to use their tap water for any potable purpose without boiling it first. The wa-ter coming to them from the Allegheny River, which runs alongside the town, wasn't adequately cleaned by the Emlenton Water Company—the Pennsylvania Department of Environmental Protection found intestinal parasites that the little utility's filtration process wasn't removing.

The utility, then owned by Jeff and Kathy Foley, who had purchased it from a previous couple a decade earlier, never did fix the problem. The state of Pennsylvania ultimately revoked the operating license of the owner of Emlenton Water Company, and forced its sale in order to get the water from the hundred-year-old plant properly treated. The boil order lasted until January 27, 2009—273 days of boiling water for the daily needs of cooking, washing fruits and vegetables, drinking, and toothbrushing, a period long enough to conceive and give birth to a full-term child.

And in January 2010, the city of Jackson, Mississippi, went without water for an entire week after a fourteen-day freezing spell fractured the city's water mains. The unrelenting cold spell froze the Yazoo clay on which Jackson sits, fracturing the pipes, and the pipes themselves also froze.

Except for the Northridge, California, earthquake in 1993, it may be an event unprecedented in modern U.S. water history. In the space of roughly eighty hours, Jackson suffered 154 water-main breaks. Jackson is 107 square miles, meaning there was on average more than one water-main break in every square mile of the city. Not only were the city's 175,000 residents put under a boil-water order, many of them literally had no water.[11]

What did no water mean in Jackson? Well, unlike Galveston, which was evacuated before Ike, everyone was home in Jackson.

Jackson's schools were closed for a week, as were three universities, including Jackson State. Jackson is the capital of Mississippi, and state offices were also closed for the week. Police headquarters was temporarily relocated.

"If we had ice on the ground, people would be much more understanding," Jackson mayor Harvey Johnson said. "We have a disaster. It's just not one you can see."[12]

Almost as remarkable as the scope of the water crisis in Jackson was the fact that it received literally no attention in the media. The *New York Times*, CNN, Fox News, NPR—none did a story on Jackson's crippled water system. Indeed, not one news outlet of any kind outside Mississippi did a story on the capital city of a U.S. state whose water system completely shut down for a week. (Jackson's water crisis started Monday, January 11; the devastating Haitian earthquake happened on Tuesday.)

It's easy to be sympathetic with what happened to Galveston's water system during Hurricane Ike, while silently shrugging. Our cities aren't, in fact, likely to be overwhelmed by a twelve-foot, hurricane-driven storm surge.[13] But while water system disasters, or even significant water system failures, remain remarkably rare, it's important to take a lesson from how they play out.

Atlanta's water crisis was really a function of twenty years of refusing to consider water as a limiting factor in growth. As Atlanta's reservoir fell by one foot per week in the midst of the drought, Roy Barnes, who had been Georgia's governor from 1999 to 2003, told reporters, "Los Angeles added 1 million people without increasing their water supply. And if Los Angeles can do it, I'll tell you, Georgia can."[14] Meanwhile, with crisis staring them in the face, Atlanta and Georgia officials didn't actually have a worst-case plan. They bet Atlanta wouldn't run out of water, and in the end, this time, it didn't. But what if the drought across the southern United States wasn't a one-time event, but prelude to a permanent shift in rainfall?

Over the course of nine months, the state of Pennsylvania couldn't manage to force a tiny water utility serving just nine hundred people to clean up its act. Pennsylvania literally had to wrest the utility from its owner and hand it off to another company to get the problem corrected.

And in Jackson, the problem wasn't a hurricane, it was a freeze—but the result was the same. The system collapsed.

Indeed, all four of these water failures—Galveston, Atlanta, Emlenton, Jackson—are the result of a man-made water system colliding with a natural system, and crumpling, or coming close. Especially in Atlanta and Emlenton, the problems were magnified by a refusal to see our relationship to water clearly, and by a refusal to pay attention to how the water supply itself is changing—in terms of climate change, and in terms of what's in our water.

Modern municipal water systems in the developed world are an unlikely combination: They are so reliable and so apparently robust they seem to have become almost part of the natural system itself, easily taken for granted in both their complexity and their cost. And they are oddly brittle—when something overwhelming happens, the water systems have no resilience. There was nothing to do in Jackson, Mississippi, but dig up every single one of 154 water-main breaks, jump down into the water- and mud-filled holes, and fix the pipes.

In Galveston, even Brandon Wade, then the deputy city manager and Eric Wilson's boss, was stunned by what happened to the waterworks.

In an interview nine months after the hurricane, he said, "I'll tell you that the most horrifying thing I heard during the entire storm was, as the sun was coming up and the storm was trying to pass, someone coming to tell me that we had no water pressure in the place that we were staying.

"It was shocking to me, and to be quite honest with you, I was briefly in denial. I was wondering what it was that the hotel had done to cause them to lose water, because I just really didn't expect that we would have lost water." [15]

It turned out that natural gas, the backup energy supply everyone had relied on—the city's newspaper, too, had a natural-gas-fired emergency generator—was the wrong choice. Along Galveston's waterfront, the Balinese Room, a famous nightclub that once hosted Frank Sinatra, along with the Hooters restaurant, had been washed away, leaving large natural gas lines torn open. The utility had no choice but to cut off service until it could close off individual gas leaks.

Wilson managed to get a diesel generator onto the island on Sunday, which allowed him to fire up two motors and pumps at the dry Airport Pump Station. That brought water back to the San Luis Resort on Sunday night, forty-eight hours after Ike knocked it out.

Five days later—a week after Ike arrived—you could drive the cleared streets of the city, you could shop at the oceanfront Kroger (which had its own water purification system—something the level 1 trauma hospital did not); you could shop at the oceanfront Wal-Mart, which had been flooded from front to back and sanitized from front to back before reopening; you could kick back with drinks, surrounded by recovery workers, at the San Luis Resort's pool bar, the H_2O lounge.

But Galveston's H_2O system was still a wreck, and it was, in fact, the main problem preventing city officials from reopening Galveston, which had literally been closed to residents since the hurricane, with checkpoints staffed by state troopers, allowing in only recovery workers. Impatient, angry residents had begun to sneak back to their festering homes using boats.

The pressure to get the water flowing again wasn't just intense, it was quite blunt. At an all-hands city staff meeting of about seventy-five people at the San Luis Resort on Saturday morning—one week to the hour from when Wilson headed out into the storm in his dump truck—city manager Steve LeBlanc turned to Wilson and said, "Eric, man, I need you to come through. Whatever parts you need to get those pump stations working, let's call the president. . . . We've got to make sure that our water and waste-water work when people come back. That is *the* central element.

"And you've got that message loud and clear, Eric."

It wasn't a question, actually. It was a decree.

As LeBlanc moved on to talk about an emergency no-burning ordinance, Wilson slipped out of the banquet room. Although he and the city utility staff of seventy had been working twenty-one-hour days for the previous week, tangible progress was mighty thin.

Thirtieth Street's backup motors—the natural gas ones—were running, helping supply limited water pressure to the seawall hotels and businesses.

And that was it. No sewage was being treated—whatever made it to the plant spilled through several settling tanks and out into the bay.

The eight drowned water pump motors—four at 30th Street, four at 59th Street—were all still bolted to their concrete pads, their shafts frozen in place. Thirtieth Street Station had been cleaned up; 59th Street had been pumped, but the motors still sat in a four-inch layer of pudding-like mud.

The residents would return in ninety-four hours, and Wilson would need every minute.

DESPITE THE TICKING CLOCK, despite the pressing needs of human biology, despite the list of major and minor-but-urgent things that need to get done—find motors, find connector parts, get old motors winched out, get new motors winched in, get new wiring, find electricity to make it all go, get a sewage pump truck waved through the checkpoint where an overzealous state trooper is holding it—there are few moments of high drama in the struggle to bring Galveston's water system back to life.

In fact, the really striking thing, hour after hour, is how almost nothing works right, no effort is rewarded the first time, but everyone simply pushes ahead.

One bit of good news is that right inside 30th Street Station are a pair of beautiful, brand-new electric motors with pumps attached. They are fire-engine red, and a crew from a construction company arrives in early afternoon to pick them up and trailer them to 59th Street, where it is hoped they can be installed as is.

The motors are tucked behind a locked chain-link fence, inside a storage area, inside the building. No one has a clue who has the key to the lock. This doesn't slow the guys from Boyer Construction down for a moment—they simply take a wrench and disassemble the hinges on the locked gate.

The gleaming motors do not come out to the loading dock easily, but after 45 minutes of sweaty, noisy, backbreaking work with crowbars, chains, and a forklift, the motors are on a trailer.

In what seems like a moment of some significance for the drowned city, the motors head off to the stucco blockhouse of 59th Street through the streets of Galveston in a slow parade of three vehicles. The Boyer Construction guys—they were on contract to Galveston when Ike struck, in the process of building a brand-new 30th Street Pump Station, just a block from the old one—back the trailer with the motors up to the side of 59th Street Station along a gravel driveway. They laid the driveway the previous day in the saturated mud around the pump station, to provide a place for heavy equipment. They used gravel from a nearby concrete company, which said they could take it, since it had been soaked in seawater and rendered useless for making concrete.

As soon as the motor-pump combos are sitting outside the building, something unfortunate becomes clear: The water pumps have extensions to allow them to connect to the water pipes they are pumping. The new pumps have flanges 34½ inches long. Fifty-ninth Street's pumps have flanges 46½ inches long. The new pumps are one foot of steel pipe from being able to connect to the pipes they need to pump.

A few phone calls. No pipe extensions that match can be quickly found.

This is exactly what happens, right along, hour after hour, at every waterworks location. A plan of action is tackled, an unexpected problem derails it, the plan is reconfigured, and work pushes forward.

No pump extensions? No matter. A crew sets to work unbolting the red motors from the red pumps. They won't replace the pumps—they'll just replace the motors.

Inside 59th Street, a second crew of men wades into the mud and starts unbolting two of the drowned motors from the pumps, and from their concrete mounting pads.

It takes the rest of the day to winch the old motors out—each weighs hundreds of pounds—and winch the new motors in, using chains and manual winches. The good news: The new motors have mounting holes that match the old motors perfectly.

The next day, Sunday, back at 30th Street Station, the navy arrives, without fanfare but with an unmistakable air of determined competence. A Humvee, a flatbed truck, and a nifty self-propelled crane wheel into the yard, along with a half-dozen Seabees. While Ike was still in the Gulf of Mexico, President Bush dispatched the USS *Nassau* from Virginia to be on station to provide recovery help as soon as the storm passed. The *Nassau* is an amphibious assault ship that looks like a small aircraft carrier and carries both attack jets and helicopters. It is anchored eight miles off Galveston—you can see it on the horizon—sending heavy equipment and a hundred sailors and Seabees to the beach each day to do whatever is most urgent.

Two brand-new motors—dull turquoise—have been delivered to the loading dock at 30th Street. The mission is for James Wooten, a Seabee mechanic, to use the X-Boom self-propelled crane to get the motors into place in the motor pits inside the big motor room.

The X-Boom has the nimbleness of a crane you might find in a video game. It sits on four oversize tires, and the whole body of the crane can tilt left and right, giving it a kind of joystick maneuverability.

Wooten gingerly lifts the first motor off the loading dock. The motor is going into a motor pit that is right through the front door of 30th Street Station. So Wooten rolls the X-Boom crane, with the motor swinging from the boom, around the front, lines up with the main door, and takes it right up the stately front stairs, rolling up one stair at a time like a creature from a *Bionicle* movie. The X-Boom's tires are taller than the balustrades, and

Wooten uses a sure, smooth touch. He has two inches of clearance on either side.

He needs a spotter to get the motor in, and over the wall of the motor pit, without the boom taking out the top of the doorframe. The metal handrail mounted in the concrete wall around the motor pit has been cut away in anticipation.

Once the motor is hanging over the pit, a small problem is evident. The motor is two feet to the right of the mounting pad it needs to go on. But Wooten can't jockey the boom sideways without destroying 30th Street's stairs and main entrance. Some consultation, some radio calls. The Galveston city staff summons an A-frame—simply a big metal stand with a winch—that can support the motor. Wooten lowers it to the floor. The city crew will use the A-frame to wrestle the big motor the last two feet into its proper spot.

The navy has offered another critical service. Motors and pumps don't connect to each other directly—they need custom-fit connector plates, which allow all kinds of motors to drive all kinds of pumps. The connector plates for the new motors are different from those for the drowned motors—and the navy has sent ashore two machinists to take measurements, helicopter back to the *Nassau*, cut the plates in the big ship's machine shop, and chopper the connectors back to 30th Street.

About a dozen sailors spend the afternoon working alongside city crews to get the motors in place and hooked to the electrical supply. On the *Nassau*, the machinists cut the new connectors. At 59th Street, meanwhile, the Boyer staff starts stringing new electrical service—starting with an enormous electrical panel drilled into a wall—so the new red motors can be fired up.

"Things are coming together quite nicely," says Wilson. "If we get power, we'll be in nice shape."

Indeed, the small motors Wooten lowered into place in 30th Street are hooked up (with the help of one of Galveston's traffic signal electricians), their connector plates fit perfectly, and they start water spinning into the system Tuesday morning, twenty-four hours before the people return.

They help, but even with the pumps at the Airport station, Wilson needs his big motors at both 30th Street and 59th Street to provide pressure when a whole city of people starts turning on faucets.

At the wastewater treatment plant, teams of men with squeegees, rakes, shovels, and huge vacuum hoses get the layers of black sand off the bottom of the tanks at the sewage treatment plant, and with twelve hours to go, it, too, is limping back into service.

But the big motors can't quite get going. With just ten hours to go before the residents return, two things go wrong at once at 59th Street. The generators brought in aren't big enough to start the two motors and keep them running. And as Wilson and the folks from Boyer Construction are trying to figure out why, they discover that the red motors installed there with such hope have slightly different specifications than the old ones. The horsepower is the same, but the motors have to turn faster to generate it. They'll need a special coupling to work at all.

Says a supervisor from Boyer, exhausted after many frustrating eighteen-hour days in a row, "These babies are definitely not plug and play."

And back at 30th Street, where the two smaller motors are pumping water, it turns out that the navy machinists have custom-cut the connectors for the big motors incorrectly. They don't match. A fresh set will have to be cut—but the *Nassau* has sailed for home port. Wilson will tap a shipyard on Galveston Island.

In the midst of the tumult, one of Wilson's phones rings (he's carrying three, plus a two-way radio). It's his big boss, city manager Steve LeBlanc. "He called to tell me that the sprinklers are suddenly running on the grass at a shopping mall." As the water system has slowly begun to come back to life, the city staff is discovering not just leaks, and sprinklers running that shouldn't be, but that many people who lost water service have left a tap or two open in their homes, so they'd know when the water came back on.

"Before I got into the business, I would have left the tap on too," says Wilson. He dispatches a city truck to valve off the shopping center sprinklers.

What he does not do is tell his boss that in just the last hour, several major problems have developed.

Plan W, at least. "We'll be fine," says Wilson. "It may be five a.m."— just an hour before residents stream back in—"but we'll be fine."

~~~

IT HASN'T TAKEN US LONG to get used to indoor, hot-and-cold running water, to water service we never have to think about. In 1940, 45 percent of the U.S. population lived in homes without complete indoor plumbing, and in 1950, more than one-third of U.S. homes still lacked indoor plumbing—including ten states where a stunning 60 percent or more of the homes didn't have it.[16] In both 1960 and 1970, the United States had a startling benchmark: During the ten years when the United States made it to the Moon, more homes had televisions than had complete indoor plumbing. In 1960, 83 percent of homes had plumbing, and 87 percent had TVs. In 1970, 93 percent of homes had plumbing, and 95 percent had TVs.[17]

What we take for granted isn't the water itself, of course, so much as the work, and the money, necessary to provide instant, safe water. It takes at least $29 billion a year in the United States just to keep up with the deteriorating water pipes and aging water treatment plants. The typical American family spends about $34 a month on its water utility bill—$408 a year. But the water system—the pipes, pumps, and treatment tanks—needs $260 per family, per year, in capital spending just to prevent things from corroding and aging into uselessness.[18]

And that doesn't count what it costs to improve the quality of drinking water and sewage treatment as scientists wrestle with the danger of micropollutants; it doesn't cover the cost of increasing demand for water; it doesn't account for the costs of grappling with water scarcity, which is often hugely expensive.

It costs $200 a foot to lay replacement water pipe under a four-lane road—that's $1 million a mile. Las Vegas is building a single new water intake pipe, just to make sure it can still draw water from the rapidly shrinking Lake Mead, that will cost $700 million. Five Australian cities are spending a total of $13 billion to build desalination plants so they can continue to provide drinking water to their residents in the face of disappearing rainfall.[19]

The cost of no water is hard to imagine or measure in advance, but crippling in real time. The state capital of Mississippi was brought to a standstill for a week because its water system shut down. Galveston, Texas, suffered all kinds of devastation from Hurricane Ike, but it was closed to its own citizens for eleven days. That's how long it took Eric Wilson and the city staff, FEMA, the navy, and private contractors—all working flat out—to get the water system patched together enough to let everyone back on, with the assurance that, while their homes might have been turned into festering messes, at least they could flush their toilets.

When people rolled onto Galveston Island that Wednesday morning, the century-old water pumping station at 30th Street had two of its four pumps running, with new motors installed and connected and tweaked.

The Airport Pump Station, undamaged except for its power supply, was running full tilt off an extra-large portable generator.

The pump station at 59th Street—where new motors couldn't quite connect to old pumps, where the generators on hand weren't quite big enough—remained silent as Galvestonians streamed back along Broadway, just a half-mile away. But by late Wednesday night, seventeen hours after the island reopened, Wilson and the men from Boyer had managed to get 59th Street online too.

Which was a good thing, because the next morning the power failed at 30th Street.

The long-term transformation of the Galveston water system to survive the next storm—almost incredibly, Hurricane Ike was only a category 2 when it washed over Galveston—is well into the planning, if not the building. The new 30th Street Pump Station, already under construction when Ike hit, was given an extra foot of protective wall in the wake of the hurricane. The Airport Pump Station is going to get a diesel generator, elevated well above potential storm surges, with its own fuel tank. No more reliance on natural gas for emergencies.

Fifty-ninth Street—the stucco building that was turned into a water tank instead of a water pump station—is still practically held together with bubble gum and baling wire. The electrical supply is zip-tied to the balcony railings. FEMA only funds one disaster repair for a facility; Wilson wants 59th Street torn down and completely replaced.

"I want to make it what I'm referring to as bulletproof," he says. "Built to the five-hundred-year storm elevation, and to survive 190-mile-an-hour wind." That would mean that the finished floor would be about where the roof is now. That one new pump station would cost $21 million—$350 for each resident of Galveston.

The sewage treatment plant, too, is functioning but is a total loss. The state of Texas has told Wilson any restoration of the plant has to comply with rules that went into force just two weeks before Ike struck—the plant itself was built in 1950. So the wastewater treatment plant will have to be rebuilt one section at a time, while still treating Galveston's dirty water— rebuilt to modern standards and to withstand another storm surge without being inundated.

And there are small things that aren't quite so small.

Virtually every fire hydrant in Galveston was marinated in seawater, and many are rusting in place. There are 1,400 on the island, and a new one costs $2,800 to buy and install. That's $4 million, just for fire hydrants. "We're caught in a FEMA endless loop on that one," says Wilson. Eighteen months after a storm that laid four to eight feet of water across the entire island—fire hydrants included—FEMA is insisting on certification from the fire department of the damage to each hydrant. Which means each one has to be inspected.

As for the water itself, Galveston gets its water from the Brazos River, on the mainland, supplied to it by a water utility called the Gulf Coast Water Authority, which cleans it and puts it in the pipe that runs across the railroad bridge to the city, where it gets a final polishing.

So the chances were that, once Wilson got water pressure back into the system thirty-six hours after Ike passed, anyone receiving that water, in- cluding both the San Luis Resort and the Comfort Inn, was getting clean water. There was a chance of contamination from leaking sewer pipes, there was a chance of contamination from the water mains themselves hav- ing been damaged somehow in the storm. And there was no way to know.

It took nine days after people got back to Galveston—the three-week anniversary of the storm—to do the testing necessary to lift the boil-water order. As to whether the water was safe before then, during the crash ef- fort to get the water flowing again, Wilson could only say, officially, that it wasn't.

"But I'm brushing my teeth with it."

5

The Money in the Pipes

When you start to think like we think, you don't see water in the
pipes. You see dollar signs.

—*Eric Berliner,*
IBM water manager in Burlington, Vermont

IN THE RANGELAND OF AUSTRALIA, sheep get frightfully dirty. They roam
the outback among all manner of plants, trees, and scrub, they loll in the
dirt, they sleep on the ground, they roll in their own poop. They only
shower if it happens to rain.

So when Australian sheep get sheared—and Australia is still the largest producer of wool in the world[1]—the fresh wool is grubby. A specialized industry exists to clean it before it can be woven into everything from
haute couture to bedding and carpeting. Wool scouring is as gritty and
demanding as the name suggests, and it is a water-intensive business.

Michell Wool has been washing wool in Australia since 1870, and the
big Michell Wool scouring plant in Salisbury, a suburb north of Adelaide,
uses about a megaliter of water a day—a million liters of water, about what
750 families use.

Just-sheared wool arrives strapped into heavy, bulging bales, chest-
high, bristling with grass, sticks, dirt, burrs. Raw wool is called greasy
wool, because in addition to dirt, the wool is coated with the sheep's natural protection, lanolin.

The machine that cleans the wool is a long line of connected stainless-steel tanks and conveyors, called a scour, which stretches more than half a football field. Wool gets unbaled and dumped into a tank at one end of the scour. It is washed in cold water, lightly agitated, wrung out, and moved up belts into successive tanks, pronged along gently to avoid damaging the fibers and ultimately washed in water that is 150°F, hot enough to dissolve off the lanolin.

The inside of the Michell scouring factory smells like a farm—a rich odor of sheep, dirt, the outdoors, and wet wool. The clean but uncarded wool itself feels like wet cotton. Michell uses a smartly counterintuitive washing technique. The cleanest water pours in at the end of the scouring process, onto the cleanest wool, and the water moves backward along the scour as the wool moves forward, so that the dirtiest water is used to rinse the dirtiest wool.

From January to July, high season, Michell runs the scours twenty-four hours a day, seven days a week. But wool is no longer as popular as it once was for apparel, and the Salisbury plant uses only a quarter of its floor space; each of the two scouring lines is running at about two-thirds of capacity. Michell wool remains well known—Armani used it in the spring 2009–2010 collection—and the firm remains not just one of the largest wool scourers in Australia but part of Australian lore. "Every ship that leaves Australia probably has wool from Michells on it," the *Sydney Sun-Herald* quoted an Australian wool official saying in 1989.[2]

It is possible to know exactly how dirty the wool coming into Michell is—weigh it before it's scoured, weigh it after. On average, the yield from raw wool is 55 percent—100 pounds of greasy wool yields 55 pounds of usable wool and 45 pounds of dirt, debris, poop, and lanolin. Greasy wool is almost half dirt. Hence the importance of the water—each pound of wool requires 3.6 gallons of wash water to get clean, almost twice what your home washing machine uses.[3]

It takes a lot of water to scour wool. And although Salisbury gets just eighteen inches of rain a year, less than Flagstaff, Arizona, and sits in South Australia, Australia's driest state, until just a few years ago, Michell was washing all its wool in the same water Salisburians were using to shower and cook—drinking water. When you see how dirty the wool is, when you see how dirty the water gets, it seems absurd to be washing greasy wool in tap water.

"Back in the 1980s, we were using in excess of a gigaliter of mains water a year"—a billion liters, three times what they use today—"and we asked ourselves, is that a sensible place to be?" says David Michell, comanaging director of Michell, a fifth-generation member of the founding family. "We're concerned about the environment, and also about making sure we don't price ourselves out of the market." Outside of farms, Michell is one of the largest single users of water in the state of South Australia, and the owners worried about a coming time when water scarcity became so serious that wool washing was competing with residential water use—in terms of price, or for adequate supply, or both. "If there is no water," says Michell, "there is no business for us. Water is a strategic issue. We started looking for a Plan B."

David Michell and his colleagues were feeling the first tickles of something most of us are utterly unfamiliar with: water insecurity. Just because the big supply pipe from utility SA Water was coming into the plant, just because Michell was spending A$1 million (one million Australian dollars) a year on water, didn't mean that in a serious drought its supply wouldn't be sharply limited, or the price wouldn't rise, or both. The tickle of water insecurity turned out to be almost scarily prescient.

SA Water, South Australia's statewide water utility, wasn't interested in providing Michell with any kind of alternative water supply. "We went to SA Water," says Michell, "and they said, 'Just keep buying water.'"

As it happened, the city of Salisbury was worrying about water, too, but from the other end: how to dispose of storm water runoff more effectively, storm water that it was collecting in drains and culverts and piping untreated into the ocean six miles west, along a sensitive stretch of coastal mangroves and sea grass.

And so the town of Salisbury started a kind of upstart water utility, and Michell Wool became its biggest customer. In the 1990s, Salisbury started collecting some of its storm water, and routing it for filtration into wetlands and reed beds—some natural, some created by the city. Salisbury then injected the water into a limestone aquifer that happens to sit directly below the city, creating a bubble of fresh water in the aquifer. That bubble is also a reservoir of reasonably clean water that's good for all kinds of purposes—watering ball fields, parks, and schools, irrigating commercial nurseries, even piping into toilets, and of course, washing wool.

"The basic idea with storm water, with urban runoff, is, Get it out of sight, get it away from my house and away from my foundation as fast as possible," says Steve Hains, who has been Salisbury's city manager for two decades. "We said, We've got the water in our pipes already—let's do something with it."

Salisbury now has fifty-three water-filtering wetlands, covering 740 acres, collecting about 8 gigaliters of water a year that it injects down into the aquifer. With no further treatment, it pumps 2 gigaliters a year of the water back out in purple pipes to customers who can use it instead of mains water from SA Water. Purple pipes have become the global standard for water that is not potable, but is clean enough for other routine use, everything from gardening, landscaping, and toilet flushing to wool scouring. The water supply pipes are purple so there is no confusing them with potable water supply pipes.

Michell Wool takes 15 percent of Salisbury's purple-pipe water, so the wool scouring now gets done not in drinking water but in cleaned-up rainwater that once would have polluted the Indian Ocean. Michell Wool started taking purple-pipe water in 2003, after investing A$1 million to jumpstart Salisbury's reclamation effort, and just as Australia's Big Dry drought began to take unrelenting hold. The drought has been especially deep and devastating in the area supplied by the Murray River, from which Adelaide's water comes, and from which Michell's water used to come.

Michell pays two-thirds less per gallon for purple-pipe water compared with what it would pay for tap water from SA Water. It has helped develop a second source of water, which, while it isn't climate independent (storm water runoff requires rainstorms), also isn't subject to the kinds of competition and politics that affect mains water. And in stepping away as one of the largest customers of SA Water just as Australia's drought has gotten progressively worse, Michell not only helped its own business but also took pressure off SA Water's crisis-level supply problems for the rest of its customers.

"Now our water is about one-third the price it would otherwise be," says Michell. "We got in early, we spent $1 million, and we helped Salisbury get going."

Just by worrying, by starting to think differently about its water needs and its water supply, Michell Wool and Salisbury have created a virtuous water cycle whose benefits seem astonishingly simple and self-reinforcing. Michell Wool has improved its competitive position by using a less costly version of its main resource, and it has increased its security by finding a new source of water (Salisbury's fresh-water "bubble" stored in the aquifer now equals about four years' supply for its purple-pipe customers). Michell Wool has also been inspired to examine all kinds of other processes involving its water. It uses about 40 percent less water per pound of wool scoured than it used to, and the company now reclaims heat from dirty wash water, using it to heat clean water, cutting the cost of making hot water almost in half.

The city of Salisbury has taken a waste product that it had to spend money to manage and dispose of—storm water—and turned it into an asset that brings in about A$1.6 million a year, a city "business unit" that manager Hains aims to grow substantially. Salisbury makes money cleaning up polluted water, and the mangroves of Barker Inlet on the Indian Ocean don't struggle to survive against urban runoff. The purple-pipe system reduces demand for a resource that has been so scarce in South Australia during the drought that SA Water is furiously building a desalination plant whose capacity was doubled even while its first phase was under construction.

Inside Michell's wool scouring plant, streams of water flow everywhere—clean water, muddy water, foaming water headed into drains, greasy water headed to have the lanolin skimmed from it. The storm-water project has had another remarkable value inside Michell and in the city of Salisbury—perhaps the most important long-term value. It's creating what Bruce Naumann, Salisbury's water manager, calls "fit-for-purpose water." You use water of a quality and a cleanliness that's good enough for the task at hand. In fact, Salisbury is home to a large residential development called Mawson Lakes, where every one of the 4,500 homes, and every business, has purple-pipe water, along with potable water. In Mawson Lakes, the purple-pipe water is a mix of Salisbury's filtered storm water and SA Water's treated wastewater, and it's used for toilets and for outdoor watering. The hose spigots mounted on people's outside walls in Mawson Lakes are bright purple. During the depth of the Australian drought, Mawson Lakes residents could still water their outdoor plants and gardens when no one else could—because they were using reuse water.

Upon reflection, it is absurd for drought-ravaged Australia to wash wool in drinking water. In fact, almost regardless of resources, it's crazy to use drinking water for things like watering soccer fields or flushing toilets. It's just what we've gotten used to.

Says Salisbury manager Hains, "We are changing the relationship between people and their water."

If there is one truly arresting sign that our relationship to water is about to shift in fundamental ways, it comes not from the world of science, or climatology, not from United Nations officials or the people who run water utilities or the aid workers desperately trying to get water to people in developing countries. No, the most unequivocal signal about what's happening with water comes from the people like those who run Michell Wool, in a quiet suburb north of Adelaide, it comes from Monsanto, the agri-conglomerate, and from Royal Caribbean, the cruise-ship company, it comes from Coca-Cola and Campbell Soup and Intel, from Levi Strauss and IBM, from GE and MGM Resorts.

They all have that same tickle of anxiety—in corporate terms—about water security that Michell Wool had a decade ago. For many companies, it's much more than a tickle. Companies that live in the world of water every day are worried about the quality of their water, the adequacy of their supply, the long-term security of their water, and they aren't waiting for someone else to alert them, or for someone else to tackle the problem. For business, water management is fast becoming a key strategic tool. Companies are starting to gather the kind of information that lets them measure not just their water use, and their water costs, but their water efficiency, their water productivity—how much work they get from a gallon of water, how much revenue, how much profit.

~~~~

IN DESIGNING ARIA, one of the signature hotel-casinos of its $8.5 billion Las Vegas development called CityCenter, the staff at MGM Resorts International went looking for a showerhead that was both low-flow—the goal was 2 gallons per minute or less, down from the typical 2.5 gallons per minute—and also provided the indulgent shower experience that guests in a luxury hotel want. Aria is a soaring, spacious, upscale hotel with 4,004 rooms, built in a desert community that gets just four inches of rain a year. That small change in one design element in a new hotel—how much water per minute the showerhead uses—can save millions of gallons of water a year.[4] But the folks at MGM Resorts couldn't find a showerhead that combined the flow they wanted, the shower experience their guests would insist on, and the design flair the Aria's high-end interiors required.

"I've heard it a thousand times," says Cindy Ortega, MGM Resorts senior vice president for energy and environmental services. " 'If I'm going to pay $400 a night, I should be luxuriating in the shower.' Yes, I've heard exactly that. Along with the plushness of the carpet and the fiber content of the pillows." CityCenter, as a development, aimed to build unusually environmentally sensitive buildings, and the goal of Aria, says Ortega, is, in part, "to dispel the notion that there is a trade-off between luxury and environmental impact. We thought we could get past the idea that luxury means big crown moldings and plush-plushy carpet."[5]

So, working with the Delta Faucet Company, through months of prototype showerheads tested in other MGM Resorts hotels like the Bellagio, and in the homes of MGM Resorts managers, Ortega and the Aria's staff designed their own showerhead. Their creation, when the hotel opened in December 2009, was bolder than anyone expected at the start: A square, flat-faced, mirrored showerhead with just four holes. And it had a flow rate not of 2 gallons per minute but 1.5 gallons per minute. The Aria showerhead could easily save the hotel, and the residents of Las Vegas, two thousand gallons of water an hour, twenty-four hours a day—enough water saved every day to supply all the needs of 140 homes in Las Vegas.[6]

In a completely different corner of the world, agricultural conglomerate Monsanto is doing exactly the same thing—trying to develop a new line of products for its customers that require less water. Monsanto is spending tens of millions of dollars a year developing drought-tolerant varieties of crops—plants whose genes have been tweaked so, biologically, they make better use of less water. Because as climate shifts, people in newly dry areas will still need to grow food. And because Monsanto sells $4 billion a year worth of seeds—it sells more seed than any other company in the world[7]—it wants to make sure it has products to sell to farmers, to nations, even when they're struggling through drought. In the summer of 2009, Monsanto opened a 155-acre water utilization learning center, a working farm in Gothenburg, Nebraska, that is an R&D facility to test drought-resistant crops in the dirt. If a crop typically takes twenty inches of irrigation water during a growing season, can you get the same harvest with eighteen inches of water? Or seventeen inches? And when in the growing cycle do you water less?

"We believe that by 2030, we can double the yield for many crops, compared to the year 2000," says Robert Fraley, the chief technology officer at Monsanto, who is helping drive the effort to find and test genes that do things like increase the efficiency of a plant's roots to absorb water, or protect sensitive parts of the plant from heat. "Literally, ten years ago, this didn't exist as an option, the possibility to confer drought or thermal protection. It's really exciting." Monsanto hopes to have drought-tolerant corn seed available in 2012.

Many of the biggest companies in the world are starting out simply trying to understand and manage their own water use, rather than producing water-related products.

Right from the Web sites of Coca-Cola, Intel, GE, and IBM, you can find out how much water each of those companies uses each year, often in stunning detail. Intel lists not only total water use but water use broken down by each of the company's manufacturing plants around the world, including what each factory's source of water is—the names of the rivers and aquifers Intel is tapping. You can easily figure that Intel isn't doing that well on its water goals, either in the big picture or based on water productivity. In 2009, the most recent year for which Intel has provided detailed numbers, the company used 19 percent more water than it did in 2005, but Intel's revenue actually fell 10 percent in that time, in part because of the recession.

So one gallon of water used by Intel in 2005 generated $5.74 in revenue and $1.29 in profit; in 2009, a gallon of water generated only $4.37 in revenue and 55 cents in profit. In terms of water, Intel's profitability fell 57 percent per gallon used. That's a measure you don't see very often, even on a Bloomberg terminal. Intel also reports renewed determination to hit its goal for water productivity—to reduce water per chip below 2007 levels by 2012.[8]

Coca-Cola, whose reputation has been doubly stung by controversy over its withdrawals of groundwater in India and by a backlash against its growing Dasani and VitaminWater bottled-water business, has vowed, in the words of CEO Muhtar Kent, that by 2020 Coke will become "the first major global corporation where we will be water neutral."

Almost all of Coke's products end up as pee—Coke's customers don't need more than a few hours to close the loop in the water cycle on the soft drinks and water they consume—and it's not quite clear what a "water neutral" Coca-Cola will look like. But the company is gathering, analyzing, and revealing cascades of water data.

According to its figures, Coke sells enough servings of drinks to indeed "buy the world a Coke." The company can buy every single person in the world more than fifty cans of Coke or Sprite a year. The company's global use of water is staggering—Coke uses enough water in its global operations each day to provide all the water necessary for an American metro area of 1.5 million people.[9]

Viewed from one perspective, in fact, Coke's business is really a water processing operation. As a company, Coke is dwarfed by IBM and GE, which together have nine times the revenue of Coke, and themselves use billions of gallons of water in industrial manufacturing. Yet Coke uses three times the volume of water used by IBM and GE together. The industrial giants use 11 ounces of water to generate $1 of revenue. Coke needs 333 ounces of water to generate $1 of revenue.[10]

Coke says that that every liter of beverage it manufactures and sells requires 2.43 liters of water—1 liter for the drink, and an additional 1.43 liters of manufacturing, cleaning, and process water. The good news is that unlike Intel, that represents a 9 percent improvement over 2004. Between 2004 and 2008, Coke cut the amount of process water per liter of drink by eight ounces. Sounds small, except when you multiply it by Coke's relentless popularity—the company serves up a drink to 67 million people *an hour.* Its improved water efficiency between 2004 and 2008 saved 8 billion gallons of water in 2008.[11]

It's easy to be charmed and hypnotized by all the details. But, in fact, the details are both beside the point—and the whole point. This kind of water reporting is amazing because it's totally voluntary, it's all new, and it is, quite literally, a window on the future. These companies aren't metering their water use with such precision to satisfy their curiosity or to amuse us. They're doing it because they want to use less water, because they think they may soon have no choice but to use less, because they've discovered that simply measuring water use quickly leads to managing it better, and because some of them see that the very act of measuring and managing water use is becoming a huge business in itself.

Even for companies, like Coke, that are utterly water dependent, thinking about water strategically, in detail, is new. In Coke's 2002 annual report—the so-called 10-K filing with the SEC required of all public companies, which Coke submitted in March 2003—there is a typical section on Coke's business operations. Under the heading "Raw Materials," the first sentence is: "The principal raw material used by our Company's business . . . is high-fructose corn syrup, a form of sugar." The word "water" does not appear in the "raw materials" explanation of Coke's business operations, as detailed in 2003, and the filing does not mention water supplies, water scarcity, water effluent, or water quality even once.

In Coke's 2009 annual 10-K filing, submitted in February 2010, the "Raw Materials" section begins this way: "Water is a main ingredient in substantially all our products. While historically we have not experienced significant water supply difficulties, water is a limited resource in many parts of the world and our Company recognizes water availability, quality and the sustainability of that natural resource for both our operations and also the communities where we operate as one of the key challenges facing our business."

Three pages deeper in the 2009 annual report, in the section titled "Risk Factors," Coke says that water faces "unprecedented challenges from overexploitation, increasing pollution, poor management and climate change. As demand for water continues to increase around the world, and as water becomes scarcer and the quality of available water deteriorates, our system may incur increasing production costs or face capacity constraints which could adversely affect our profitability or net operating revenues in the long run."[12]

In 2003, the main ingredient in Coke's products isn't mentioned. In 2010, water supplies around the world are one of the key challenges Coke faces, so much so that Coke is warning that access to water could impact its cost of doing business. Just to be clear: Water didn't change between 2003 and 2010—what changed was Coke's appreciation of water, Coke's understanding of water.

Coke is hardly alone. Although water supplies are critical in semiconductor manufacturing, in Intel's 2002 10-K filing, just as in Coke's, water was not mentioned as a risk to its business. By 2009, it was included.[13]

This water focus isn't trendy green consciousness, or corporate altruism, although in the case of Coke, it is vitally important PR. It's business. Any water that the showers in the luxury rooms in the Aria hotel don't spray on guests is water that MGM Resorts doesn't have to buy from Las Vegas, water it doesn't have to heat, then pump up to rooms, and then pay to have cleaned as wastewater. Similarly, the 8 billion gallons of water that Coke's bottling operations didn't use in 2008 was 8 billion gallons that Coke didn't have to buy or pump out of rivers or aquifers, clean to food-manufacturing standards, and then dispose of. Companies are realizing that the water bill includes the electric bill, the natural gas and heating oil bill, the chemical treatment and filtration bills. Reduce your water use 9 percent, and you reduce a cascade of costs alongside the water bill.

That is the sense in which, in this instance, business is actually ahead of politics, and ahead of popular awareness. When a company that cleans dirty wool in Australia, a company that hosts gamblers in Las Vegas, a company that makes microchips in Ireland and Israel, and the company that sells the most popular drinks in the world in more countries than belong to the UN—when four such different organizations, in such different geographies and lines of work, all agree that a major shift is under way in something as basic as our relationship to water, when they don't just agree, but change their behavior, that's something the rest of us should pay attention to.

In the last decade, business has discovered water as both a startling vulnerability and an opportunity to reduce costs and turn water itself into a business. No less a sage than Warren Buffett has quietly realized how the water landscape is changing. In 2009, Buffett's company, Berkshire Hathaway, became the largest shareholder in Nalco, a water services, treatment, and equipment company that has no public profile but twelve thousand employees and nearly $4 billion in revenue.

Sometimes the water consciousness percolating all over the world of commerce results in efforts that are less revealing than they are slightly silly. Levi Strauss, the apparel company, has been worrying about water for more than a decade. Since the 1990s, Levi has tried to get the outsourced global factories that make its jeans to treat their wastewater, in order to reduce pollution. In 2009, Levi released a painstaking "life cycle analysis" of the water involved in a single product—a pair of men's 501 medium-stonewashed jeans. Levi wanted to track water use during the life of a pair of blue jeans. Levi, which now only designs and markets jeans, outsourcing all the sewing, discovered that that single pair of blue jeans required an astonishing 919 gallons of water during its lifetime. Levi attributed 450 gallons of water (49 percent) to growing the 2.2 pounds of cotton the jeans required; it charged 416 gallons (45 percent) of water to our washing the jeans once we bought them, leaving just 53 gallons (6 percent) for which Levi itself was directly responsible. In a moment of humility, Levi assumed the jeans would last only two years (104 washes)—otherwise our water burden as customers could easily have been much greater. One of the company's great water-saving insights from this project is that the jeans would "use" a lot less water if we, the wearers, would replace our top-loading washing machines with more water-efficient front-loading ones.[14]

This kind of water analysis is so new, in fact, that a certain amount of silliness is inevitable, even desirable. No one has ever asked the kinds of questions that we are starting to ask about water and how we use it. Perfectly reasonable questions will sometimes bring silly answers; apparently silly questions will sometimes result in wonderful changes in perspective.

Cruise ships are fascinating water laboratories, because while floating on an unlimited cushion of water, they must be water self-sufficient, either tanking up on potable water in port or using fuel to run onboard desalination and purification systems. Every toilet flush, every cup of coffee, every shower, every ice cube, uses water that had to be ordered and accounted for. One of the great symbols of indulgence on cruise ships is the dining, and nothing captures the onboard culinary culture quite like their prodigious buffet lines, offering dozens and dozens of items, often available fourteen hours each day to provide anytime dining. Those buffet displays require literally tons of ice on each ship each day. For this ice, water has to be made or loaded onboard, ice makers have to run nonstop, ice beds must be laid out and replenished, and meltwater must be drained into the ship's water treatment system, where energy has to be used to clean it before it's released back into the ocean.

In 2008, the vice president of culinary operations for Royal Caribbean's high-end Celebrity ships, Jacques Van Staden, suggested substituting superchilled river rock for ice on the buffet lines. Van Staden had come to Celebrity from Las Vegas, and he thought in addition to saving water, the river rock would look better. It had a high-fashion flair—distinctive black rock instead of prosaic clear ice.

"This was the heyday of super-high fuel prices," says Scott Steenrod, director of food and beverage operations for Celebrity. "As Jacques and I talked about it, we knew this would save a lot of energy as well. We tried it on one ship. We knew immediately that [the rock] was equally effective at keeping the food chilled. And people liked it—it looks good."

In fact, testing showed that the smooth black river rock actually held cold longer than ice. Now, on all nine of Celebrity's megaships, the river rock has replaced ice for cold food on the main buffet line at breakfast, lunch, and dinner. Each ship has two sets of 1,500 pounds of rock—one set of rocks clean and being chilled, one set out on the buffet line. The rock is easily sanitized—kitchen staff take it from the buffet line and run it right through the standard dishwashers on sheet trays. The rock is chilled in belowdecks cold rooms that are already in use, and wheeled out to support the buffets as easily as ice.

Each Celebrity ship used to make 7,500 pounds of ice a day, just to support that one buffet line. So each of the nine ships is saving 2.7 million pounds of ice-making a year, ice that requires 330,000 gallons of water to be made, frozen, and then treated and pumped back overboard.

From one perspective, on ships using more than a million gallons of water each a week, the rocks-for-ice swap is trivial. It comes to saving about two gallons of water per passenger per cruise.[15]

On the other hand, it is a small stroke of genius. Royal Caribbean has eliminated a whole category of water use, reducing its costs while improving both the environment and the cruise experience the company is trying to offer. "We were able to turn off one ice machine completely on each ship," says Steenrod. "We literally put a sign on it that says, 'Not in Use.' It's off at the circuit breaker." And of course, on a cruise ship, every bit of electricity has to be generated by burning fuel, so unplugging an ice maker that used to run 24 hours a day saves real fuel and smokestack emissions, however modest.

More than that, the rocks-for-ice swap represents exactly the kind of mind-flip that a smart-water culture requires. Not just, How can we use less water? but, What are we using water for? Water, it turns out, has the capacity to inspire creativity about how we use it.

AT THE IBM MICROCHIP PLANT in Burlington, Vermont, the factory where they make the ultra-pure water necessary to produce semiconductors, the staff knows a lot about its water.

For the ultra-pure water—the exotic liquid that is so clean it isn't safe to drink, so clean it requires its own separate factory inside the microchip factory—the water staff measures eighty characteristics all the time, in real time. Just for a moment, see how many characteristics of water you could imagine measuring—temperature, flow rate, pressure, pH, clarity. Unless you're a chemist, good luck getting to ten.

But beyond the ultra-pure water system, IBM Burlington has created an internal nervous system to monitor and gather data about its water across the whole facility. The plant's pumps, tanks, and pipes are wired with five thousand electronic sensors, which each gather about 1 data point a second. The water staff at IBM Burlington gathers 400 million data points about the factory's water every single day.[16]

It's hard to know where you'd even begin to use that much data effectively. If you wanted to, you could just sit at your computer in IBM Burington and keep up in real time: you'd only need to observe and interpret a stream of 300,000 data points a minute. For comparison, the double-deck stock ticker streaming along the bottom of CNBC provides 52 data points a minute.

Eric Berliner, one of the water and environmental managers of IBM Burlington, is giving a tour of the Central Utilities Plant, the place where water is heated, chilled, pumped, and cleaned to the point that only microchips can drink it. The plant hums twenty-four hours a day with the sound of pumps moving fat pipes of water; it has the musty smell that comes from water and metal pipes in contact for years. Berliner stops in an alleyway deep inside the plant and nods toward the ceiling. It's hard to absorb the array of piping overhead. Perhaps six distinct layers of pipe, crossing over each other, some as big as a person's waist, some no bigger than your wrist. Many have labels—"Hot Water," "Chilled Water"—with arrows pointing in the direction the water is flowing.

"When you start to think like we think," Berliner says, his eyes tracing the pipes, "you don't see water in the pipes. You see dollar signs."

The water bill at IBM Burlington, just to get 3.2 million gallons a day into the plant, is $100,000 a month. And that's not the important cost. The water staff turns plain municipal water into a product—actually, into a portfolio of products, depending on whether someone is mixing high-tech chemicals or running air-conditioning chillers or supplying water fountains and coffeemakers. IBM's utility plant creates nine custom varieties of water—and that's where the real money goes, for chemicals, filters, energy for pumps and boilers and UV disinfection, for staff on duty 24 hours a day, 365 days a year. Each brand of water costs four or five or ten times the cost of the raw water itself.

A few years ago, Janette Bombardier, site operations manager in Burlington, and her staff had a revelation: Water is so important that although it seems far removed from the final product—the computer chips—it could actually be a competitive advantage. "We've moved from being a facility that makes chips for IBM products to a facility that makes chips directly for the consumer market. We make cell phone chips, we make chips for printers, for TVs, for cameras and GPS systems. We go head-to-head with other fabricators in the Far East."

We don't think about the cost of the water necessary to make our cordless phones or DVRs or Sony PlayStations. But you can bet that IBM and its chip competitors think about it—there is hardly a more relentlessly price-cutting arena than computer chips.

From Bombardier's perspective, if she and her staff can find ways to use less water, and to make water more smartly, she's directly reducing the cost of IBM's chips. Wringing expensive water out of the process helps the giant stay nimble.

"All the issues with water, with energy, with the increasing cost to produce water and move water," says Bombardier, "that's always inches from my nose."

The daily water bill at IBM Burlington, including energy and chemicals, is $10,959. Most of the water used each day—2.2 million gallons—becomes ultra-pure water, the most expensive kind. Of the $10,959 bill, $9,300 a day goes to make ultra-pure water. That's the big target for the water staff. Not much point in worrying about how much water the toilets use, when 85 percent of each day's cost is in the ultra-pure water.

That, in fact, is the first lesson from IBM Burlington. It's not about saving water per se—it's about understanding how you use water; where the costs are, and reducing them; where the value is, and preserving that.

In that sense, IBM Burlington's water factory is just like a Celebrity cruise-ship buffet line or the shower of a Las Vegas hotel. You still need the qualities that the water is providing—you want to rethink your use of ice without leaving the chicken salad lukewarm, you want to reduce the amount of water the hotel bathroom consumes while preserving an indulgent shower experience. But if in those examples Royal Caribbean and MGM Resorts are working with an inspired idea and good instincts about their customers, IBM Burlington is working from the analytics—from the billions of bits of information it gathers about water, sifting for patterns, trends, for bulges of wasted energy that aren't being harnessed. That, in fact, is part of IBM's business: teaching people to sift huge quantities of data for important insight, and then selling them the computers and the software to do it themselves.

In the ultra-pure water factory, though, as on the buffet line, it's the mind-flip about water that gets you started. You have to take a step back and look at the water cycle as a whole. "One of the most innovative things we've done," says Bombardier, "is we take the energy the water inherently has in it, and we use it for other purposes." Or, as her deputy Eric Berliner put it, Everywhere you see water flowing in pipes, think dollar signs.

Water comes into IBM Burlington cold from Lake Champlain and the Champlain Water District. It's so cold, in fact, that it has to be warmed up before they can turn it into ultra-pure water. Meanwhile, the factory has thirteen massive, two-story-tall chillers using huge quantities of electricity to produce cold water, even in winter.

If it seems stunningly obvious to connect these two problems—well, not really. There was coldness in the incoming water that for most of its fifty years, IBM Burlington wasn't quite smart enough to use—in fact, the coldness was undesirable; IBM spent money getting rid of it. In another part of the 750-acre campus, water had heat in it that was undesirable, and IBM spent money getting rid of that. In most companies, in most organizations, though, there wouldn't be much of a pipeline connecting the specialty department that creates ultra-pure water with the everyday engineering department that is running the air-conditioning systems.

What IBM Burlington's engineers have done isn't nearly as glamorous, or as comprehensible, as substituting cold river rock for ice. But it is, in fact, exactly the same concept. In a plant that already has something like eighteen plumbing systems—from steam to a segregated fire-sprinkler system—they've created three fresh loops of water, to capture cold and heat where it is and use it where it's needed. The cold incoming water, for instance, is routed to areas that need chilling. It provides "free" cold, and in the process, it gets warmed up, also "free," so it's ready to be ultrapurified.

IBM Burlington also now uses cold outside air—which is abundant in Burlington, where the average high in December, January, and February is never above freezing—to make cold water in winter, instead of using its big chillers.

All of this saves water, and it saves all the things water requires to do its jobs. These kinds of projects are daunting enough that IBM Burlington uses computer models to track water, temperature, and energy to make sure its ideas are going to work.

And the result? Between 2000 and 2009, IBM Burlington cut its water use 29 percent—that saved the factory $740,000 a year in water bills. But here's where the magic of water really kicks in. Cutting water use by $740,000 is saving $600,000 in chemical and filtration costs each year. It is saving $2.3 million in electricity and energy costs.

For every $1 that IBM Burlington cuts its basic water bill, it saves $4 more in chemical, electricity, and energy costs.

By the end of 2009, production of chips at IBM Burlington was up 30 percent compared with 2000. So over the course of nine years, the water staff had cut water use 29 percent, saving $3.6 million a year—while the facility was actually increasing its output by a third. The result: Between 2000 and 2009, "water productivity" at the plant very nearly doubled. A thousand gallons of water in 2009 produced 80 percent more chips than a thousand gallons of water in 2000.

So it isn't just that Janette Bombardier's team saved $3.6 million a year by being smarter about water. If they'd done nothing, increased production of chips would have actually raised the cost of water by perhaps $2 million in 2009. The real savings—out there in a world where even $4.99 birthday cards contain computer chips—the real savings is $5 million a year.[17]

"We did fifty things to get there," says Bombardier. "Angles of usage, treatment, energy capture, using less pump capacity, capturing internal pressure that comes with the water in the line—fifty different things."

As IBM has discovered, the measuring alone creates an imperative for curiosity and innovation, for changing behavior—just like when you keep track of every calorie you eat, you start cutting back, just like when there's a real-time miles-per-gallon number on a car's dashboard, you can't help but drive in such a way as to keep the mpg number high.

"We are never done," says Bombardier. "We are never out of ideas."

Water consciousness has a kind of infectious quality, an upward spiral in which better water management spins off all kinds of benefits that reinforce the original impulse to think about water. For IBM, the real inspiration from Burlington has been far more dramatic than simply saving water and money. Burlington has helped IBM change the way it thinks about itself. IBM, the computing company, is creating a whole business around water. IBM wants to do for its customers—for companies, for cities, for utilities, for whole natural ecosystems—what it has done in IBM Burlington.

IBM's leap seems bemusing on the face of it. Why would the world's legendary computer company go into the water business? The answer is really both simple and brilliant. In most places, in the United States and the rest of the world, water is not smart. Traffic signals have intelligence, highways have intelligence, the electric grid has intelligence, the cell phone network, the cable TV network, heck, even Wal-Mart's long-haul trucks are connected on an intelligent network. Water's network typically moves only water, not any information about the water. Even at the simplest level, for instance, most water meters are still read, not automatically but manually, with someone striding along and popping open your water-meter cover.

"Water is not really measured and monitored in a way that allows you to manage it," says Sharon Nunes, a vice president at IBM in charge of the company's Big Green Innovations effort. Her job is to create businesses for IBM out of the exploding world of sustainability. "We think there is a big business opportunity around managing water. Water is not disappearing. But as it becomes more scarce in more areas, it becomes critical to manage it better."

IBM, in fact, wants to do for water what Apple's iTunes has done for music. At the simplest level, iTunes is just what the corporate IT types would call a "dashboard" for managing your music. You can see what you've got, you can see what's out there, you can see how much it costs, you can see what you've bought, you can even see what other people are buying. iTunes is a music ecosystem—Apple doesn't know anything in particular about music, except how you might want to use it, display it, arrange it, analyze it. iTunes offers you a "smart music" system.

That's exactly what IBM wants to offer for water users. What IBM can do is lay down a nervous system of water sensors, feeding an array of computers, loaded with analytical software that lets you see and understand your water—whether you're running a microchip factory, as IBM does, or a sprawling university, or a sewage treatment plant, or trying to understand the hydrodynamics of a whole bay. IBM wants to offer a "dashboard" of water intelligence, a way of grasping your whole water ecosystem. That kind of intelligence has transformed the world of music—for anyone who listens to music, for music companies, and for the artists themselves. (One crucial difference, of course, is that iTunes is a closed system, valuable but hermetic; water is the original open-source system.)

IBM, in short, wants to usher in the era of what it calls "smart water." That's what it has created with its five thousand sensors and its 400 million data points a day in Burlington: smart water. Not just the kind of information that lets it use less water here and there, the kind of information that lets it take the qualities inherent in the water it is using, and shift those qualities around to where it needs them.

In March 2009, IBM formally announced the creation of a water management services business unit, along with a list of pilot customers and projects, including a sensor system to monitor Ireland's Galway Bay, a similar system to model and monitor New York's Hudson River, and a contract to create an "end to end" smart-water utility for the island nation of Malta.

The conventional estimate is that around the world, water is a $400 billion-a-year business—that's four times the size of IBM's annual revenue, but it includes everything from digging up worn-out water pipes to building billion-dollar desalination plants. IBM says the information technology part of water, the smart-water market, could be $15 billion or $20 billion a year.

For the moment, water seems to be inspiring not just a mind-flip at IBM but also a burst of creativity and cross-pollination that is a reminder of how spartan water technology really is, despite a hundred years of modern water systems. The century-long golden age of water has made the water world complacent. There aren't many areas of modern life in the developed world where thousands of staff people routinely maintain vital technology that is forty or fifty or even a hundred years old. But that's the standard in water. Even lightbulbs—evolving from incandescents to compact fluorescents—have made more progress than most water technology.

In the spring of 2010, IBM vice president Sharon Nunes announced a partnership with a Saudi Arabian research center to develop a new, inexpensive desalination system that could be powered by solar energy. In the Middle East, of course, where the whole region needs to manage its fresh water with an eyedropper, finding ways to use the sun to make cheap drinking water is a near obsession. What was remarkable about the IBM announcement is that the project relies on combining two unrelated areas of IBM's technology portfolio: microprocessor technology (in a new kind of solar panel) and nanotechnology (in a new kind of desalination filter), in the service of a third business—making clean water, a business IBM wasn't in just four years ago.[18]

If water is going to get smart, or more to the point, if we're going to get smart about water, that's the kind of convergence, the kind of cross-disciplinary leaps, that are going to be required.

IBM, in fact, seems to be betting that it can learn about the water business even while it is teaching its customers about their water. Most of its early water projects include partners with deep experience managing or understanding the water part of water systems. "There are very few water experts in IBM," says Cameron Brooks, who is part of Nunes's team in charge of building IBM's water business. "For the moment, we're trying to bring the capabilities we already have to this new area, to figure out how to make a difference."

IBM's favorite example of smart-water effectiveness, in fact, is its own Burlington semiconductor plant. "We bring the institutional expertise in how to do this," says Janette Bombardier. Her senior water system manager, Jeff Chapman, has been tapped dozens of times for sales presentations around the world, including as far afield as Singapore, to explain what it means to look at your pipes as if they have dollar signs flowing by. "Jeff is helping to create a strategy for the whole corporation," says Bombardier. And what is Chapman's hourly consulting rate to IBM's sales operation? Bombardier smiles. "I give him up for free, until there's a real contract signed."

THE LARGEST BOTTLED-WATER FACTORY in North America is located on the outskirts of Hollis, Maine. In the back of the plant stretches the staging area for finished product: 24 million bottles of Poland Spring water. As far as the eye can see, there are pallets double-stacked with half-pint bottles, half-liters, liters, Aquapods for school lunches, and 2.5-gallon jugs for the refrigerator.

Really, it is a lake of Poland Spring water, conveniently celled off in plastic, extending across six acres, eight feet high. A week ago, the lake was still underground; within five days, it will all be gone, to supermarkets and convenience stores across the northeastern United States, replaced by another lake's worth of bottles.

Looking at the piles of water, you can have only one thought: Americans sure are thirsty.

Bottled water has become the indispensable prop in our lives and our culture. It starts the day in lunch boxes; it goes to every meeting, lecture hall, and soccer match; it's in our cubicles at work; in the cup holder of the treadmill at the gym; and it's rattling around half-finished on the floor of every minivan in America. FIJI Water shows up on the ABC show *Brothers & Sisters*; Poland Spring cameos routinely on NBC's *The Office*. Every hotel room offers bottled water for sale, alongside the increasingly ignored ice bucket and drinking glasses. At Whole Foods, the emporium devoted to sustainable food, bottled water is the No. 2 item by units sold.[19]

Thirty years ago, bottled water barely existed as a business in the United States. In 2009, we spent $21 billion on bottled water, more on Poland Spring, FIJI Water, Evian, Aquafina, and Dasani than we spent buying iPhones, iPods, and all the music and apps we loaded on them.[20]

Indeed, if there can be said to be a "business of water" in the United States, a business with which we are both familiar and utterly at ease, it is the business of bottled water. For most Americans, bottled water is the one spot in their daily lives where water and commerce routinely intersect.

And bottled water certainly represents a water mind-flip. A generation of American adults raised on tap water and water fountains, we now drink a billion bottles of water a week, and we're raising a generation that views tap water with disdain and water fountains with suspicion. We now drink more gallons of bottled water than milk. Touchy about the cost of everything from the monthly cable TV bill to microwave ovens, we've nonetheless acclimated ourselves to paying good money—two or three or four times the cost of gasoline—for a product we have always gotten, and can still get, for free, from taps in our homes.

It would be easy to regard bottled water as a bemusing sidelight to the world of water. While the volumes of bottled water are staggering when considered as a consumer product—we drink eighteen times as much bottled water per person today as we did in 1976, seventeen bottles a month for every man, woman, and child in the country—the amount of actual water involved is trivial. In 2009, a total of 8.4 billion gallons of water was sold on store shelves in the United States. That's 27 gallons of water per person, per year. If you take one bath all year, you use almost twice that. In fact, U.S. water systems leak about 7 billion gallons of water a day—so the water pipes supplying our homes leak more drinking water in thirty hours than we buy at stores in a year.[21]

But bottled water isn't a curiosity. In some ways, it isn't just the water business we are most familiar with, bottled water is the water phenomenon of our times. The water business has remained stagnant for forty years, even as dramatic innovation has swept through almost every other arena you can point to, from the engines in our cars, to how we cook, to the "technology" embedded in our Kleenex. If you were to look for water innovations that people could point to, could appreciate, in the last forty years, well, the availability of chilled water from the island nation of Fiji in a sexy square bottle in the cooler of your corner 7-Eleven—that would be one of the standouts. Bottled water has certainly turned out to be more popular, and more prized, than, say, the dual-flush toilet.

Walk into the water aisle at any large suburban supermarket—an aisle that didn't even exist thirty years ago—and you'll find water from three or four continents, water from glaciers (they're melting anyway). You'll find water with added oxygen. IBM notwithstanding, you'll find bottles of SmartWater, with added electrolytes to "one-up Ma Nature." There's certainly been more creativity in bottled water in the last three decades than there has been from the makers of the lowly water fountain.

The problem is that bottled water is a wacky, funhouse-mirror version of the real world of water. Bottled water subtly corrodes our confidence in tap water, creating the illusion that bottled water is somehow safer, or better, or healthier. In fact, tap water is much more tightly regulated and monitored than bottled water.[22] The three largest brands of bottled water—Nestlé Pure Life, Coke's Dasani, and Pepsi's Aquafina, which together make up 20 percent of all bottled water—are nothing but municipal tap water, repurified and packaged up for our convenience. Bottled water undermines our financial and civic commitment to a reliable public water system. Why accept an increase in the water bill, why vote yes on a water-system bond issue, when you can always get your water at the supermarket? Indeed, bottled water offers a kind of spring-fed vision of water security, a vision that turns out to be a mirage the moment we need to depend on it. When one of the main aqueducts feeding Boston's water system failed in May 2010, 2 million people in the Boston metro area were ordered to boil their water before using it, not just to cook but even to wash their dishes. Supermarkets across Boston sold out of bottled water within hours after the boil-water order went into effect. And Boston's system didn't fail to deliver water—everyone still had water pressure for toilets and showers; the water just wasn't dependably clean enough to consume during the water-main break.[23]

Americans spent $21 billion on bottled water in 2009. It doesn't seem like an astonishing sum of money—about $65 per person, $1.25 a week. But in the context of water, $21 billion is huge.

Consider, for instance, what Americans spend for all the water delivered to their homes—350 gallons per family per day, 365 days a year. The water bill comes to about $412 a year. Which means we spend $46 billion a year on all the household water we use all year long—to run the morning shower, to boil the pasta, to water the lawn. As a nation, we spend $46 billion for a year's water, always on, whenever we need it. And we spend another $21 billion—almost half as much—for bottled water, for an amount of water that wouldn't get us through eight hours of water use at home on any given day.

But there's an even more arresting comparison. We spend about $29 billion a year maintaining our entire water system in the United States—the drinking water treatment plants, the pump stations, the pipes in the ground, the wastewater treatment plants.[24]

So as a nation, we spend very nearly as much on water delivered in small crushable plastic bottles as we do on sustaining the entire water system of the country.

When we buy a bottle of water, of course, what we're often buying is the bottle itself, as much as the water. We're buying the convenience—a bottle at the 7-Eleven isn't the same product as tap water, any more than a cup of coffee at Starbucks is the same as a cup of coffee from the coffee-maker on your kitchen counter. But we're also buying the artful story the water companies tell us about the water: where it comes from, how healthy it is, what it says about us.

Bottled water, in that sense, is often simply an indulgence. The problem is that it is not a benign indulgence. We're moving 1 billion bottles of water around a week in ships, trains, and trucks in the United States alone. That's a weekly convoy equivalent to 37,800 18-wheelers delivering water. (Water weighs 8.33 pounds a gallon. It's so heavy you can't fill an 18-wheeler with bottled water—you have to leave empty space.)

Meanwhile, of course, one out of six people in the world has no dependable, safe drinking water. The global economy has contrived to deny the most fundamental element of life to 1 billion people, while delivering to us in the developed world an array of water "varieties" from around the globe, not one of which we actually need. That tension is only complicated by the fact that if, as a nation, we suddenly decided not to purchase the lake of Poland Spring water in Hollis, Maine, none of that water would find its way to people who really are thirsty.

The chilled plastic bottle of water in the convenience-store cooler is the perfect symbol of this moment in our relationship to water. As a business, bottled water stands triumphant. The variety available is staggering—it's as hard to pick the right water as it is to pick the right toothpaste, or the right laundry detergent. There's water from the French Alps, from the Italian Alps, and water from a spring in Indiana packaged in a striking glass wine bottle. There's water touting qualities that seem a bit redundant—skinny water, life water, zero-calorie water—and water with added antioxidants, or vitamins, or a zest of raspberries, which seems to be drifting away from actually being water.

Evian packages its famous French mineral water in an aerosol mister so you can use Evian to refresh your face as conveniently as you use it to refresh your thirst. The Evian facial mister may offer the most expensive water routinely available at retail: the 3.3 tablespoons in a palm-size aluminum can cost $5.50. At that price, a single half-liter bottle of Evian would go for $55—$427 a gallon.

Bottled water is the final flowering of the old water culture. Nothing says indulgence, in fact, like paying for something you don't need to pay for, like paying for something you don't need. Superficially, it looks like a somewhat silly triumph for capitalism—look what really smart, creative people can do with something as utterly pedestrian as water. In fact, it's a reminder of exactly the opposite—the market has created very persuasive solutions for water problems that don't exist, while failing to find any solutions for real water problems.

As the first decade of the twenty-first century came to a close, though, bottled water was taking a beating in terms of its public image. The cities of San Francisco and Seattle and the New York city council all banned the purchase of bottled water with city money except in the case of emergencies.[25] There was a growing awareness of bottled water's environmental impact. We only recycle 27 percent of the plastic bottles in which our water comes—so in the United States alone we're throwing away 36 billion plastic water bottles a year, 115 discarded water bottles for each of us.[26]

In fact, the amount of bottled water sold—which grew by 8.6 percent in 2004, by 10.8 percent in 2005, by 9.5 percent in 2006—peaked in 2007. Whether because of the recession, or because of bottled-water skepticism, sales fell in 2008 by 1 percent, and fell again in 2009 by 2.7 percent. The slippage was small—less for bottled water than the drop in beverage sales overall. But the real significance was this: 2008 was the first year bottled-water sales in the United States had fallen since 1976, the first pause in thirty-two years of nonstop growth. Said the Beverage Marketing Corporation, in its analysis of the 2009 bottled-water sales: "While not specifically measured for this study, tap water was likely one of the winners in 2009, driven by cost-conscious consumers."[27]

It is, of course, just bottled water, not high on the list of modern sins one might routinely commit—hardly comparable in impact to driving a Hummer or driving a Prius while texting. Carbonated soda, the only category of drink in the United States that outsells bottled water, is hardly better for you, or for the environment—and it, too, is mostly water in a bottle. In a country where one-third of us are obese, we could clearly use a little more water and a little less soda.

Is the bottled-water aisle at the supermarket really different from, say, the cookie aisle, where the varieties of Oreos alone now number at least ten?

It *is* different, in this way: The variety and intensity of the bottled-water business isn't a signal that the water economy is flourishing with creativity. It's a signal that the water economy has malfunctioned. The bottled-water aisle is less a sign of intelligent life in the water business than of water illiteracy—mostly on the part of us, the customers. No matter what the ads say, bottled water is not, in fact, smart water.

If bottled water has been the water phenomenon of the last thirty years, FIJI Water has been the phenomenon of the bottled-water boom, carefully cultivating celebrity customers and fashionable outlets to become the No. 1 imported bottled water in the United States, besting Perrier and Evian.[28] Much to the company's delight, Barack Obama and his family were photographed drinking FIJI Water on election night. Obama has since been reported to drink FIJI Water during his daily workouts.[29]

FIJI Water is a miniature miracle of the modern global economy—water from an aquifer on the isolated north coast of Fiji's main island, bottled in a state-of-the-art factory that fills and packs more than a million bottles of water a day, water that then makes its way by truck, cargo container, ship, and even the Panama Canal, to the hippest clubs and restaurants in Los Angeles (5,520 miles from Fiji) and Miami Beach (7,480 miles from Fiji). Meanwhile, more than half the residents of the nation of Fiji do not themselves have safe, reliable drinking water. Which means it is easier for the typical American living in Beverly Hills or Miami or Manhattan to get a drink of safe, pure, refreshing Fijian water than it is for most people in Fiji.

IN 2008, GE STARTED AIRING a TV commercial that was beautiful, original, and a little eerie. Scored to a haunting version of Creedence Clearwater Revival's "Have You Ever Seen the Rain?" the video takes place in the clouds, where it turns out that a silent staff is at work, powering nature's water cycle. Everyone wears white, including white hard hats and white hooded jumpsuits. Water evaporates into the clouds in white buckets. A huge white bellows turns the water into cloud vapor. Water flows through the old-fashioned wringers once used to launder clothes—tended by women in white jumpsuits—presumably to get cleaned. A white-goateed technician samples the water with a test tube, then examines it closely. Long, meandering lines of white-suited workers form bucket brigades, filling enormous white watering cans, which, as thunder rumbles, are tipped over to bring on the rain.

The lighting and design perfectly capture that mood before a storm when the thunderheads are building but the sunlight is still streaming through them with heightened brightness.

The commercial's opening scene plays in sync with the opening line of the song. "Someone told me long ago, there's a calm before the storm. / I know, it's been coming for some time."

A first-time viewer would be captured but, for almost the entire commercial, would have no idea what it was about. Right at the end, as the watering can rocks over, a narrator offers twenty words. "Just as nature reuses water, GE water technologies turn billions of gallons into clean water every year. Rain or shine."

There is a brief moment—two seconds—showing water bubbling in the tanks of a GE water treatment plant. As the GE "ecomagination" logo comes up on the screen, vocalist Juju Stulbach sings the song's final line: "Coming down on a sunny day."[30]

It would be hard to pack more into forty-five seconds—more vivid imagery, more unspoken foreshadowing about the future, more potent reassurance of water well cared for—than GE's "Clouds" commercial does. The world of water may seem calm now, but it's the calm before the storm. Nature does a super job cleaning and recycling water, but so does GE. Clouds are great, but nothing beats a battalion of white-suited technicians, who are respectful enough of the purity of water to cover their hair and wear white gloves. In fact, as the narrator points out, GE water technologies have one tiny edge over nature—GE provides water, rain or shine. Water, from GE, "coming down on a sunny day."

The giant industrial conglomerate—it makes diesel locomotives in Erie, Pennsylvania, jet engines in Cincinnati, Ohio, wind turbines in Pensacola, Florida—has an ambitious water division, GE Water, with eight thousand employees at fifty manufacturing facilities worldwide and revenue of about $2.5 billion.[31] GE Water cleans water for a Virginia coal mine to reuse; GE Water built the largest desalination plant in Africa, in Algiers; GE Water created a small wastewater purification plant that produces 172,000 gallons a day of reuse water to keep the fairways and greens at Pennant Hills Golf Club in Sydney, Australia, lush right through Australia's brutal drought. (Australians refer to it as the country's first "sewer mining" facility; the golf club literally taps a nearby municipal sewer line as the source for the water its GE-supplied facility cleans up.)

China's Lake Taihu—source of drinking water for 30 million people in the Wuxi metropolitan area—suffered an algae bloom in 2007 so catastrophic that it covered 70 percent of the lake's surface and cut off drinking water supplies to at least 4 million people. Wuxi subsequently chose GE Water to provide the technology necessary to clean up the heavily polluted lake, which is larger than Florida's Lake Okeechobee.[32]

Around the world, most of the water treatment systems in use in Coca-Cola bottling plants come from GE Water.

GE Water even has a "mobile water" business unit that puts a complete water treatment plant inside an 18-wheel truck trailer or a cargo container. It has 800 of the self-contained, deployable mobile water units, which each make about 1 million gallons of clean water a day, for use in emergencies or when a factory needs to take its own water system off-line for maintenance. In an emergency, GE says, it can get a mobile water unit to most places in the United States in two or three hours.

Whether you're trying to turn seawater into drinking water for your city, clean up your coal-seam water before it pollutes nearby streams, reuse sewer water for irrigation, or remedy decades of pollution that's already gotten away from you, GE Water stands ready.

Says Jeff Fulgham, chief marketing officer for GE Water, "The beauty of the GE portfolio is, we can treat it all."

The astonishing thing is, in 1999, GE wasn't in the water business. Moving with quiet speed, GE has assembled its water division by spending $4 billion to buy up five existing water companies. Now GE can make the rain fall, and scrub it clean if it turns out to be acid rain. (For the truly conspiracy-minded, there is an odd bit of history from the GE corporate attic. Although GE only stepped into the water business in 1999, it was a GE researcher, Vincent Schaefer, who first developed the technology of cloud seeding in 1946, using a freezer, a cloud of his own breath, and dry ice. GE says the company has no involvement in the cloud-seeding business now, although that history does cast the "Clouds" TV commercial in a slightly different light.[33])

The new GE business is busy, but it hasn't grown as fast as CEO Jeffrey Immelt would like. GE wants water to quickly be a $10 billion-a-year business, but it turns out that many companies are skeptical about spending money on water when there is no urgent pressure—financial, governmental, or scarcity—to do so. "Customers aren't feeling a cost for their water," says Fulgham, "so they're reluctant to spend money to improve their situation."

Immelt, at least publicly, has urged patience. "People see me investing in the water business. It's financially only so-so. You take four steps forward, three back. But they hear me constantly saying, 'Don't touch it. Someday this is going to be a really great business.'"[34]

It's a funny moment in the world of water—big companies, water-dependent companies, companies with a particular risk or a particular sensitivity are ahead of the rest of us in worrying about water. Companies are cracking open their own understanding of how to use water, matching the right quality of water to the right need, learning how to reuse water or capture the qualities of water, like heat and pressure, that have often been discarded.

That's good in all kinds of ways—it's good because innovative companies are already trying to find solutions to problems the rest of us don't know exist, problems that could become widespread; it's good because it's a clear signal to the rest of us to start paying attention to water; it's good in the simplest terms of all: When the water crises start to break out more routinely, at least someone will be ready.

But it should also make us nervous. One CEO of a small, water-related company has been watching GE's move into water with a touch of wariness. "It's like a *New Yorker* cartoon," he says. "The world is one man, dying of thirst, crawling on his hands and knees through the desert. Just up ahead stands a smiling guy in a suit, holding the last glass of water, available for a fee. That's GE Water."

Except for air and water, in fact, we pay for almost everything else in life that is essential; we entrust everything, from electricity to hospitals, to private companies. Private companies hold our money and, these days, manage our friendships and our love lives. But just as water technology has not evolved dramatically in the last fifty years, neither has water law or water regulation or water oversight. There is a strand of altruism running through all water management efforts—people at Michell Wool and Coke and GE drink water every day too. But it's also true that you don't pay $4 billion to create a business, and then complain publicly about its financial performance (as Immelt did about GE Water at an investors conference in December 2009) out of altruism.[35]

GE technology does, in fact, turn billions of gallons of water into clean water every year. But GE is also the company whose PCBs poured into the upper Hudson River, creating a pollution problem that stretches 197 miles, to the southern tip of Manhattan, and that GE hasn't managed to fix since that stretch of Hudson River was declared a Superfund site by the EPA in 1984. Over a quarter-century of negotiations, foot-dragging, and preliminary work, two things are true: GE hasn't yet removed the PCBs from the Hudson's water and riverbed, and the company that spent $4 billion in seven years to create a clean-water business has spent not even $1 billion on its Hudson River efforts, which won't be finished at the earliest before 2016.[36]

So while it's important that companies are leaping into the water business, while it's impressive that they are providing leadership and pursuing innovation and even a vision for the future, it's also vital not to let business get so far ahead that we cede the future of water to commercial interests.

The companies themselves are discovering that our complicated attitudes about water mean that the rational solution, the logical fix, even if presented with style, may not work very well in the face of the very personal connection we have to water.

The new square showerheads for the 4,004 rooms at MGM Resorts'

Aria hotel in Las Vegas, for instance, were so dramatic and successful elsewhere that the Delta Faucet Company for months featured them on the home page of its Web site. The showerhead takes advantage of precisely the sort of cross-disciplinary technology that the whole world of water is waiting for. The goal was to get strong pressure, and to retain warmth in the shower spray, while using dramatically less water.

"The technology came over from the automotive industry," says Paul Patton, a senior product development manager with Delta. "It came from car windshield washers. Remember ten years ago, the washer jets would be narrow, or weak, or they'd shoot over the top of your car? Today you get a nice even fan of water that goes on the window.

"That's the 'fluidics,'" says Patton. "It's really not so different from the requirements for a showerhead—the windshield washer doesn't have much pressure or much water." Delta's high-tech showerheads don't have any moving parts inside. The water follows a kind of raceway through the showerhead—gaining force the way you do on a waterslide—and being broken into larger drops than typical, to retain heat further from the showerhead—so you feel like more water is hitting you. The four holes are misleading—with the new internal design, they manage to create a wide, warm spray zone.

There's only one problem. Within a month of Aria's opening, guests were complaining about the showerhead. "We were having mixed reviews from it," says MGM Resorts senior vice president Cindy Ortega. Yes, luxury hotel guests in Las Vegas will take the trouble to grouse about the showerheads in their rooms, particularly if they don't feel like the showerheads do a good job of rinsing off soap and shampoo.

Just weeks after the hotel opened, Ortega says, "I went and spent the night at the Aria for the purpose of testing the showerhead. To see what it's like, again. I have to say, I thought the shower was really great. I like it. I think it's really a love-it-or-hate-it kind of thing. That's people's reaction." [37]

The Aria opened in December 2009. By April, the showerheads had all been replaced with a less cutting-edge, less distinctive, Delta model—classic round shape, forty-five holes instead of four, with a standard cone-shaped spray. The water use is pretty good: 2 gallons per minute, not as good as 1.5 gallons per minute, but better than 2.5. "With the showerhead, you know, we were reaching," says Ortega, "and sometimes you reach too far."

One thing that's often oddly missing from the conversation about the business of water is the price of the water itself. The companies that are taking water seriously at this point have something at risk—their inability to function without reliable water, their reputation if they squander or damage local water supplies. Or they see an opportunity in convincing other businesses to try to understand their water risk.

The problem is that the water itself is free, or next to free. So making an argument to spend money in order to save water, or better manage water, on the basis of the cost of the water itself makes no sense. It is often cheaper to simply take fresh source water than to purify and reuse the water you've already got.

Although we don't often notice it, every gallon of water we use has an economic value—the value of whatever we can actually do with that water, whether it's boil a pot of rice, or grow an acre of wheat, or make a microchip.

In fact, we typically behave—in our homes, our schools, our companies and organizations—as if the opposite were true: We act as if clean, on-demand water had zero economic value. Especially in the developed world, the economic value inherent in the water is hidden under a cloak of invisibility, because although the water has indispensable usefulness, it rarely has a price.

That's really what we mean when we shrug and say that people take water for granted—we take it for granted because good water is basically free, so we can afford to take it for granted.

What's really interesting about the business of water is that people who start to take the economic value of water seriously immediately start to use water differently, and also to think about it differently.

6

The Yuck Factor

The "yuck factor" is a deeply ingrained psychological thing.
—*Alan Kleinschmidt,*
manager of water operations,
Toowoomba, Australia

THERE ARE MANY WAYS TO DIE of thirst in Australia.

In January 2008, which is the middle of the Australian summer, a young married couple traveling through the Tanami Desert, three hundred miles northwest of Alice Springs, with the husband's uncle, gradually gave all their drinking water to the leaking radiator of their Mitsubishi Pajero SUV.

It is hard to overstate, or overimagine, the heat and isolation of areas like the desert the trio were driving through. Daytime temperatures were 106°F. Even the most ordinary road maps of Australia have blunt red-print warnings directing drivers to register their travel through the outback with police agencies, so searches can be mounted quickly if they are overdue, and advising them to carry ample extra water and fuel.

When the trio did not arrive as scheduled in Nyirripi, police began searching. They found the seventy-year-old uncle walking along the road, and they found the disabled Pajero. The uncle told police they had been without water for two days after the SUV gave out. As police continued searching for the young couple, they left water bottles along the road, in case the pair might find them. The next morning, police found the thirty-one-year-old woman, barely alive by the side of the road. They used her footprints to backtrack, and found her thirty-four-year-old husband, already dead, three hundred feet from a dry waterhole. His wife also died, at a medical clinic.[1]

In January 2007, the *Australian* reported the deaths of two men with a few restrained paragraphs that seem to contain a novel's worth of storytelling, irony, and regret:

> He could meld into the harsh Australian landscape like few others, but it was a simple flat tyre that caused the death of legendary bushman and celebrated Aboriginal artist Nyakul Dawson on an isolated outback track.
>
> The 69-year-old and his nephew, Jarman Woods, 45, died on Dog Fence Road, on the western extremities of the Nullarbor Plain, just days before heavy rains drenched the desert country they had crossed without incident countless times.
>
> Accustomed to the remote backblocks of western and central Australia, Dawson and Woods did not even take the necessities normally packed into the back of any car on such a potentially dangerous journey. . . .
>
> Dawson's body was discovered 15 meters from [their] early-model Toyota Land Cruiser, lying under a blanket. It appears Woods covered his uncle, then wandered off for help, only to perish a little more than 3 kilometers south of the broken-down car. He had died under a makeshift shelter.
>
> Police suspect that the pair got into trouble after they left the Aboriginal community of Coonana when the vehicle's back-left tyre went flat and the men drove on until the wheel studs snapped off. It is possible they did this because their spare tyre was also flat.[2]

They drove until the wheel studs snapped off; they walked until they could literally no longer stand the heat; then they lay down and died of dehydration.

Later that same summer, in March 2007, the drought had become so severe that camels in Australia's Northern Territory were dying of thirst.

In April 2005, two migrants heading north through Western Australia to catch up with fruit-picking season broke down in their thirty-one-year-old Land Rover. According to police, the men had no extra fuel, no tools, no two-way radio, no detailed map, and just twenty liters of water. The men trekked nine miles, round-trip, out from their vehicle in search of water, and back. A detailed map would have shown them a working well five miles from their vehicle, in the opposite direction they had walked. Again, the desolation of the area where the men died is startling. The men, found beneath the Land Rover, along with their dog, had been dead a week before a jackaroo—a ranch hand—driving along the road found the Land Rover. There had been no other drivers for a week. The jackaroo had to travel a day before he reached a community where he could summon police.[3]

But Australia is so wild and so dangerously dry, it is possible to die of thirst even when you have cell phone coverage. In a devastating event that triggered national headlines and a coroner's inquest, a Sydney schoolboy out for a three-day hike with two friends in the Blue Mountains, just a few miles from Sydney's western suburbs, in the summer heat of December 2006, got separated from his friends. Despite seven increasingly desperate calls to Sydney's emergency services—triple-0 in Australia—no one was dispatched to rescue seventeen-year-old David Iredale. In the calls, operators insistently asked Iredale for a street address they could send an ambulance to. At the inquest thirty months later, his calls were so harrowing that his parents left the courtroom to avoid hearing the tapes. Although experienced hikers, he and his schoolmates only carried four liters of water each, expecting to find water in the wilderness rivers, some of which Australia's extraordinary drought had dried up. Before he died, Iredale told the operators he had gone seventeen hours without water, in hundred-degree heat. It took searchers eight days to find his body.[4]

Of course, at some level, there is only one way to die of thirst, in Australia or anywhere: to run out of water. What connects the recent dehydration deaths in Australia is that each started with a small misjudgment, a minor instance of poor planning, an all-too-human moment of overconfidence. Most of the time, such moments aren't fatal. But running out of water is like slipping off the edge of a cliff—it's hard to be saved.

Australia is a country where the cities themselves have begun running out of water during the last ten years—a place that has discovered how quickly an elaborate system for gathering and providing water can become inadequate, or even irrelevant. Australia is also living through something completely new—the division and damage that sudden water scarcity can do to the shared sense of values, to the politics, of a community.

Back when the people of Toowoomba, in the state of Queensland, thought things were bad—when it hadn't rained in five years, when the city's reservoirs were down to 34 percent full—they came together in their desperation to try a drought-ending strategy people have tried for five thousand years.

They turned their eyes and their voices heavenward, and they prayed for rain.

On Thursday evening, April 22, 2005, hundreds of people packed the soaring, cathedral-like sanctuary of St. Luke's Church, including Ian Macfarlane, the local member of Australia's national parliament, community leaders, and ministers from a half-dozen churches.

Reverend Herman Ruyters asked the crowd if they had brought their umbrellas.

City councilor Joe Ramia said the levels in the city reservoirs were extremely worrying to the council. "The dams we have built seem to me to be in the wrong place."

The Reverend Edgar Mayer, of Living Grace Church, offered God a two-for-one deal. "If it is our sins that caused the drought," he prayed, "then please wash our sins away with your rain." [5]

That evening's prayers were not answered—not then, not by the following April, or the next, or the next, not even four Aprils later, by April 2009. It was impossible to know on that Thursday night, but Toowoomba's water troubles hadn't even begun.

Toowoomba, which deservedly calls itself the Garden City, is the kind of place Americans often seem nostalgic for. It is a classic small town, with a prosperous main boulevard—Margaret Street—lined with boutiques, theaters, cafés, clubs, and restaurants serving everything from pub food to Indian and Thai. Downtown is surrounded by a grid of neat, well-kept homes that seem transplanted from 1960s Fort Lauderdale—one-story bungalows, with low rooflines and yards planted with plenty of flowers. Toowoomba is friendly, and despite the fact that Australia's decade-long dry spell has turned much of the landscape from green to khaki, it takes fierce pride in its annual Carnival of Flowers, started in 1949. Toowoomba is regarded as a conservative corner of Australia—more like Birmingham, Alabama, than Austin, Texas—and it's not unusual for people here to go to church on Sunday, a habit that is considered a curiosity in the rest of Australia. In 2008, Toowoomba was named Australia's "tidiest town."[6]

Toowoomba is snug without being insular. It has a university—the University of South Queensland—and has become home to a handful of sophisticated private boarding schools where students from rural families can get a better secondary education than in the outback. Half the high school students in Toowoomba are from out of town.

The most curious thing about Toowoomba, in fact, is its name—which is only odd to outsiders. Many smaller towns in Australia have Aboriginal names—Toowoomba seems to have been derived from *tawampa*, an Aboriginal word for swamp.[7]

In Australia, in fact, Toowoomba is not an unusual place. The rural landscape here has not been homogenized by big-box stores, and even the smallest towns have thriving main streets with locally owned toy stores, restaurants, and women's dress shops. Nor is Toowoomba, with a population of about 120,000, considered small by Australian standards. Indeed, except for the capital at Canberra, Toowoomba is Australia's largest inland city.[8]

Toowoomba perches atop the Great Dividing Range, a line of low mountains that runs up Australia's east coast, as the Appalachians do along the east coast of the United States. It is one of Australia's highest cities—with an elevation of 2,100 feet. The mountain views are panoramic. One of Toowoomba's most distinctive features is one of absence. It is the rare major Australian city with no dramatic water in its geography. It has no river, no bay, no lakes, no oceanfront. Dousing rainfalls, captured in three city reservoirs, historically provided the water Toowoomba needed.

But by the time of the April 2005 prayer service, Toowoomba was already distressingly short of water—watering lawns was already forbidden, and the steep, normally submerged sides of its reservoirs were overgrown with grass. Toowoomba needed to find some water.

Rosemary Morley, a cheerful, matronly, lifelong Toowoomban, remembers precisely the moment when she learned that the city had, in fact, found a large, renewable source of water. It was the moment she began to view Toowoomba water with suspicion, and the moment when that suspicion took over her life.

"We had a ladies' club," says Morley. "We met once a month"—about forty women, ages fifty to seventy. For the May 2005 meeting, the speaker was Dianne Thorley, Toowoomba's popular mayor. Mayor Di, as she was known, had just won a second term with 67 percent of the vote. Mayor Di is not exactly a women's club–style politician. She looks like a female rugby coach, and she talks like the foreman of an Australian copper mine—punctuating her speech with "shit" rather than the watered-down "bloody." A self-made woman—she started out cooking in pubs, eventually starting her own catering company in Toowoomba—Mayor Di has no college, but at one point in a varied work life she did work in an abattoir.[9]

That afternoon at the ladies' club, Mayor Di talked with gusto about the amazing new source of water the city had discovered, the answer to the devastating lack of rain, a kind of perpetual fountain of renewable water. Toowoomba was going to recycle its wastewater back into drinking water, using advanced technology. Right there at the monthly meeting of her ladies' club, Rosemary Morley learned that the best hope for Toowoomba's water future was in her own toilet, as she flushed it.

It was Mayor Di's first public discussion of the sophisticated recycling plan that she and the city's water managers had been working on for six months. She talked with the enthusiasm of someone unveiling the solution for an intractable problem, a solution so obviously persuasive it will carry everyone along.

"She held forth for an hour," says Morley. "She was so animated, she was so excited about it. 'You're all going to drink from the sewer!'"

The mayor made it clear the decision had been made: Recycled water was the future, the only way to refill Toowoomba's reservoirs.

The reception among the women's club? "The ladies in that room were dumbfounded," says Morley. Herself included. Morley is a grandmotherly sort. She and her husband run a house-painting business ("I'm the office end of things"). She left school at age fourteen, but she was also the first woman to be elected president of Toowoomba's chamber of commerce. Her approachable manner disguises an unyielding resolve, and a firmly held worldview. Rosemary Morley grew up drinking rainwater, and her house is set up so she and her husband can still do that if there is enough rain to fill her home's water tanks. She is deeply suspicious of both government officials and scientists; drinking sewage sounded like an idea that could only come from a conspiracy of those two groups.

"I came home from that meeting and my reaction was, How can you go forward with a project like that without running it by people? I thought, This is such a sneaky thing. There must be something about it that's funny."

What was happening in Toowoomba—a steady disappearance of the water supply that had been unfailing for a century, a desperate search for a way to replace that water—was not only not unusual, it was typical. Water scarcity was turning into Australia's most urgent issue, from Brisbane in the east to Perth in the west.

The change in Australia is both simple and startling.

Australia has built an entire way of life that assumes a certain availability of water—from the way homes are laid out and the way people spend their free time, to the way the nation raises its food and runs its cities. For a hundred years, those water levels, those water assumptions, were unthinkingly reliable. The Toowoomban economy requires that level of water, the Australian economy requires that level of water—the farmers and the factories, the backyard flower beds and the swimming pools. Every economy in the developed world, in fact, operates exactly the same way—that's why both Atlanta and Las Vegas are so nervous about falling reservoir levels. Water is not simple to supply in the first place, but replacing the supply everyone has built their lives around is much more difficult.

And Australia's water has disappeared, with stunning speed and almost unbelievable thoroughness. In the last ten years, the rainfall that fills Australia's rivers, its reservoirs, and its aquifers has simply not come. Australians refer to the last decade as the "Big Dry."[10]

The change is so dramatic that the man in charge of supplying water to Toowoomba, Kevin Flanagan, knows every time it has rained for the last twenty years. "March 1999 was the last time our dams were full," he says. "We've had just ten significant rain events in the last twenty years; they account for all the filling of the dams." None of those rainfalls came between 1999 and 2009.

"The last ten years have been an absolute horror—nonstop downslope on the capacity of all three dams."

Australia has had to remake habits and priorities to adjust to a new water reality. There's no reason the same shifts in water availability can't overtake anyplace on Earth.

That's why understanding the experience of Toowoomba, and of Australia, is so important. They are a window to the future for the United States and the rest of the developed world.

Australians have had to do three things simultaneously—tackle the complicated engineering and public policy issue of replacing vital water that is suddenly absent, while persuading people to live with less water; and develop a politics of water, a way of making expensive, high-stakes decisions about a topic that has historically been left to engineers and water bureaucrats.

The complexity of the technical decisions makes the politics of water difficult; the politics of water has often frustrated leaders' ability to make the water decisions they thought were best. The story of the effort to solve Toowoomba's water scarcity problem perfectly illustrates that as challenging as the technical issues are, the politics of water is both more challenging and ultimately more important.

Because the politics of water involves two issues that most developed countries have little experience with, but that turn out to be highly emotional: who gets water, and what kind of water they get. Despite the utterly carefree attitude most people in the developed world have about their daily water, we turn out to have complicated feelings about water, and to hold fiercely to them.

Although Mayor Di didn't realize it, the battle over Toowoomba's water future, which would dominate the town for the next sixteen months, started at Rosemary Morley's women's club meeting, with the mayor's cheerful enthusiasm for drinking recycled wastewater.

What Mayor Di didn't appreciate that day in May 2005 was that she was introducing a whole new way of thinking about water. She wasn't being "sneaky"—to use Morley's word—in the least. But Mayor Di didn't seem to grasp that people might have different attitudes about water, and about what kind of water is wholesome.

The months-long debate over Toowoomba's water-rescue plan fractured the town, the way high-pressure groundwater fractures rock, unleashing anger, distrust, and contempt that haven't faded four years later. And even in a country that now debates water policy as readily as economic policy, the water politics of Toowoomba became a national spectacle in Australia, entangling even the prime minister's office.

Mention the town, and Australians smile and shake their heads—oh, those wacky Toowoombans. But there is no cartoony comfort in the story of Toowoomba, either for Australia or for the rest of us. It is a complicated cautionary tale that illustrates how water abundance can smooth over very different, often conflicting, views of water.

In retrospect, that was all evident at that first women's club encounter—evident, if not obvious. Mayor Di's rough-hewn glee at the prospect of using technological alchemy to turn dirty water into drinking water, running smack into the utter disbelief of a room of ordinary women, who couldn't quite absorb the idea that their city would pump the sewage right back into the reservoir, no matter how many fancy filters it went through first.

Di Thorley went on from the women's club to the formal announcement of what was called Toowoomba's Water Futures project ("keep our future flowing"), officially unveiled July 1. The story in the next day's Toowoomba *Chronicle* newspaper was headlined with stark simplicity: "The Plan to Save Our City." [11]

Waterwise, things got grimmer in Toowoomba through mid-2005. Dam levels fell below 30 percent, water restrictions banning any use of outdoor hoses took hold, Bunnings—an Australian version of Home Depot—sold out of watering cans, and Toowoomba empowered four water cops to enter the backyards of residents, without warrants, if the cops suspected violations of water rules. The main worry seemed to be that Mayor Di's recycling plan wouldn't move fast enough. It wasn't due to come online until 2009—four years away, with the dams down to a three-year supply of water.[12]

Rosemary Morley, meanwhile, was so appalled by the details she read from Mayor Di's formal unveiling of the recycling project that she immediately set about organizing opposition, forming a group called CADS— Citizens Against Drinking Sewage (or Citizens Against Drinking Shit, depending on the audience). What especially aggravated Morley was Mayor Di's insistence that the decision was made, that the city's staff, scientists, and leadership had considered all the water options, and this was the only viable one.

"No consultation, no debate," says Morley. "That's like waving a red flag in front of a bull."

THE SPEED WITH WHICH Toowoomba's water problems came upon it is almost hard to grasp—its reservoirs overflowed in March 1999 and were still 80 percent full at the end of 2001. Less than four years later, the city was praying for rain. Its vulnerability is all the more striking because it seemed relatively invulnerable—with regular rain, and three reservoirs scaled to hold ample supply.

Toowoomba has an unusual water challenge. Because it sits at the top of a mountain ridge, all of its water runs away from it. The city, in fact, sits at the headwaters of Australia's great river, the Murray. There is an aquifer beneath Toowoomba, but the aquifer is falling and use by cities is sharply regulated. Although it's relatively small, Toowoomba owns all its own waterworks—the reservoirs, pumping stations, and water treatment plants.

Starting in 2006, after the dams fell below 20 percent full and level 5 restrictions were triggered, it was forbidden in Toowoomba to use city water to do any outside watering—even sprinkling an urn of flowers with a watering can.

In the office of Kevin Flanagan, Toowoomba's director of water and wastewater services, there is a tall potted umbrella plant, its stems and leaves dried brown and crackly. The plant's big plastic pot bears a hand-written sign: "Level 5 water restrictions are in force in this office. Walking the talk."

"That plant's been dead almost ten years," Flanagan says with just a trace of a smile. He figures to get a new one when it's once again legal to water your lawn with city water.

Kev Flanagan is smart and a bit impatient, a technocrat. At fifty-six, he's youthful and energetic. A bluff Irish Catholic—one of seven children, married father of four, grandfather of three—Flanagan is an engineer, a pipes-and-pumps man. He's been in the water business thirty years, and he knows how to watch the rain, the reservoirs, the aquifers. His job as director of water services for Toowoomba is rarely a position of public renown.

Whatever his critics think of him—the opponents of drinking recycled water took to referring to Flanagan derisively as "Kevvie"—Flanagan conveys a visceral sense that it is his responsibility to make sure water always flows when Toowoombans turn on the kitchen tap.

"I invented the idea of bringing recycling to Toowoomba," says Flanagan. Given the level of scorn that the plan, and Flanagan personally, have been subjected to in the last four years, insisting that the idea is his is a small act of courage.

"It was November 2004. There hadn't been any significant rain for five years. I was talking to a coal mine about supplying them water. They wanted water from our wastewater treatment plant. I thought, Shit, we have nine thousand megaliters coming from the wastewater treatment plant [a year]—if we can clean it up for the coal mine, why can't we clean it up for us?

"At that moment, our supply was grim. I was worried. The options were very limited." [13]

Of course, cleaning water so it can be used to wash coal is very different from cleaning water so it can be used to rinse your toothbrush. Flanagan envisioned taking the city's wastewater—everything collected by the sewers—through a series of filtration and disinfection steps similar to those used, for instance, by the folks at IBM's Burlington microchip plant.

The water would go through Toowoomba's regular wastewater treatment plant, Wetalla. The effluent from Wetalla—which now goes into a creek, and eventually to the Murray River, where farmers use it for irrigation—would instead be routed to a new advanced wastewater treatment plant (AWTP).

It would go through ultrafiltration, which takes out particles and bacteria; it would then go through reverse osmosis, the process by which seawater is turned into drinking water around the world. Reverse osmosis removes almost everything else—viruses, pharmaceutical residues, salts, and minerals. The water would have been blasted with UV radiation for a final dose of disinfection. Toowoomba's water purification factory would then have pumped its product, what is called "six-star water" in Australia, on to Cooby Dam. Six-star water is so clean it actually has to have some minerals added back at the end of the process so it doesn't leach nutrients from the bodies of people who drink it.

The blend of Cooby Dam and recycled water would eventually have been pumped to Toowoomba's Mount Kynoch drinking water plant, where it would be put through conventional drinking water treatment, including another set of filters and chlorination. The whole cycle is called indirect potable reuse (IPR)—cleaning the wastewater back to drinking-water cleanliness, then mixing it with the routine water supply.

The purified water would have been far cleaner than the reservoir it was pumped into. Cooby's surface area is 750 acres when it's full, and it drains an area of sixty square miles, the size of Washington, DC. Every kangaroo or koala that pees or poops in the forest around Cooby has its waste washed into the reservoir, assuming it rains. The debris from every decomposing possum carcass is eventually flushed into either the ground or the reservoir. Cooby is simply a lake, open to every bird flying overhead, every insect (dead or alive), every cascade of leaves from trees, everything washing off lawns and farm fields and nearby roads, from sheep dung and pesticides to leaking motor oil and the tiny particles as car tires wear away.

More remarkable still, the recycled sewer water coming out of the new water factory would have been cleaner than the water coming out of Toowoomba's taps—simply because the technology at Mount Kynoch wasn't designed to turn koala pee into ultra-pure water, as the proposed AWTP would. People who were nervous about the recycled water should have been really nervous about their routine Toowoomba tap water.

Kev Flanagan knew all that, of course. Water professionals know the good news about water: You can't really hurt it. These days, you can take water that is as dirty as you could possibly imagine, and clean it to whatever level of purity desired. There are no technical issues, just questions of effort, energy use, and expense.

Convincing people that the water really is clean is much more difficult, and much less scientific. This is the "toilet to tap" conundrum.

The condoms flushed away, the stagnant water from the vase of roses that stayed too long, the washing machine water from the dog's bath towels, the sour milk poured down the kitchen drain, the deceased goldfish given a toilet-bowl funeral—you can clean all that out of the water, no problem. But no matter how crystalline the water itself, you can't filter away the images of where it comes from.

As with many other things in modern society, we are more comfortable in ignorance. We don't really want a vivid picture of where our hamburger comes from, we don't want to meet the Bangladeshi who made our cheap blue-jean skirt.

We want a comforting mental and physical distance between the last time our water was dirty and the moment we use it to stir up a pitcher of iced tea. It's easy for water professionals who live every day of their careers with the reality that while there is plenty of pure water, there is no fresh water—our water was *Tyrannosaurus rex* pee and dirty snow at some point, because there is no other water. For ordinary people, though, our consciousness of water doesn't even include a willful forgetting about its source, as it does with the hamburger. We really don't know where our water comes from, just that it needs to be "fresh" when we fill the ice cube trays.

Flanagan says now that he understood this: "I was fully aware of the issue [for residents] of the safety of the water, and of the need to handle that right."

Even Mayor Di, who would ultimately lead the recycling campaign with unvarnished enthusiasm, started out skeptical of Flanagan's solution.

"She wasn't negative," says Flanagan. "She was cognizant of the fact we live on top of a mountain. Any water source to come to Toowoomba is very difficult to get, so if you've got your own water here already, why not recycle it and reuse it? She was skeptical of it. We had to prove it to her."

No one in Australia was then using the kind of water recycling system Flanagan envisioned. So Flanagan and Thorley together visited cities in the United States with experience recycling water. They went to Fairfax County, Virginia, a suburb of Washington, DC, where treated effluent has been flowing into the Occoquan Reservoir, and back into the water supply of one million people, for thirty years. They visited Orange County, California, where the world's largest wastewater recycling system was then under construction—Orange County's plant now produces 70 million gallons of highly purified, recycled water every day, ten times Toowoomba's total daily use.

Orange County designed an elaborate public education process while its plant was being built—and it uses precisely the technology Toowoomba proposed, except for one final element. Orange County pumps its finished water into an underground aquifer, rather than a reservoir. The water comes back into Orange County's supply by tapping the same aquifer. Water professionals typically refer to the final steps in making water palatable for people as "polishing"; in terms of the mental polishing recycled water requires, it turns out people like their water to disappear into the ground for a while, or into a huge river, rather than simply sit in a quiet reservoir.

Flanagan and Thorley also visited Singapore's world-famous NEWater operations. Singapore uses the phrase "used water" instead of sewage, and brands the purified wastewater as NEWater. In a subtle but smart positioning move, most of the earliest NEWater customers are high-purity users like technology factories, whose cleanliness standards are far more stringent than those of the average thirsty jogger. Singapore is introducing NEWater into the drinking water supply very gently—1 percent of daily drinking water comes from recycled "used" water, set to rise to 2.5 percent by 2011.[14]

"My thirtieth wedding anniversary was January 4, 2005," says Flanagan. "I was looking at recycled sewage plants." But the visits did the trick. Flanagan came back confident that his idea would work. And Mayor Di, says Flanagan, having seen the technology in action, came back gung-ho.

The story in the Toowoomba *Chronicle* about Mayor Di's "plan to save our city" included not a word of skepticism about the technology or a sentence of dissent from anyone in town. Indeed, leaders lined up to support the idea. In the course of applying for federal funding for the A$68 million (US$56 million) water factory, Toowoomba's city council approved

Flanagan's plan 9–0. Peter Beattie, then governor of Queensland, endorsed the plan. Most significantly, Ian Macfarlane, the member of parliament from the area that includes Toowoomba, enthusiastically endorsed water reuse, reassuring voters that "this water is . . . going to be better quality than what we're currently drinking."

In most places, that would have been it. Toowoomba's funding application would have been approved, design and construction contracts would have been let, the water plant would have been built, the water would have been tested, and by the end of the decade Toowoomba would have had a climate-independent source of water—along with a new tourist attraction, because the water factory was planned to have a center where visitors could watch water being purified just as Singapore's NEWater facilities do.

Instead, Toowoomba was heading into a whirlpool of a year unlike any in its history.

Clive Berghofer, the richest man in Toowoomba, the 118th-richest person in all of Australia, dropped out of school at age thirteen.[15] "I could barely read or write," he says with a lopsided smile of perverse pride. He went on to become a real estate developer, and has built huge swaths of Toowoomba. He claims to have named four hundred streets in the subdivisions he's rolled out, including a short lane called Berghofer Street.

Berghofer, who is now a cantankerous seventy-four years old, spent almost two decades on the Toowoomba city council, the last ten years of that as mayor. He, too, worried about possible water scarcity back in the mid-1980s—he presided over construction of the city's largest reservoir, Cressbrook, almost twice as large as the older two reservoirs combined. He is so well known that Toowoomba's newspaper can refer to him in headlines simply as "Clive." Five recreation venues in Toowoomba bear Berghofer's name, as does the intensive care ward of the local hospital.[16] He is easily Toowoomba's most influential resident, and he shares with Mayor Di a willingness to speak his mind.

Within weeks of the announcement of Thorley and Flanagan's recycled water plan, Berghofer came out against it. "It'll kill this city," he said. "It's an absolute disaster."[17]

Berghofer, a tough-as-rocks Australian, gets squirmy about recycled water.

"If I die, they run all those fluids through me, they drain me out before they put me in the coffin—that stuff goes right into the sewer. Hospitals, funeral homes—all that material goes right into the sewer.

"They claim you can't get all the hormones out of the water. We don't know what that will do. It might take two generations to find out. The fish are all turning female.

"It's like smoking. It's bad for you, but you're not going to die when you take the next cigarette. You might never die of smoking. But we know it is bad for you.

"Why would you drink recycled water if you didn't have to?"

Berghofer dubbed supporters of the city's Water Futures plan "sewage sippers." He said his beloved Garden City would become known nationwide as "Poowoomba" and "Shit City." Opponents of water recycling took out newspaper ads telling readers, "You deserve fresh water" and urging residents to "think for yourself." Resistance to drinking recycled water is known in the water trade as "the yuck factor." Berghofer, and Rosemary Morley's CADS group, experienced the yuck, and also plumbed it.

Mayor Di, Kev Flanagan, and the supporters of the recycled water technology were deftly outmaneuvered by Rosemary Morley and her opponents, not once but twice. The opponents were better at the politics, and better at the populism.

The very first public CADS meeting organized by Morley featured as its main guest speaker a well-known Aussie antirecycling character named Laurie Jones, a plumber from Queensland's Sunshine Coast. Jones has no formal scientific or water treatment training, except, as he says, "I am a licensed plumber and drainer and I know what goes into a sewer."[18]

Jones has focused his online research on micropollutants in wastewater—hormone and pharmaceutical residue left over after basic sewage treatment, how those chemicals feminize fish and might feminize Toowoomba's men if they were forced to drink recycled water.

That meeting, which drew five hundred Toowoombans, including both Clive Berghofer and Mayor Di, was presided over by Snow Manners, a patrician entrepreneur who became Morley's partner in dedicated opposition to the recycling plan.

"Laurie Jones had a lot of extreme stories," says Manners. "He told us [that the men] would grow breasts, that our testicles would fall off." Not a particularly scientific approach, but an undeniably memorable one.

Asked at the meeting if the recycling plan was open for debate or reconsideration, Mayor Di told the crowd it was "nonnegotiable." She was booed.[19]

"It got boisterous," says Manners.

The images, the questions, the language of the opponents were so vivid that three years later, with almost no prompting, ordinary Toowoombans volunteer them back. "The hospital washes blood down the drain," says the waitress at one of Toowoomba's Italian restaurants before serving dinner. "I don't want to drink that."

In the wake of that first CADS meeting where Toowoomba's testicles were put on notice, Rosemary Morley organized a petition drive to bring the recycling plan to a halt. By early October, just seven weeks after the public meeting, she had collected 7,048 signatures on petitions opposing adding recycled water to the dams, more than 10 percent of Toowoomba's electorate.[20]

Morley didn't present those petitions to Mayor Di—she took them straight to Ian Macfarlane, the federal member of parliament who just two months earlier had called the recycling plan "bold," and said he was "proud to push its merits." In the face of the petitions, and what he said was an overwhelming volume of calls and letters to his office, he pirouetted, meeting with Mayor Di to tell her, "Given information I have obtained from independent sources, I am currently unable to support recycled effluent being discharged into Cooby Dam."

Months of behind-the-scenes political jockeying ensued; what happened next took official Toowoomba by stunned surprise. The following March, the Australian federal government said it would pay for the new advanced water purification plant—but only if the people of Toowoomba endorsed it in a referendum. That kind of contingent federal funding was unheard-of in a century of Australian history. What's more, it's not clear

that any community anywhere has ever voted on whether to include purified wastewater in its drinking water supply—certainly residents in Singapore, Fairfax County, and Orange County never did.

The referendum provided political cover. Water is one of the top two
or three political issues across Australia—depending on recent rainfall and
the economy. Rather than be accused of forcing Toowoombans to accept
poo-water, the national leadership would let Toowoombans themselves
decide.

And so what should have been a rational, careful conversation became
a raucous political campaign. "We knew we were pretty well stuffed once
the referendum was in place," says Kev Flanagan.

The referendum was announced in late March, the vote was set for
Saturday, July 29, 2006—the recycled water campaign lasted all of April,
May, June, and July.

The city produced a full-color, forty-page book explaining the reasoning of the recycling project, with graphics of molecules and filter barriers, pages of text explaining the technology, photos of water in every
possible mood, and many pictures of children. Clive Berghofer paid to
have an eight-page newspaper written, printed, and mailed to every home
in Toowoomba with the front-page headline "Clive Says 'NO,'" above a
photo of a churning tank of raw brown sewage, with the caption "Is this
our city's future?"

Three members of Toowoomba's city staff did nothing for those
months but explain and promote the water plan—Mayor Di, Flanagan,
and Flanagan's chief deputy, Alan Kleinschmidt. Flanagan and Kleinschmidt were relieved of their routine duties to concentrate on the referendum full-time.

Mayor Di, says Flanagan, lived up to her reputation. "She was rough
as guts," he says. "She was a fist on the desk. 'You're going to drink it, and
if you don't like it, you can drink bottled water!' That was her style. Sometimes I had to say to her, 'Di, today you're a diplomat.'"

The science itself was hard—hard to present, hard to grasp, hard to
be quickly comfortable with. Proponents started out saying that the purified water would have nothing left in it—the barriers were finer and finer,
and in the end, only water molecules would be able to dance through. But

while that's mostly true, it's not quite true. The purified water does have infinitesimal amounts of things in it—some that slip through the barriers, some that are the result of the purification process itself, but in quantities so small they are measured in parts per trillion.

For instance, some tested samples of sewage water purified in Queensland using the Toowoomba process contained a bit of acetaminophen—the active ingredient in Tylenol. How much? If 685 people drank two liters of that recycled water every single day of their lives, and all 685 lived to be a hundred, then on their hundredth birthday, the day they had together drunk 50 million liters of water, they would have at last consumed a single adult dose of Tylenol—collectively.

So the water in question didn't contain no acetaminophen, but an infant running a fever gets more in two doses than you'd get in a hundred years of drinking recycled water.

Some tested samples of the purified water contained bisphenol A (BPA), one of the most controversial industrial chemicals, a substance whose risks to health are unclear. How much bisphenol A was in the water? The analysis found 10 nanograms per liter. That's not nothing. But in the course of a normal day, a typical person consumes 2,300 nanograms from all kinds of other sources—food containers, plastics—so the tiny amount in drinking water doesn't change your exposure to BPA. Drinking 20 liters (5 gallons) a day of recycled water, which is impossible, would only add 10 percent to your BPA exposure.[21]

In any meaningful measure of reality, there was nothing in the purified water but water. In scientific terms, there were some molecules of other stuff.

But the difference between "nothing" and "virtually nothing" is the difference between security and anxiety. In a heated political campaign, it's the difference between trust and suspicion.

"I heard things like, 'A nurse has told me they can't get the hepatitis C virus or the AIDS virus out of the water,'" says Flanagan, shaking his head. Both HIV and hep C virus particles are five to ten times larger than the pores in the very first filtration barrier; they are 500 to 1,000 times larger than the holes in a reverse-osmosis membrane. For those worried about blood, hospitals in Toowoomba are forbidden to flush bloody waste into the routine sewer system—but even if they did, blood is organic. The first wave of routine sewage treatment destroys blood. And blood cells are 60 times the size of the HIV virus—60,000 times the size of the pores in the filters.

But what if the nurse your friend knows is right, and the water officials are just saying whatever is needed to get their water recycling plant? Toowoomba became a swirl of urban legend that was moving far faster than any ordinary person could keep up.

"There are some people who are genuinely afraid. They can't get their heads around the idea that it is possible to drink our effluent," says Peter Swannell, an engineer who had been president of the University of South Queensland in Toowoomba. Swannell was a vigorous public supporter of the recycled water plan. "It does require education to get used to it." Of Berghofer, Swannell says, "The guy is unshakeable in his misunderstandings."

The unnerving thing, in public policy terms, is that the opposition campaign was as much about perception as about facts—and in that sense, the debate, the campaign itself, became a kind of negative self-fulfilling prophecy.

Clive Berghofer seemed to relish saying that people would come to know Toowoomba as "Poowoomba" or "Shit City." But those terms rarely, if ever, occurred in the media unless he was the one using them or inspiring someone else to use them.

"The fear of recycled water is reasonable," said Berghofer, "not because it would hurt you—it might hurt you. . . . Is it safe? It's all a matter of perception." In fact, whether recycled water is safe is not a matter of perception. It's a matter of science, a matter of reality.

The opponents had one final political masterstroke. As the campaign wound toward election day, the opposition ads and flyers positioned the no vote as a way for Toowoombans to keep their water options open.

"No gives you options," shouted one print advertisement. "Yes = drinking recycled sewerage. No = putting all options, including recycling, on the table."

It was, truly, brilliant. No was a better version of yes than yes.

What the supporters of recycling failed to do was two things: They failed to enumerate the costs, even the dangers, of voting no with anything like the vividness with which their opponents painted the costs of voting yes. And they failed to find a good-humored way of pointing out to people that if you insist on thinking of it this way, every drop of water on Earth is "poo-water," because the water has been around longer than life itself, so every creature that has ever lived on Earth has done its version of pooping into that water, whether it comes from Cooby Dam or the springs at San Pellegrino.

It's all recycled water—it's just a question how big a gap in time and space and imagination has opened.

In the end, Alan Kleinschmidt and Snow Manners had a very similar thought, from very different perspectives.

"The 'yuck factor' is a deeply ingrained psychological thing," says Kleinschmidt. "As a water industry, we're partly to blame for that. All of the advances we've made in the last hundred years in public health are really about separating the water and the sewage.

"Now we're telling them to forget all that. That's hard.

"Of course, we're not really telling them to forget it—we're moving the separation to the molecular level. But still . . ."

Snow Manners is blunter. "The reason Toowoomba erupted was, they were attacking the fundamental values here," he says. "They came up with a proposal that attacked the core values of Western suburban life: You go to the sewage plant to get your water."

It drizzled most of the day before the election—but not enough to run in the gutters. Rosemary Morley predicted that the rain would get the opponents 5 percent more votes.

They didn't need it. In what was perhaps the first ballot of its kind anywhere in the world, Toowoombans voted against drinking purified recycled water, 62 percent to 38 percent.

In defeat, Mayor Di got it right. "The community didn't trust their mayor and their council," she said. "They believed that I would put their families and themselves at risk."

Clive Berghofer merely harrumphed. "I'd like to have seen [the vote] a bit higher than what it is, to prove that we're not idiots in Toowoomba."

Toowoombans may not have been idiots, but their acrimonious debate and vote provided no answer, no deliverance. It provided no water. After the votes were counted, Toowoombans were simply a year deeper into their water crisis. By the end of 2006, the year of the referendum, Toowoomba's dam levels had fallen to half what they were at the time of the prayer service in April 2005. Kev Flanagan still had to solve the problem of how to get water to Toowoomba.

TOOWOOMBA'S PROBLEMS are simply a miniature version of the crisis every major city in Australia has gone through in the last decade (except Darwin, on the relatively lush north coast).

The details of how the water is disappearing vary, and the details of the politics vary, but every big city has run dramatically short of water—and all at the same time: Brisbane, Sydney, Melbourne, Adelaide, and Perth.

The result has been an urgency about water issues in Australia, and an immediacy. Every state in Australia now has a cabinet-level minister for water or climate change—or both. Australia's federal government also has a cabinet minister for water, the same rank as the minister for defense. And in the last four years, Water Minister Penny Wong has gotten as much media attention as the defense minister.[22] Water is a topic of daily conversation, debate, worry, and speculation. Water scarcity, water restrictions, and plans for finding more water are written about in Australia's major newspapers virtually every day. The weather pages feature color charts and graphics that show how much water residents have used—total water in the last twenty-four hours, per capita use, how much water remains in a city's reservoirs. The figures are updated each day.

Australian officials at all levels are scrambling to find new sources of water and frequently promoting desalination and recycling because, while expensive, they are "climate independent," their success doesn't depend on rain. Officials are also trying to instill new habits of water use. The changes are coming fast, because there is no choice. And the consequences are being felt in every corner of Australia—not least because of the astonishing cost that no one anticipated, a cost to try to rescue the water status quo and the advanced economy it supports.

Australia's state and federal governments have embarked on a A$30 billion effort to reimagine and rebuild the nation's water system. The A\$30 billion is far more dramatic than it first appears. For while Australia is physically the size of the United States, it has only about as many people as Florida. Imagine if the United States consisted of only the people who lived in Florida—and the rest of the country was empty. That's Australia. The country's spending on water projects comes to A\$1,500 for every Australian—the equivalent, in the United States, of \$400 billion, half Barack Obama's entire economic stimulus program. Just on water.[23]

This is what water scarcity looks like, this is what the impact of climate change looks like, in a developed country where the economy, the per capita GDP, even the movies, music, and pop culture, are very much like that in the United States. There is no debate about the "reality" of climate change in Australia. It is as real as 2009's global economic crisis. In Australia, climate change, like the economic downturn, is big, moving across the whole landscape, impossible to either ignore or redirect. And climate change, like the economic downturn, can be startlingly immediate, even intimate.

In Perth and Melbourne, in Adelaide and the farm towns along the Murray River, the adults may hope that it will one day rain again, but in all those places, by 2009, kids entering their teens had never known a year when it rained like it did in the twentieth century.

Australia has always had a watery spirit, despite a climate of legendary dryness. Eighty-five percent of Australians live within a strip just thirty-one miles wide along the country's coastlines. Half the nation's homes are eight miles or less from the ocean.[24]

Riversides, bays, waterfronts, and beachfronts are so present in the daily life and the daily landscape of Australia's major cities—in Perth, in Adelaide, in Melbourne, Sydney, Brisbane—that it can be easy to forget how precariously dry Australia is. Australians live at the water's edge precisely because most of the country is so parched.

Australia, whose land area is almost the same as that of the continental United States, is by far the driest inhabited continent on Earth. Rainfall across Australia is highly variable, but averages just eighteen inches, less than falls in Flagstaff, Arizona. What makes Australia so truly dry compared with other places, though, is a topography and a climate that mean very little of that rain ends up as runoff in Australia's rivers. The dry landscape absorbs or evaporates rainfall quickly and thoroughly. Just 11 percent of Australian rainfall becomes runoff. The average rainfall in North America is only 20 percent higher than in Australia, but the runoff is 52 percent, so five times as much rain ends up as runoff. In wet times, the combined flow of all Australia's rivers is just half the flow of the Mississippi River alone.[25]

After a century and a half living in such a climate, Australians understood their relationship to rain, water, heat, and drought. Or they thought they did. In the last decade, the Big Dry has laid hold of the whole continent, withering an already arid country. Today, whether suburbanite or farmer, politician or factory manager, water scarcity touches the life of every Australian every day in some way.

Perth, Australia, sits at the far western edge of the country—in about the spot San Francisco does in the United States, and with the same spirit of independence and informality. With a population of 1.5 million, Perth is capital of Western Australia, a single sprawling state that is bigger than the United Kingdom, France, Spain, Germany, and Italy combined.[26] The Swan River winds through downtown Perth, wide and placid, with a miles-long greenway along its banks crowded with runners, skaters, strollers, and bikers. Perth is just eleven miles upstream from the Indian Ocean and the artsy, relaxed beachfront town of Fremantle.

In the last thirty years, Perth's average annual rainfall has fallen by 20 percent. But because dry ground drinks up sparse rainfall more readily, the 20 percent drop in precipitation has led to a 75 percent drop in the water running into Perth's dams. Since 1980, Perth's population has doubled; the water flowing into its reservoirs has dropped to a quarter what the city depended on.[27]

Even as Perth scrambles to build its second desalination plant, officials have sought to permanently alter the community's water culture. Perth's residents are famous for their English-style gardens—the city has vast carpets of suburban single-family homes where, like Toowoomba, the cultural devotion to gardening is at odds with the Mediterranean climate. Now residents are only allowed to water their gardens two days a week, and always before 9 a.m. or after 6 p.m. Sprinklers are banned during Australia's winter months, from June 1 to September 1. The watering rules aren't drought restrictions—they are permanent.

It was Perth's water problems that inspired the renowned Australian scientist and author Tim Flannery to predict, in June 2004, that Perth could become uninhabitable. "You don't want to lead people to despair, and it's not too late, but there is a possibility Perth will become Western civilisation's first ghost metropolis," Flannery said. "I think the chances are that, in the next fifty years, Perth will face challenges it simply won't be able to overcome." In Perth's major newspaper, that prediction appeared under the headline "Perth Will Die, Says Top Scientist." [28]

The Big Dry hasn't just imposed an era of water austerity. Scarcity has quickly shown that Australia's water systems weren't set up to decide who should get water when there wasn't enough for everyone—the Big Dry was quickly causing water conflict.

Adelaide, the bustling commercial center that sits in the middle of Australia's southern coast, depends for most of its water on the Murray River, which runs east of the city. The lack of rain had reduced water in the Murray River so dramatically that Adelaide's core drinking water supply was at "extreme risk," as a 2009 report put it. The condition of the Murray also created friction between the thousands of rural farmers upstream who tap the river to grow Australia's food, and Adelaide's urban residents, a conflict that isn't just about water, but quickly becomes about culture, politics, and economics. Whose community, whose work, is more important? Water scarcity that comes on quickly often seems like a zero-sum game—water that farmers use is water that doesn't make it to Adelaide's spigots—and the rivalry over water quickly becomes an argument about priorities, values, and lifestyles.

The water restrictions imposed in Adelaide cut water use enough that residents saved the equivalent of the output of an entire desalination plant, which would have cost A$1 billion. Still, Adelaide's officials are so worried about the future of both the rain and the river Murray that they not only had a desalination plant under way in 2009, they decided to double its size before construction on the first stage had even begun. [29]

In Melbourne, the daily water culture has been revolutionized. Melbourne is a hip and cosmopolitan metropolis of three million people; it sits along Port Phillip Bay, and the Yarra River arcs through the middle of the city. You can run and bike for miles along seaside and waterside paths, and some of the city's most popular neighborhoods are right on the edge of the bay, with an air of 1960s Miami Beach. Despite the daily presence of water in the city landscape, the total water in Melbourne's ten reservoirs in mid-2009 was lower than it had ever been—down to sixteen months of drinking water, after twelve years of low rainfall. Residents are exhorted to "Target 155"—use only 155 liters of water per day per person, 41 gallons, about what an American would use with a single bath.

If you own a swimming pool in Melbourne, starting in 2008 you could only legally fill it with drinking water using a bucket, and then just to keep the pool from cracking and popping out of the ground. Hose filling was forbidden. It became popular for high-income people to truck in water from outside Melbourne to fill their pools. A seven-thousand-gallon tanker cost a swimming pool owner A$1,600 (US$1,350)—23 cents a gallon. A backyard pool requires two tankers to fill.[30]

But in the twelve years of the Big Dry, Melbourne residents have transformed their individual water use. They've gone from using 358 liters per day—95 gallons, typical for someone in the United States—to 143 liters per day (38 gallons). Melbourne residents have changed their daily routines in ways that have allowed them to eliminate 60 percent of their water use.[31]

To try to guarantee a climate-independent supply of water, Melbourne, too, has laid plans for a desalination plant, Australia's biggest, at a staggering cost of A$3.5 billion (US$3 billion). Said the premier of Victoria, the state of which Melbourne is the capital, "We don't want to be a 'pray for rain' government."[32]

For Australians, the most potent symbol of the Big Dry is the emptying of Australia's great river, the Murray, which winds through the heart of southeastern Australia. The Mighty Murray, as it is known, occupies a place in the Australian imagination akin to that of the Mississippi for Americans. The Murray is a character and a setting for Australian stories; its early exploration and its use by paddle wheelers to move people and goods are a central part of Australian history. The Murray River runs for 1,591 miles, and with its main tributary, the Darling River, the Murray-Darling Basin drains one out of every seven acres on the continent. In the

Big Dry, the Murray River has gotten so low that in 2007, 2008, and 2009, it simply ceased to flow.

More than simply a symbol, the Murray is a vital economic engine for Australia. It waters farmland that produces 30 percent of the country's agriculture—a basin productive enough, by itself, to feed every Australian. But the combined total flow in 2007, 2008, and 2009 was little more than half what a single typical year is.[33]

The results for Australia's breadbasket have been devastating, both in terms of the amount of food produced, and socially. More than 100,000 fruit trees in the region have been allowed to die because there is no water to sustain them. Australians used to chart their nation's economic success by counting sheep—"Australia rides on the sheep's back," the old saying goes. The number of sheep and lambs being raised in 2009 was the lowest since 1920. The rice harvest in 2009—Australians grow a huge amount of rice, most of it in the Murray basin—was just 5 percent of the typical year. And 2009 was three times the size of 2008, when the rice harvest was 1.6 percent of normal.[34]

Australians fear that a whole way of life is being hollowed out by the Murray basin drought—farmers are lining up to sell their water rights back to the federal government, effectively cashing out of farming, and depopulating farm towns north of Melbourne and east of Adelaide. Without irrigation rights, a piece of farmland is worth almost nothing.

Everything about life in Australia—business, leisure, food production—assumes a level of water availability that was backed by a hundred years of history and infrastructure. The dams are in certain places for a reason; the farms are in certain places for the same reason: there was water there. In the last ten years, shifting rainfall patterns have seemed to make a mockery of those choices. The water in many of Australia's reservoirs is so low that the construction roads used to build the dams are now visible.

Indeed, in 2010, even as rain in some parts of Australia allowed a slight easing of water restrictions, it was clear that in terms of water—use of water, perceptions of water—life has been permanently altered. In the state of South Australia, home to Adelaide, Water Minister Paul Caica said he fully expected to make the drought-inspired ban on lawn sprinklers permanent.

"I used to like running through sprinklers as a kid," he said. "My kids liked running through sprinklers. But what we've got to realize is . . . we all have a responsibility to use that water effectively, and [sprinklers] are not an effective way of using water."[35]

It would be as if the governor of Texas had announced that he never expected to see lawn sprinklers legal in that state again.

"The message of our experience is very clear," says Ross Young, who spent eighteen years with Melbourne's water utility and now heads an industry association of Australia's large water utilities. "Once climate change hits, it hits at a pace and a level of severity that no one ever predicted in the climate models."

Supplying clean water has for a century been a fairly straightforward engineering task in the developed world. But as the political fight over the water rescue plan in the city of Toowoomba shows, the politics of water scarcity often turn out to be surprising, emotional, and confounding.

THE TUMULT OF TOOWOOMBA's yearlong water battle obscures something important, something that connects Toowoomba not just to its much bigger sister cities with water troubles in Australia, but to cities around the world that are nonchalant about water now, but may not be for long.

Toowoomba lacked nothing it needed to tackle water scarcity.

It had good water resources before the Big Dry took hold. It had an energetic water staff that anticipated the problems and searched for solutions. It had a community that was strong and stable, smart and cohesive. It had plenty of money—the federal and state governments stood ready to offer tens of millions of dollars for a well-argued water rescue plan.

What Toowoomba didn't have was a way of talking calmly and thoughtfully about water. Toowoomba's public officials really didn't grasp that water is so personal that tinkering with people's drinking water unleashes emotions that the normal political process, and the normal political leadership, are not equipped to handle. Recycling sewage into drinking water tapped something primal and disquieting for a lot of Toowoombans. Toowoomba was totally unprepared for the surprising, and surprisingly caustic, politics of water.

For Americans, for anyone with a secure supply of water they never question, the fractiousness of water politics can seem unexpected, even bemusing. But Toowoomba isn't some oddity—the intensity of water as a public policy issue in the developed world is not to be underestimated. Water issues are often a combustible combination of two things that don't typically mix—something immediate, like taxes, that touches your life every day; and something intimate, like abortion, that taps your deepest beliefs. Water may be mostly ignored, but when it becomes important, it often ends up being about emotion as much as science or rational policy-making. When water becomes a crisis, when it cascades into politics, our responses are hard to predict, and hard to manage.

Of course, subjecting such technical decisions to a referendum brings democracy into an arena where it is not at its most helpful. We did not vote on the stringing of power lines or the laying of water mains, we do not vote on even major zoning decisions or the routes of highways or the locations of their interchanges. We elect officials to make those decisions, and if we don't like their choices, we elect new officials.

For ordinary people, the molecular science behind the drinkability of purified wastewater is a leap of faith. You can look at all the multicolored diagrams you want, you can hold reverse-osmosis membranes in your hands, you can listen closely as scientists and engineers explain how UV radiation disinfects anything that slips through. You can read reports reassuring you of the purity of the water, and you can taste the water itself. But in the end, none of that is firsthand experience of the water's safety. You have to decide you believe what you're reading and hearing. Drinking any water, but particularly water you might be leery of, requires belief that it's okay.

What happened in Toowoomba points to a question about water that, in the developed world, we haven't had to worry about in a hundred years, but is coming back with fresh urgency: How clean is my tap water anyway?

It's not quite as easy a question to answer as it would seem, and it's not quite as easy to answer as it once was. The recycled water Toowoombans rejected would have been cleaner than the water coming out of their kitchen faucets every day—the purified wastewater would have been cleaner than the tap water anywhere in Australia or the United States. But the tone of the conversation in Toowoomba meant that the science couldn't compete with the emotion.

That's the larger point: If we're going to manage our water well, and if we're going to avoid water debates that are emotional, even scary, but not helpful, we need more than good science about how clean our water is. We need a way of talking about what's in the water that gets beyond alarming headlines about Prozac and Tylenol and birth control pills lurking in the reservoir.

In fact, the debate in Toowoomba really points to a new question about water, what might even be called a modern question: What exactly is clean water? What does it mean to say that water is clean?

Americans are already both suspicious of and nervous about the quality of their tap water. We love to talk about its taste, about whether it's really safe. In a 2009 Gallup Poll, Americans were read a list of eight environmental problems. The No. 1 issue they worried about "a great deal" was "pollution of drinking water": 59 percent worried about drinking water "a great deal"; another 25 percent worried about it "a fair amount."[36] (It is precisely this worry that drives the $21-billion-a-year bottled water business in the United States.)

The worry might seem irrational. In the United States and most of the developed world, our tap water is remarkably safe. Clean tap water has contributed significantly to the dramatic increase in our life spans. But two things have changed recently that have shaken confidence in tap water. First, scientists have developed the ability to test for contaminants that are present in only the tiniest amounts—at the level of 1 or 2 or 3 parts per trillion.

What is 1 part per trillion?

Think of it in terms of your own income. Most of us will not earn $10 million in our entire working lives—you would have to earn $100,000 for each of 100 years.

Now, look around you for a penny. You probably don't have to look beyond your pocket, the bottom of your purse, the drawer in your desk. Pennies are everywhere.

If you earned $10 million in your lifetime, 1 part per *billion* of that income would be a single penny, out of all the pennies that rattle through your life. So the new ability to test for substances at a concentration of 1 part per trillion is the same as the ability to find a single penny out of a lifetime of $10 million in earnings, not for one person but for 1,000 people. And not just to find a penny, but to find a single specific penny.

Put another way, it's the ability to zero in on one specific second out of 1 trillion seconds, 1 second out of 31,689 years.

That kind of detection capability is simply astounding. Something that appears as a couple parts per trillion is in the water, but only barely.

That testing ability has allowed us to discover a whole wave of things that we never realized are in our water—almost like the discovery, one hundred years ago, of the bacteriological pollutants in water.

The result this time is a little tricky: Some of the things we're "discovering" have been in the water all along; others are relatively new. The water itself isn't dirtier. We're just seeing the water we once thought was clean in a new way.

That we couldn't detect the "dirt" ten years ago doesn't mean it wasn't there, and doesn't mean it wasn't damaging the environment or human health. The tricky part is that the opposite is also true: The fact that we can detect the substances, their very presence in the water, doesn't mean they *are* harmful, or even significant. Just because we suddenly realize there's stuff in the water we didn't know was there before doesn't mean we have to take it out.

We actually don't know. That is, we don't know how clean the water needs to be.

One reason we don't know has to do with the second big change in our water supplies: what we're putting into the water. We've started to wash substances into our wastewater, and into our lakes, reservoirs, and rivers, that simply didn't exist a hundred or even fifty years ago.

Tens of millions of people now take maintenance pharmaceuticals every day: antidepressants and cholesterol-lowering medicine join birth control pills and blood-pressure medicine, along with somewhat rarer cancer and organ-transplant drugs. The residues of those medicines end up in our urine and in our wastewater. Farms, mines, and gas-drilling operations use all kinds of exotic chemicals, some regulated and some not, that end up in wastewater. And all the products of modern life—from shampoos and detergents to the fire-retardant chemicals that infuse our children's pajamas—are depositing a faint rainbow of contamination in our rivers, lakes, and reservoirs. Because of the way water works, of course, that means those substances are also starting to appear in the raw water that utilities rely on to supply our tap water.

What we don't know is if these micropollutants in our water are hurting us, and how.

"It is the difference between what's detectable and what's dangerous," says Shane Snyder, a toxicologist who is codirector of the Arizona Laboratory for Emerging Contaminants at the University of Arizona. Snyder was one of the scientists who helped first discover the micropollutants in U.S. water supplies, and until August 2010, he ran an unusual research and development laboratory for Patricia Mulroy's Southern Nevada Water Authority in Las Vegas.

We have the ability to pollute our water in ways that are new, and we have an ability to detect those pollutants that is new—but our ability to understand their impact on the environment and our own long-term health is seriously lagging.

And, of course, what you don't understand you can't effectively regulate. In the United States, the Safe Drinking Water Act, which is updated periodically, requires utilities to test for 91 contaminants, but it was written 35 years ago. Big U.S. water utilities routinely find hundreds of unregulated chemicals in their water supplies, albeit in minuscule amounts. Water systems haven't kept up with modern technology. They haven't kept up with our ability to ask questions about what's in our water, and to figure out what it might be doing to us and how best to neutralize it.

The amounts of most such substances are almost unimaginably tiny, the equivalent, as Snyder puts it, of a single grain of sand in an Olympic-size swimming pool. But many of the chemicals are what's called "endocrine disruptors"—that is, in animals, and potentially in humans, they have the ability to act like hormones. The very nature of hormones is that a tiny amount can have a significant impact.

The superficially easy solution is simply to filter the substances out of the water. And in purely technical terms, we can of course filter our tap water to whatever level of cleanliness we desire. But to take out substances that appear only in the parts-per-billion and parts-per-trillion range would, with current technology, double or triple or quadruple the cost of cleaning the water. It would dramatically increase the power required ("we don't have enough power plants in the country to do that," says Snyder). And it would be premature, if not foolish, on at least two counts. First, 95 percent of the water that utilities provide isn't used for drinking or cooking, it's used to flush the toilet, fill the bathtub, and water the lawn. That water doesn't need to be ultra-purified. Second, we don't actually know which substances are hurting us, so we'd be spending enormous amounts of scarce money on a public health effort that might not, in the end, improve public health.

For the moment, that kind of purity should be reserved for drinking water, and the way to achieve it is to filter the water you actually drink—at the tap.

That doesn't require any technological breakthrough. The most ordinary of water filters, an activated-charcoal filter like those found in Brita pitchers, PUR faucet attachments, or the cartridges built into refrigerators, "are incredible at taking this stuff out," says Snyder. "Almost anything in the water binds to the charcoal—the chemicals, the pharmaceuticals, the disinfection byproducts." For anyone worried about the quality of his or her tap water, the filters offer an easy and inexpensive margin of reassurance. The only issue, says Snyder, is how long the filters last. They should be replaced every couple months, because their effectiveness fades.

But a Brita pitcher isn't any way to manage the micropollutants in the long term. Understanding them, and finding a way of explaining their impact to people without scientific training, is one of the critical issues of water policy in the next decade. More and more we will want to reuse our wastewater, because as water scarcity grows, the wastewater that utilities already have will be one of our easiest and least expensive "sources" of water. But that wastewater is precisely where the micropollutants turn up first.

So if we're going to reuse wastewater successfully—whether as gray water for things like irrigating athletic fields and golf courses, or as a source of fresh drinking water—we need to understand how clean it has to be, to know how to clean it, and to be able to make people comfortable that it is, in fact, clean.

Shane Snyder, whose institute is doing exactly the kind of research the water industry needs, says that rather than try to filter drinking water, it may be much smarter to find ways of removing the pollutants from our wastewater before we release it back into lakes, rivers, and aquifers. The micropollutants, he points out, are not just an issue for people. "It's not a drinking water problem," he says, "it's an environmental problem. Let's take it out of the wastewater, not just because of human health, but because of environmental health."

An aggressive and open effort to understand what's in the water in developed countries, where it's coming from, and what impact it has, should not only bolster confidence in tap water, it should make possible all kinds of innovative water reuse without getting tangled in the debate Toowoomba put itself through. The very concern that Americans express about the safety of their drinking water should mean there is support for making sure water supplies are well taken care of.[37]

Even Toowoomba's Mayor Di started out skeptical about drinking recycled wastewater. But she gave herself the time, along with hands-on visits to actual water purification facilities, to get comfortable with purified wastewater. What she didn't do was give her constituents the same space to get comfortable—she wanted them to take the water she was giving them on faith.

Water itself doesn't respond much to the power of belief. "In the last eleven years, we have not had a lot of rain," says Rosemary Morley. "The city also hasn't run dry of water. Droughts do end, and when the drought breaks, our reservoirs will fill up, and we'll be fine."

"One of these days," says Clive Berghofer, "it will rain."

Actually, maybe not.

Three years after the referendum, there had still been almost no rain, and Toowoomba's reservoirs had fallen to 90 percent empty. The three reservoirs looked stark, scary. A vast reservoir that is 90 percent empty is kind of a spooky place—like standing on the side of an empty Olympic-size swimming pool. Vaguely ominous, even dangerous. An empty reservoir is so big, and so empty, that your immediate thought is, That's not right—they should fill that thing up.

In fact, after years of uncertainty, in late 2009, Toowoomba was rescued. Twelve hours a day, from dawn to dusk, construction crews furiously laid twenty-four miles of pipe that now connect Toowoomba's independent water system to the sprawling water grid of the state of Queensland—the water system that supplies Brisbane and its surrounding communities.

To get the pipeline in the ground on a tightly orchestrated nine-month schedule, five construction gangs worked on separate stretches simultaneously. Each segment of the Toowoomba pipeline is a fat black steel pipe forty-four feet long—longer than a typical American yellow school bus. The pipes are lined with a thin layer of black concrete, and the interior diameter is thirty inches. There are 3,600 segments of pipe to be buried and sealed—beneath fields, through the heart of a small town, along a gorge. The pipeline goes dramatically uphill—rising 754 feet from Queensland's Wivenhoe reservoir to Toowoomba's Cressbrook reservoir, the height of a sixty-story skyscraper. It's not a trivial push. When the pumps kick on each day, the water sitting in the twenty-four miles of pipeline angled up the mountains weighs 38 million pounds. (The grade is significant enough that the pipe enters Cressbrook at the base of the dam, not over the side, to save thirty meters of elevation.)

The pipeline is designed to pump 14,000 megaliters of water up to Toowoomba each year, enough water just from the pipeline to supply all of Toowoomba's present needs. With additional pumps, it could supply 18,000 megaliters.

The pipeline is costing A$187 million (US$156 million)—about A$8 million per mile, with Toowoomba's water customers paying A$75 million. But the advanced water purification plant voters rejected would have cost A$68 million. And because the feds and the state would have helped pay for the recycling system, Toowoomba would only have had to foot one-third of that cost—A$23 million. So the most immediate cost of the referendum is A$52 million in extra costs to sustain the flow of water to Toowoomba. That comes to A$433 per person in Toowoomba—it's why the city's water bills will double over the next several years.

A pipeline from Wivenhoe was one of the ideas considered before Flanagan and the council settled on recycling. It was rejected, in part, on the basis of cost.

Sending the water from Wivenhoe to Cressbrook requires significant amounts of energy. Each day, electric pumps will push 60 million pounds of water up the mountains, equivalent to pushing a train loaded with 400,000 people up the hill. The electricity is expensive enough that the pumps will operate mostly overnight, to take advantage of off-peak utility rates.[38]

So the Toowoomba pipeline has two costs that would have been smaller with the recycling plant: the electricity to move all that water is a new operational cost that is permanently added to water bills, on top of the one-time capital cost (the recycling plant would have required electricity too, one-third less than the pipeline).

Finally, in an astonishing twist that no one could have imagined before Toowoomba's tumult of the last four years, the new pipeline will inevitably bring to Toowoomba's water mains, and Toowoomba's kitchen faucets, exactly what it has fought to avoid: purified, recycled wastewater. While Toowoomba was debating whether to build an advanced wastewater treatment plant, the big water system on the coast, the South East Queensland Water Grid, spent A$2.5 billion to build three such facilities; together they can purify ten times as much water each day as Toowoomba uses. Those three plants are up and running, using the technology Kev Flanagan proposed, and the highly purified water they produce is being supplied to power plants. The utility Seqwater isn't using any of it in drinking water reservoirs yet. But Seqwater has been very clear: When Wivenhoe Dam falls below 40 percent full, the recycled water will be piped to Wivenhoe, and from there to Toowoomba.

Meanwhile, Kev Flanagan quietly went ahead and built a new water purification plant anyway, so he could sell Toowoomba's purified wastewater to the coal mine that originally inspired the idea to use the water closer to home. The new plant doesn't use reverse osmosis, so it doesn't produce water of drinking-water quality. But Toowoomba has a twenty-eight-year contract to supply the coal mine with purified water. Right now, the coal mine is buying A$8.5 million worth of wastewater from Toowoomba a year, so after repaying the A$14 million cost of the plant, that contract will yield at least A$200 million for Toowoomba.

"It's very nice. It gives me a bit of satisfaction," says Flanagan. "I'm not the dumbest engineer around."

Mayor Di Thorley served out the remaining twenty months of her second term after the referendum, then bought a tavern in Tasmania, Australia's island state, 1,400 miles from Toowoomba. The first big newspaper story about Thorley running her own pub opens with her breaking up a fight between two male patrons.

Snow Manners, the cerebral opponent who ran Rosemary Morley's first public meeting about recycled wastewater, says that, upon reflection, he thinks he could have successfully persuaded Toowoomba to accept the idea—that he could have done what Mayor Di and Kev Flanagan failed to do in the face of his opposition.

"I could have sold it," Manners says, with slight smile. "I would have used a gradual process. I would have put it in fountains and had goldfish swimming in it, with water lilies."

7

Who Stopped the Rain?

I think the days of big water are gone.
—*Laurie Arthur, Australian rice farmer*

LAURIE ARTHUR jounces around his farm in a white 2006 4WD Toyota Land Cruiser that is dusted with grit to the windowsills, equally dirty inside and outside, the way only farm vehicles get dirty. Across the front end is bolted a grille to reduce the danger from hitting kangaroos. Arthur killed one the previous night, with his wife, Deb, in the passenger seat, on the way to a Mother's Day dinner. During the workday, three dogs keep Arthur company from the rear cargo compartment. Three antennas wang from the front bumper—he's got radios and a dash-mounted cell phone, to be able to stay in touch across a lot of distance. A box of rifle cartridges is jammed in a crevice of the front seat (the guns are locked up in the house); a lot of tools are in easy reach across the backseat.

Arthur is a rice farmer in the basin of Australia's Murray River, with 10,450 acres of fields in the wide-open rangeland called the Riverina. At the extremes, he has fields 40 kilometers (25 miles) from each other; his nearest neighbor, who also happens to be his best friend, is fellow farmer Nick Lowing, 20 kilometers down the road; the Thai restaurant where he, Deb, and their daughter Lauren had Mother's Day dinner is 70 kilometers from home; the nearest movie theater is 160 kilometers away.

Arthur's farm neighborhood is crisscrossed with well-maintained dirt roads, and Arthur cruises at 100 kilometers per hour (just over 60 mph), often flanked by squads of kangaroos, who move with the ease of dolphins alongside a boat.

Arthur points out a wedge-tailed eagle, gliding low above the scrub. Native to Australia, the black-feathered raptors are as big as bald eagles, the largest birds of prey in Australia, and are often seen feasting on road-killed kangaroo. Farmers in the Murray basin grumble about them, because they have a reputation for taking young lambs. Arthur is more tolerant.

"I think they just get the lambs that are dead already," says Arthur. "Sure, they do kill the occasional lamb that's not dead. But they do it with such style."

Although the little town of Moulamein, near Arthur's farm, is about 150 miles from the edge of the true Australian outback, and all this land has been ranched and farmed for 150 years, this is the Australian bush. The land is arid, flat, and empty; it is no place to underestimate nature.

No one knows that better than Laurie Arthur.

Normally, he's a rice farmer on land that is fertile but dry. His water comes from irrigation canals, supplied by the Murray River.

Now he's trying to be a rice farmer in the Big Dry. The irrigation canals are dry, the Murray River itself is dry.

We wheel into an empty brown field called Jurassic Park—Australian farmers call their fields "paddocks," and Arthur names all his paddocks. "The previous farmer wasn't that attentive," he says. "His excavator broke down here, and when we bought this, it was a jungle rather than a well-tended field. So we called it Jurassic Park."

We climb out, and we're looking out across 150 acres of dirt—a wide space that stretches halfway to the horizon, a single field that has a border two miles around. It is 1 percent of Arthur's farmland. The ground is rough and uneven—you wouldn't drive anything but a piece of farm equipment into Jurassic Park. The soil is claylike; when you lift a chunk, it has a surprising heft.

Even empty, the tilled field makes you want to draw in a deep breath. Arthur, who lives in the modern farming world of digital specificity— his fields laser-leveled, his tractors GPS-guided, his crop yields computer-mapped—loves simply being outside. His eyes take in the sweep of Jurassic Park. "It's beautiful country, eh?" He smiles.

"I sure would like to be pulling six hundred tons of rice off this field right now," Arthur says. Looking at the empty, sandpaper-dry field underneath a cloudless sky is like looking at an idle factory: bad news. "Normally, you'd never catch me at this time of year. I'd be on a combine. For the second year in a row, I'm growing basically nothing."

Farmers grow cotton here, and oranges, almonds, apples, wheat, and miles and miles of grapevines to supply the worldwide taste for Australian wine. The Murray River basin is dairyland too; to grow grass for milk cows, farmers routinely flood their paddocks.

Arthur, like all the farmers for hundreds of miles around, is an irrigator—rain is essential, but the rain needs to fall to fill the Murray River. Water comes into the river from reservoirs five hundred miles away, and then to Arthur's fields in irrigation channels designed to rely on gravity flow to get the water delivered.

Standing at the edge of Jurassic Park, the obvious question is, What are they thinking? Who in the world would imagine a quilt of emerald-green rice paddies here, in this semi-desert?

The Australians, of course, weren't thinking any differently than the farmers of California's Imperial Valley, which is part of the Sonoran Desert and gets just three inches of rain a year. But with water piped eighty miles from the Colorado River, it is the eleventh most productive agricultural county in the United States. The Australians weren't thinking any differently than the 2 million people of Las Vegas, they weren't thinking any differently, in fact, than the rice growers of Egypt or the operators of the two hundred golf courses in Phoenix, Arizona.[1]

If Arthur's Jurassic Park doesn't seem like a spot where an ordinary person would ever imagine growing rice, both the land and the farmers know better. The soil loves growing rice. "Our yields are the best in the world," says Arthur, "just ahead of the Egyptians." Arthur averages four tons of rice per acre—20 percent more than U.S. rice farmers. Indeed, despite the drought, in 2010, with the overall rice crop at just 20 percent of normal, Murray basin rice farmers set a world record for productivity—

growing 4.9 tons per acre, three times the world average.[2] One reason Arthur and his colleagues grow rice is, "It's very profitable."

But to grow rice, Arthur needs to turn Jurassic Park into a shallow pond, flooding it with four inches of water and keeping the rice plants diked in with four inches of water for fourteen weeks. Keeping that pond of water on Jurassic Park right through the Australian summer ultimately requires flowing in enough water to have filled the field four feet deep. Every dinner-plate serving of the medium-grain Japonica rice Arthur grows requires 14.4 gallons of water in the field.[3]

When it comes to water, Arthur, who is fifty-six years old, is all-in in a way most people never have to be in their whole lives—about anything, let alone simply water. He has A$3 million in farm equipment, and land that cost nearly A$2 million. Without rain, without water, it's all dust.

"Water is 90 percent of the value of my assets," he says. "I've been farming for twenty-eight years. I like what I do. Ultimately, if I get it wrong, I'll go broke."

In the Big Dry, Laurie Arthur is both a water baron and a water prisoner.

Arthur owns six thousand megaliters of water a year, and when he can, he uses it all. That is, Arthur, a lone Australian, owns and uses enough water to supply the entire city of Toowoomba for half a year. When he gets all six thousand megaliters, he can grow enough food in a year to feed 100,000 people, which means that with half the water the city of Toowoomba uses in a year, Arthur can raise enough food to feed the whole city for a year.[4]

Six gigaliters of water is an amount hard to imagine. A single liter of water is what one of those shapely, slightly chubby Evian bottles holds: one liter.

Six gigaliters of water is 6 billion one-liter bottles: Arthur uses enough water each year to allow him to hand almost every person on Earth a liter of Evian.

Six gigaliters of water is really a lake of water. If you had a valley, for instance, half a mile wide and 1.5 miles long, six gigaliters of water would fill it ten feet deep.

It seems like a colossal slug of water, six gigaliters.

But for Arthur, it's the right amount, and really not that much. For the land he owns—10,450 acres—it comes to just 5.5 inches of water on average across every square foot of dirt, over nine or ten months of good growing season, underneath Australia's blazing sun. Which is to say, in order to feed 100,000 people for a year, Arthur needs just 5.5 inches of water, his land, and his labor.

But Arthur only gets to buy the water he's entitled to if there's enough water in the Murray River—each year, he gets some allocation of the six thousand megaliters, based on the water available in the dams and the river. From 1991 to 1999, for nine years in a row, he got 100 percent.

Over the three years from 2007 to 2009, Arthur received enough of his water allocation to put 5.5 inches on just 950 acres. During a time when he should have planted, watered, and harvested 31,350 acres of crops—his land, three times over—he's had enough water to irrigate just 3 percent of it. It comes to one-tenth of an inch of water per acre per month. A kangaroo peeing in the right place could have changed the productivity of his fields—if there had been any reason to plant them in the first place.

"The last three years have been among the worst for water in this area in recorded history," says Arthur. "And I've written this next year off as well. But I find it hard to believe that the extremes are going to become the norm."

Even in years when he gets all six thousand megaliters, managing water is as much a part of Arthur's work as the planting or the harvesting. Arthur thinks the rain is coming back; he thinks the water is coming back. But after eight years of drought, he is too much in touch with the daily rhythm of weather, sunshine, and dry dirt not to have an almost elegiac view of the future. "I do think the halcyon days are gone," he says. "I think the days of big water are gone."

The anger and emotion in Toowoomba about how to secure the city's water supply were deeply felt and personal—and yet, while the gardens may have withered, while people may have felt like they were living in a Third World country because they had to bucket water to their outside plants, the water never for an hour stopped flowing in Toowoomba's mains. While the water shortage felt intensely personal, even urgent, it was for almost everyone a purely political conversation, even a theoretical one. No one in Toowoomba risked having his life endangered, or even substantially altered, by recycled water, or by Toowoomba's steadily emptying reservoirs.

In the Riverina region where Arthur farms, water scarcity is now the dominant fact of life, driving farmers to sell their water rights, or their land, or both, causing some to commit suicide, closing businesses that depend on farmers, slowly drying and emptying the farm towns, changing everyday life in ways big and small.

This is what happens when you do not have enough water to do your work.

Laurie Arthur's situation is not just caused by water scarcity, it is quite literally the very same water scarcity that caused Toowoomba's reservoirs to go dry, the scarcity that caused the water conflict there.

But the water problems of a big rice farmer in a vitally productive agricultural area, and how his water connects to the water needs and the economy of a huge slice of Australia, pose a much different, equally important set of questions.

Should Laurie Arthur and his colleagues be farming in an area that can't grow anything without irrigation water?

Should they be growing rice, a particularly water-intensive crop, in an arid region—and does it matter that, when watered, the soil loves to grow rice so much that Arthur and his fellow rice growers have the most productive rice fields in the whole world?

The city of Adelaide relies on the same Murray River water as Laurie Arthur. When water runs short, do cities, with their density of people and their intensity of economic activity, always trump farmers, who are small in number but use water in quantities that are hard to grasp?

The farms, of course, are no less vulnerable to economic destruction from lack of water than the cities, and the farms are no easier to restore, or replace, when the water itself returns.

But how do you weigh a single farmer, and the food he raises that can feed 100,000 people in the city, against the water needs of those very same 100,000 people?

And perhaps hardest of all, who decides?

How do you make choices that are fair when those needs are competing directly against each other for the very same water, in a very short time?

Because one of the legacies of scaling an economy to abundant water is that when the abundance disappears, it turns out we not only don't have the water, we don't have a water system that can adapt to scarcity.

Arthur is struggling against running out of water. If he does, it won't kill him, but it will surely kill his way of life.

"I do have moments when I think, Nuts, I'm doing the wrong thing." Doing the wrong thing as in, insisting on continuing to be a farmer, and a rice farmer at that.

The two chief weapons Arthur deploys against the drought are an absolutely unrelenting work ethic—he works the farm until sundown, he has three demanding organizational roles that routinely take him from the farm to Australia's cities a couple days a week, and he sits at the computer in his office until 2 a.m. five days a week—and an equally unquenchable good cheer, a kind of wide-open Aussie optimism.

"Do I believe in my heart the rain is coming back? Yes. And I believe it in my mind too."

Arthur looks like Paul Newman, especially when he smiles. He's a big man, but he moves lightly. He has a quick, dry sense of humor, and a restless curiosity, equally at home talking about "return on capital," or how to raise the occasional orphan kangaroo as a pet, or the safety rules he has for himself while flying his helicopter. On the shelf in his farm office are books on French (he speaks it), land surveying (he does his own), and rice growing.

Despite having no recent farm income, Arthur has two expensive projects in the works to blunt the damage from the lack of rain. After a five-year break, he's gone back to raising sheep—"just to make some money"— which is why he's building a sturdy woolshed the size of a three-bedroom house, to handle the two thousand sheep that need to be sheared. He's also laying a new pipeline to handle some of his irrigation water, replacing an open channel—a ditch, really—that is leaky and sloppy with water, and is fifty years old. Even though he's doing most of the work himself, with the help of one of his sons and a farmhand or two, the new woolshed is costing A\$49,000. The new pipeline is costing A\$450,000. At a time when he doesn't have the water to make his existing A\$5 million farm investment pay, Arthur is spending a fresh half-million dollars to make himself a better farmer when the water returns.

Arthur loves his seven-year-old helicopter. The cockpit is a tiny red bubble, slightly less roomy than a Smart Car, from which Arthur can get a view of his farm's condition unlike any from ground level. But he has stopped paying the insurance on the copter—nearly A$1,000 a month— "and I've put it on the market. Everything is expendable." Arthur used to take fuel deliveries at the farm five thousand gallons at a time, filling an on-site storage tank. Now he takes five hundred gallons at a time, and gases up his SUV in town. "I just don't like to have all that money tied up in gasoline."

Water scarcity always creates water consciousness, especially for those of us who, unlike Arthur, typically don't think about water at all. When water is suddenly in short supply, we not only pay much closer attention to how we use it, and how much we use, we're suddenly alert to how other people use water, and how much they use. Rather than broaden understanding—Oh, that's how much water it takes to raise rice—the result is often resentment, even conflict.

"I recently went to a dinner—it was a school event for my daughter," says Arthur. "I'm sitting there with the other parents, we're going around, each of us saying what we do.

"I said, 'I'm a rice farmer.'

"One of the other parents said to me, 'Oh, you're a water waster, then.'"

He pauses to savor the presumption of the parent. Arthur has the ease of someone who can repair a A$400,000 John Deere combine or draw his own blueprints, but who would never mention either skill.

"This fellow thought he was making a profound observation about what I do," says Arthur. "As if I might not have thought about it."

AUSTRALIA IS EXPERIENCING the first wave of water envy. No one thinks he personally overindulges in water, but everyone can see the water gluttony of the farms upstream or the cities downstream or the next-door neighbor whose automatic lawn sprinklers run even when it's been raining. Water envy seems like an all-new phenomenon.

In the developed world, water service has typically been so robust that even in places like Arizona, Nevada, and California, where much of the essential day-to-day water is imported from hundreds of miles away, there has still been such an abundance that you never had to scowl over the fence at how your neighbor was using water, except in times of extreme drought. But outbreaks of water envy, of water resentment, are going to become as common as water scarcity itself.

In a sense, water envy is like class envy, or even racial enmity. It's often the result of ignorance, of our inability to imagine how the world looks from the perspective of the poorer person, or the Hispanic person, or the person using water completely differently than the way we do.

In fact, farmers have been growing rice in southeastern Australia as long as settlers have been trying to farm the land. During the early Australian gold rush in 1860, rice was grown to help feed the laborers. A well-known Australian character named Jack Brady went to California in 1920 to see how farmers there were growing Japonica, and by 1928 Australia was self-sufficient in rice production.[5]

Two things made the rice—not to mention oranges, grapes, and dairy cows—seem reasonable. The first was the presence of the farmers in the first place. During the first half of the twentieth century Australia was eager to populate, cultivate, and green its interior. To spark settlement, Australia gave soldiers returning from the first and second world wars six-hundred-acre blocks of land free if they agreed to farm them, and to stay for at least five years.

The other thing that made growing rice seem reasonable was the ready availability of water. As disastrously dry as the last nine years have been across Australia, and for the Murray River in particular, there hasn't been a period this dry, for this long, in Australia since the start of European settlement. The Murray River and its main tributary, the Darling River, have for a century been a steady source of water for all the things Australians use water for—transportation, recreation, drinking water, and growing almost half of the country's food production.

To smooth out the season-to-season and year-to-year variations, Australians built dams and reservoirs, and much of the Murray River's flow is now controlled by Hume Dam, finished in 1936, specifically to store enough water in wet years to make certain those who relied on the river would have water even through the dry times. Hume was joined in 1970 by an even bigger reservoir, Dartmouth.

The Murray basin feels not just wide open, but lightly settled—the main roads often cut through completely empty landscape, not even accompanied by power lines—but the free-land settlement project begun ninety years ago has been hugely successful. The Murray-Darling Basin is home to almost 10 percent of Australians, 40 percent of Australian farmers, and it produces 40 percent of the nation's agricultural products by value—A$15 billion a year in produce. The water smoothing system has been so successful that what is called "consumptive water use"—farms, factories, cities, and towns—almost tripled along the Murray between 1950 and 2000.

Why not grow rice?

Irrigation water powers the economy of the whole region, which is the size of Kansas, Missouri, Iowa, Nebraska, Oklahoma, and Arkansas combined.[6] The irrigation system is a huge, permanent institution. It's the area's water circulation system, just like an interstate highway system or a power grid, with thousands of miles of canals, hundreds of employees, and tens of millions of dollars in revenue a year. Although it only has 10 percent of Australia's people, the Murray-Darling Basin uses 52 percent of the country's annual water consumption, and most of that water goes to farming.[7]

If Laurie Arthur runs out of water, he will not be alone. The irrigation system on which he and thousands of other farmers depend is running out of water, the Murray River on which the irrigation system depends is running out of water, the dams on which the Murray depends are running out of water. Ultimately, the very sky over one-seventh of Australia seems to be running dry.

And yet, in many important ways, if the dams, the river, the irrigation canals, and Arthur's livelihood all die of thirst, this, too, will be because of a series of missteps and misjudgments, poor planning and overconfidence. Arthur's struggle is a warning that in much of the developed world we have built not just flower beds and lawns but a whole lifestyle—the way our food is raised, the way our communities and our economy are organized—around an assumption of water abundance.

Laurie Arthur has created a bar graph on his computer that shows the relationship between how much water he's gotten and how much grain he's grown going back two decades. Irrigators have to pay for the water that they own the rights to (when he gets all his water, the water bill is about A$100,000 a year), and the percentage they get is based on rainfall, how much water is stored in the reservoirs, and flows along the river. For twelve straight years, starting in 1990, Arthur got 100 percent of his water every year except one. He hasn't gotten 100 percent of his water once since 2001, and in four of the last seven years the allocation has been 10 percent, 9 percent, and no water at all, twice.

It's hard to overstate how dramatic and how sudden the falloff has been. The first year he got zero water, 2006–2007, wasn't just the first time of no water for him, it was the first time his water supplier, Murray Irrigation, had provided no water in seventy years of operation.[8]

Humans have short memories in all kinds of ways, but the withering of the Murray River has made what is really a water-based economy look not just wacky or foolish but positively profligate. When it isn't raining, when the Murray isn't flowing, the whole lifestyle seems to defy common sense.

"When I go to Adelaide for meetings," says Arthur, "I hear tales of the evil rice and cotton growers. What we do seems like a crazy thing to those people."

Partly that's because Adelaide is in direct competition with the farmers for water, and water scarcity makes it quite easy to see others' intemperance. But it's also because there is a huge distance between urban Australia and rural Australia. Just one out of 125 Australians is a farmer, and even well-educated Australians often can't get their minds around the basics.[9]

An editorial in the *Age,* Melbourne's major broadsheet daily newspaper, said rice and cotton farming in the Murray River's basin "should never have been encouraged," and that such water-intensive crops "promise no more than a slow and lingering death" for farmers and farm communities. In support of its argument, the *Age* said that the rice industry used the astonishing, unsustainable quantity of 1.7 billion liters of water a year.[10]

"The question," says Arthur, "was, do we write them and tell them they made a mistake? I mean, they were off by a factor of a thousand."

Laurie Arthur himself—just one rice farmer—uses far more than

1.7 billion liters of water a year. "We [rice farmers] use 1.7 trillion liters of water. Should I write and tell them how stupid they are? Nah.

"And the *Australian*"—Australia's major national daily—"did the same thing," says Arthur. "They said it takes 22,000 liters of water to grow one kilogram of rice. Well, I use 1,100 liters of water to grow one kilogram of rice. They're only off by a factor of twenty. But who's counting?"

In fact, though, the wildly erroneous numbers don't really matter. Water scarcity causes some people to tell other people how to behave, and the numbers were really decorative, not substantive. They simply illustrate the sense among urban Australians that those Murray Valley farmers—while growing enough food to feed the whole country—simply don't have any idea how to use water.

"Thirty years ago, we were greening the land," says Arthur. "Now some people think if the water crisis doesn't kill rice growing, it has failed."

Water, of course, is a natural resource, a basic material, a commodity like petroleum or copper, lumber or wheat. But even with something as vital and contentious as oil, we rarely experience oil envy. The reason is price. We let price sort out scarcity—in some roughly satisfactory way, the oil goes to the uses that make economic sense. When the price goes up, people drive less and heat their homes less, the airlines park their less efficient jets, building owners switch to LED lighting. Water envy, though, is aggravated by the fact that water has no real price attached to it that has any market meaning. During periods of water scarcity, we don't have a good way of sorting out the best uses of water in economic terms, because water isn't priced according to supply and demand, even in local areas. And we'd all agree that some uses of water that have no immediate economic value are indispensable anyway.

Water envy isn't a problem when there's more than enough clean, cheap water. But when there suddenly isn't enough water to support the lifestyle that has been created, water envy is more than just a matter of resentment or social friction. It stands in the way of making good choices during a pressing crisis, and it can prevent a state or a country from remaking its water rules, its water infrastructure, its water economy, in a way that fairly adjusts to a future of less water, or more expensive water, or both.

In that sense, Australia's problems are a gift to the rest of the developed world. They are a warning how quickly and perilously water availability can change; they serve notice that the rules we have for giving out abundant water won't serve us well when there is no water to give out.

The city of Adelaide, 1.2 million people at the western edge of the Murray basin, requires 200 gigaliters a year of water; metropolitan Melbourne, 4 million people at the southern edge of the Murray basin, requires 400 gigaliters a year. The 39,680 farmers of the Murray basin use at least 7,000 gigaliters each year—the 39,680 irrigators use ten times the water that 5.2 million city dwellers require.[11]

No wonder there's a water shortage.

THE RIVER MURRAY has been flowing across southeast Australia for 40 million years, and it has created landscapes startling for both their beauty and their variety. At the Murray's mouth, where the river meets the ocean at the town of Goolwa, the coast is estuarine, with barrier islands, the smell of the sea, and white sand dunes reminiscent of the Outer Banks of North Carolina.

You can stand on the beach, looking south directly across the mouth of the Murray. The river comes in from the right, the sand dunes part, and the Mighty Murray meets the Indian Ocean. In the opening are huge windswept breakers. Looking south through the gap in the dunes made by the river, there is no land between you and the coast of Antarctica—just 2,500 miles of cold, unruly ocean. You can hear the waves, and it seems like a perfect place for Australia's greatest river to meet the ocean, wild and beautiful.

In fact, though, the Murray's water at Goolwa is absolutely still—a flat pond that only Australia's classic black swans seem to find inviting. The Murray River has been so low because of the Big Dry that it hasn't actually flowed into the Indian Ocean for four years. There hasn't been enough water to make a current.

As the river makes that final turn toward the sea, there is a dam across its width, almost half a mile. The dam is low and utilitarian, with a lock in the middle to allow boat traffic. You can walk out across the top of the dam, and here you discover something remarkable that you can't see from shore. Looking down on the Murray River, you can see that the water on the ocean side of the dam is higher than the water on the river side. If the dam were to suddenly disappear, or even if the navigation lock were to be opened, the Indian Ocean would pour upstream, overwhelming the Murray. The mouth of the Murray isn't a river mouth these days—it's an ocean inlet, with the fresh-water river itself protected from the sea by a wall. It is as if the Mississippi—or the Thames or the Yangtze—had run out of water.

You can follow the Murray by car from where it meets the ocean toward Laurie Arthur's farm—five hundred miles upstream—and about three hours' drive from the mouth, the Murray's personality changes dramatically. The river drifts in and out of view on the west side of the road headed north. Then, all of a sudden, the land to the west drops away, and the Murray reappears at the base of curving cliffs that run for miles and soar three hundred feet tall. The cliffs of the Murray River sit on the eastern bank—they face west. In the setting sun, the limestone walls glow terra cotta, salmon, and ivory, as if lit from within. Before it was tamed with dams and reservoirs and irrigation canals, the Murray had enough spirit to carve a stunning half-canyon that looks like something from the red-rock region of Utah. Nothing here betrays to the casual traveler a river in desperation, as is so clear in Goolwa. This stretch of river looks so different than the Murray you find at the ocean that it's hard to believe it's the same river.

The Murray River was once both free and wild. Australians tamed it to protect themselves against its occasional dramatic floods, and to harness its water to feed the country. For most of the last eighty years, this domestication has worked brilliantly—the vast productivity of the Murray Valley comes directly from being able to use the Murray itself as one vast, meandering irrigation pond. But as happens in other arenas, water abundance has camouflaged a serious problem. The half-century from the end of World War II to 2000 was uncharacteristically wet for the Murray River. During that time, farmers and cities grew to regard the abundance as typical.

"Every drop of water in the river is owned by someone," says Robyn McLeod, commissioner for water security for the state of South Australia.

"It isn't a free-flowing river at all, and the environment usually gets what's left last." Her job is to be a watchdog of water security for the state—to look at the big-picture policy issues and to lobby state government to make the best long-term decisions so South Australians have the water they need and the Murray River does too.

"Until a few years ago, Adelaide and South Australia thought we had the most secure water supply in all of Australia," says McLeod. "We didn't have dams, we had the great, mighty river. And it had been a great, mighty river. Until the last few years."

It might seem that nothing could more starkly illustrate the state of the river Murray than Laurie Arthur, and hundreds of fellow farmers, receiving literally no water to grow food in two out of the last three years.

But it's worth stepping back to understand the Murray River's water intake as a whole, to appreciate how quickly and dramatically change has swept down on the Australians—because their sense of water security in 2000 was no less than our own.

In the water year 2008–2009, the Murray River received only 1,860 gigaliters of water inflow total. The farmers alone typically take 7,000 gigaliters. The river itself needs 600 gigaliters just for what is called conveyance water—the water necessary to keep the water itself moving through the locks and dams.

The whole purpose of the Murray River's reservoirs is to act as a cushion—a water savings account—for precisely those years when the rain doesn't fall. But at the end of that same 2009 water year when the river only got 1,860 gigaliters, the reservoirs were in their ninth year of below-average rainfall. The main water storages held just 980 gigaliters total. The Big Dry had all but evaporated the Murray River's emergency water savings.[12]

Indeed, if you take from that 1,860 gigaliters the bare minimum for the river—600 gigaliters—and take another 300 gigaliters for the critical water needs of Adelaide, you're left with just 900 gigaliters, for the whole river, and all its dependents, for the whole year.

So it's not just that, as Robyn McLeod says, "every drop of water in the river is owned by someone." Accounting for Adelaide, for the smaller cities along the river's length, for the irrigators, for water evaporating and seeping into the riverbed itself and being lost from irrigation canals, it's much worse than that. There aren't enough drops for every person with a claim on the river. Every drop is owned, and some drops are owned by two different people. "Welcome to my problem," says McLeod. "It is an absolute catastrophe."

Or, as Laurie Arthur puts it, with characteristic reserve, "The river is overallocated."

It is hardly a problem peculiar to Australia. Indeed, it's fairly common—it's just that most of the time there's enough water to camouflage the shortage. The Colorado River, which supplies both the fertile farm fields of California's Imperial Valley and the gaming tables of the Las Vegas Strip, is overallocated. The Chattahoochee River, which supplies both the city of Atlanta and the oysters of Apalachicola Bay 435 miles downstream, is overallocated. The Tigris and the Euphrates, shared by Turkey and Iraq, are so overallocated that some scientists predict that the Fertile Crescent, the valley framed by those two rivers where farming was born ten thousand years ago, will dry up.[13]

The question the Big Dry has revealed for the Murray is starkly different from the water scarcity question that faced Toowoomba. Toowoomba was simply arguing about how to restore its water supply. The question from the Snowy Mountains, where the Murray River begins, past both Laurie Arthur's farm and the city of Adelaide, to the Indian Ocean, is what happens if the last hundred years have been unusually wet for the Murray, and the weather pattern in place the last nine years is really the normal pattern?

"Drought" implies a devastating water-related event, like "flood" or "blizzard," but also one that comes to an end. Robyn McLeod is one of the people whose job is to take a long-term view of how best to think about the Murray River. "I don't use the word 'drought,'" she says.

IN SEPTEMBER 1996, Mundaring Weir, Perth's oldest reservoir, drew thousands of visitors to the quiet eucalyptus forest around the lake the dam creates, to watch an unusual spectacle. Visitors could stand on a walkway across the top of the 1,000-foot-wide dam as 16 million gallons of water a minute cascaded across the top, just below their feet, and waterfalled down the face of the 130-foot-high dam. The dam hadn't overflowed in more than a decade.[14]

Perth sits all the way across the continent from the Murray River, and for Australia, Perth has served the role that Australia is serving for the rest of us. Climate change and water scarcity hit Perth hard, and came to the city five years before the problems became clear back east.

It was hard to be worried about Perth's water supply as the water was pouring over Mundaring dam, but the rains and the runoff of 1996 didn't distract one man. Jim Gill was looking at the bigger picture, and losing sleep. Gill had taken over the water system of the state of Western Australia just the year before, at age forty-nine, chosen for all kinds of good reasons, with one startling exception. "I knew nothing about water," he says.

The Water Corporation, as Western Australia's water utility is called, had plenty of people who knew water. Gill was an engineer, manager, and government technocrat who designed bridges, then built roads in Western Australia's deadly dry deserts ("It was very, very remote—you never saw a boss out there"), and at age forty-one had been handed the state's tangled railroads to run.

On the phone, Gill conveys the impression of a tough, muscular, no-nonsense Aussie engineer—a guy who could wrangle outback construction workers and lay down a ribbon of macadam to a nickel mine, who could make sense of a railroad system that in 1988 still stubbornly maintained track widths different from the railroads it connected to.

In person, Gill looks like a character from a Woody Allen movie— short, with a swirl of peppery untamed hair, a big head, big hands, and small shoulders. He looks like a college math professor.

In fact, both impressions of Gill are true, and when he was given the Water Corporation to run, Perth didn't realize how desperately it would rely on both his experience and his inexperience.

"I had no idea what was facing the community," says Gill. "What I didn't realize was, it wasn't raining anymore."

Gill used an utterly sincere naiveté to spot Perth's impending water catastrophe. Then he used a quiet political and bureaucratic jujitsu to triumph in public water politics that ran every risk of becoming as inflamed as those of Toowoomba. And he used decades of experience building roads and running railroads to steer Australia's first desalination plant to completion in just two years, on budget and on time.

Gill took over the water system of Western Australia as Perth was heading into its driest years in recorded history. As CEO of Water Corporation he was responsible for securing and delivering water from reservoirs to 2 million residents, concentrated in the city of Perth but spread across an area of astonishing breadth—a single state, and a single water utility, across a piece of land as big as Texas, New Mexico, Colorado, Utah, Arizona, Nevada, and California combined.

It was a single page in a report—a bar graph—that caused Gill to ask the question Robyn McLeod is asking about the Murray River: What's normal?

Gill started at Water Corporation in March 1995. Two months later, the utility issued a fifty-year strategic plan that had been in the works for two years, a document that laid out the predicted water needs of Western Australia through 2045, and how Water Corporation would meet those needs. It was pretty routine stuff—Western Australia is booming because its incredible mineral resources are being mined to power the global economy, and Water Corporation expected to meet the needs of the new residents and industries the way it always had, with reservoirs, and by tapping some underground aquifers.

"I looked at it as it was published," says Gill, "and I said, There's something funny going on here."

Among the items in the report was the single-page bar graph—exactly the kind you learn to draw in second grade—showing rainwater inflows to Perth's dams every year going all the way back to 1907. Each year got a vertical, and the line's height showed how much water Perth's reservoirs received. There were some stunning years in both directions—six lines in ninety years soar above 800 gigaliters in a single year; four years don't even show 100 gigaliters. The average going back to 1907 was about 330 gigaliters, and that's what Perth relied on.

What was odd, from Jim Gill's perspective, with the bar graph in front of him in 1995, was this: From 1974 to 1995, there was not one big water year. In fact, there wasn't even one "normal" year—not one year out of the previous twenty-one in a row where the reservoirs got even 300 gigaliters, let alone the average of 330. What was even odder was that in fourteen of those twenty-one years, the reservoirs didn't even receive 200 gigaliters.

The most recent twenty-one years of rain and reservoir water looked completely different from the previous twenty, and from the previous sixty years.

But Water Corporation's strategic plan didn't acknowledge the most recent twenty-one years of low water—it assumed a future of at least 330 gigaliters of rainwater runoff a year.

"I said to the staff, 'Well, surely the last twenty years is a better indicator of what will happen next year, and the following year, than the last ninety years.'

"They said to me, 'No, no, no. You don't understand. You have to take the entire average.'

"The problem was, from within [Water Corporation], we had a consensus that the best indicator of the future was the entire past," says Gill. "It turned out to be good, coming in from the outside clean like that."

Gill was new to predicting reservoir inflows. And you had to be careful, because new dams and reservoirs are expensive, hard to get approved, and hard to build.

"I was in charge of this water utility, in a state that was growing. We have a monopoly—we cover the whole state. It's a pretty big responsibility. And I was being paid to be slightly paranoiac. I was waking up at 3 a.m. after the plan was released, thinking, There's something going on here. It's just wrong. We have to do something about it."

He went back to Water Corporation's staff. "Quickly, they started agreeing with me. They said, 'Yup, things are getting dryer, and getting dryer more quickly.' Which also horrified me."

How could the outsider so easily see something that the water professionals had overlooked? And, Gill wondered, what would they do if what they were seeing on the bar chart was the new reality?

This was all just before 1996, which turned out to be a big year only in comparison with the previous twenty-two years—although the Mundaring dam overflowed, 1996 didn't quite make it up to the ninety-year average.

It was easy enough to do the math. Perth's average reservoir inflows were 338 gigaliters a year from 1911 to 1974. They were exactly half that, 177 gigaliters, from 1974 to 1996, including the flush year of 1996. Despite having seven reservoirs, Perth was using more water than was coming in.

Soon Perth wouldn't even have enough water to meet its current needs, let alone support booming growth.

In February of 1996, in the midst of the first moderately wet year in two decades, Gill convened a scientific seminar for his own staff, about climate and rainfall trends. For Gill, the most startling presentation came from a scientist from the University of Washington who had studied the relationship of redwood tree rings and stream flows, going back five hundred years.

"If you core those redwoods, and measure the growth rings, going back five hundred years, the thickness of the rings is closely correlated to stream flows," says Gill. In his presentation, the scientist first revealed fifty years of data, and there was a clear pattern. In the next slide, he revealed the prior fifty years of data. "Oh, you'd think, I see now, it's actually wetter!" says Gill. The scientist just kept adding fifty years of earlier data, slide by slide. "Every time he unveiled a bit more of the picture, you'd come to terms with it, you'd get the picture. But the picture you got was just wrong.

"By the time he'd unveiled five hundred years of data, it was clear there was no cycle at all."

And if five hundred years of data is humbling in rainfall and climate terms, well, Perth's hundred years of data is in fact representative of nothing, except what you've built your dams to expect.

That meeting helped change the conversation inside Water Corporation, and helped change the sense of urgency. Gill and his staff took the fifty-year plan and accelerated much of the "new source development" into the next five years—three new dams, a couple new treatment plants, development of deep wells to tap groundwater.

The single-page bar graph showing water flows into Perth's dams is as plain a presentation device as you can get—it is utterly without adornment or digital-era special effects. It would be exactly the kind of report page or PowerPoint slide you might flip past. But if you take the time to study it, it has an utterly arresting quality. In fact, the single-page Perth bar graph has become a staple of the Big Dry across the country. Water officials in

Adelaide, in Melbourne, in tiny Toowoomba can instantly produce versions with their own data: years running along the bottom, bars rising to show water flows. Laurie Arthur even has one for the water availability at his farm. They are all equally austere, and equally stunning.

Jim Gill has learned to use an updated version, with water flows starting in 1907 and running right up to 2008, to devastating effect, using precisely the technique he found so effective from the University of Washington professor.

Gill begins in 1995, the year he started, leaving the whole century open but covering the future years—then he slides away the cover sheet, unveiling each recent year, year by year. You don't know where you're going. You don't know how much water is coming.

You don't even need to know the real amounts of water, just watch the heights of the bars. The long-term average water flow is a line 1.5 inches tall—that's the water Perth actually needs. The tallest bar on the sheet—with a century of data—is 4 inches high, 1945. Gill's first year, 1995, is three-quarters of an inch tall. Then comes 1996, not quite 1.5 inches. Okay! Then 1997: a half-inch. 1998: less than a half-inch. Two years in a row with one-third the necessary water.

Nineteen ninety-nine and 2000 feel like a relief—both pop back up to three-quarters of an inch tall. But wait—they only look good compared with the years just before them. Together 1999 and 2000 equal just one average year going back to 1907.

"You have to plot scenarios," says Gill. "Let's look at the worst years. What if we have a string of those?"

The shift in perspective is striking. You can feel a tiny bit of what the water suppliers, the water officials, see and feel at times like this. We're grumpy about not being able to water our lawns, about our dirty cars and our short showers. This small exercise gives you a taste of the anxiety of being responsible for the empty reservoirs. How much water is coming?

Gill keeps revealing years. Two thousand and one is a stunner: barely a quarter-inch of line, a tiny stub—the lowest year of water flow in a century.

"In 2001, it was clear this was one helluva crisis. After 2001, I started asking a different question: What if that 2001 rain is repeated for two or three years in a row? What happens then? We were just going to run out of water."

Mind you, by 2002, Water Corporation had done almost everything in its fifty-year plan to "secure" more water. "We had built three new dams and there just wasn't much water in them," says Gill. "You could see the mud in the bottom."

THE POLITICS OF WATER was never far from the surface in Perth.

The city has the relaxed charm of South Florida or Southern California from a more innocent era. Downtown is compact and walkable, with a waterfront along the Swan River that is busy with people taking advantage of Perth's mild climate; suburbs of single-family bungalows with gardens in front and a patch of yard in back roll out for miles in all directions.

Even before the Big Dry, Perth's climate was more like that of Greece or Spain than the English countryside, and yet in Perth, as in Toowoomba and much of Australia, the individual English-style garden is much prized.

Sue Murphy was a senior Water Corporation manager hired by Jim Gill, and she succeeded him as CEO in 2008. She often uses Perth's English gardens as an example of how water habits that literally make no sense take hold and come to seem natural. "If we'd been settled by Mediterraneans of some kind—Greeks, Spaniards—we would not have these English gardens everywhere. We have them because life in the beginning was miserable here, and people were trying to re-create what they had back home in England," says Murphy.

It's not just the gardens. Perth's residents have installed 100,000 backyard swimming pools.

Despite the slow erosion of its water supply by 2001, the water culture in Perth was complacent. "We hadn't had water restrictions of any kind for twenty-three years," says Gill. "Perth had grown—people here are affluent. At Water Corporation, we felt all hell would break loose if we only allowed outside watering two days a week."

Gill is a better tactician than his colleagues in Toowoomba. He didn't attempt to shock the culture—he wanted simply to wake people up to the water situation. Water Corporation started advertising aggressively, getting residents tuned in to the drought, urging them to reduce their daily water use, inside and outside, in advance of restrictions. Dual-flush toilets were mandated in new construction, as were low-flow showerheads.

Inside Water Corporation, there was an air of controlled panic. Gill and his staff planned to be able to run Perth with no water at all in the reservoirs that had historically provided 70 percent of the city's supply. They planned in an emergency to use nothing but wells.

Gill and his lieutenants also established a drought war room in a second-floor conference room at Water Corporation headquarters. "We were looking at the weather forecasts every day," says Gill. "We were looking at the sunshine out there every day. I looked at the satellite images every day. I looked at the three-month look-ahead forecasts. I looked at the global circulation models. It always seemed bad."

Beyond conservation—Perth residents ultimately scaled back per person consumption enough to save 45 gigaliters a year, which is equivalent to what an entire desalination plant produces—there were really only two options for adding water that didn't depend immediately on rainfall: tapping a vast aquifer called the Yarragadee and building a desalination plant.

In typical Gill fashion, he went for both. "We wanted two solutions on the go at any one time," says Gill, "because one might fall over at any time."

Yarragadee posed political and scientific problems: Assessing the dynamics of deep aquifers is very difficult, and there wasn't much science available on the Yarragadee. Farmers and residents in the area were suspicious of tapping the aquifer to supply the entire city, worried their own wells would dry up, worried about salt water creeping in from the Indian Ocean, worried about the sustainability of sucking water out of a source that, while seemingly large, had taken 100,000 years to accumulate.

Environmentalists were equally opposed to a big desalination plant. One of the problems with desalination is that, when you take in a hundred gallons of seawater, you typically produce about forty-five gallons of pure drinking water. You're left with fifty-five gallons of water that has all the salt in it that was originally in the hundred gallons—double-concentrated brine. You aren't, in fact, adding any salt back into the ocean that wasn't there to start. But before it is diluted, that stream of brine can do a lot of damage to the ecology of the ocean where it is disposed of.

The site of Perth's proposed desalination plant—and the place where it was ultimately built—is on a bay called Cockburn Sound. So the desal plant wouldn't be sending double-concentrated brine back to be diluted by the Indian Ocean's tides and currents. The brine would go into a semi-enclosed bay. The very real worry was that the desal plant's super-salty effluent would turn Cockburn Sound into a lifeless salt sea.

Desal faced opposition for another reason, one Americans find hard to appreciate. In Perth, there is widespread agreement that the absence of rain is the result of climate change. A desal plant is a huge consumer of electricity. So there was resistance to desal based on the idea that the desal plant was only necessary because of climate change—but that building it would increase greenhouse gas emissions and so ultimately make worse the very problem it was supposedly solving.

As Perth's water crisis became more and more obvious, all kinds of ideas were floated. The level of water or engineering expertise was irrelevant. Colin Barnett, running for premier of Western Australia, backed what became known as Colin's Canal—the idea that all Perth's water problems could be solved with a big ditch. Because the unpopulated northern part of Western Australia is flush with tropical rainfall that runs into the ocean, Barnett argued that the state should simply build a canal to bring the water down to Perth. Sometimes the distances in Australia seem to confound even Australians. It is literally the equivalent of proposing to supply Las Vegas, Nevada, with water by building a canal from Niagara Falls, New York.[15]

Gill set his staff to analyzing whether the city could really tap, and then rely on, the Yarragadee Aquifer, which was his first choice, and the first choice of many at Water Corporation. He was skeptical of desalination—it wasn't just expensive and energy intensive, many recent desalination plants had proved infuriatingly difficult to get built and running properly. The largest desal plant in the United States, in Tampa Bay, was just half the size of the one being considered for Perth, and Tampa Bay's plant ran 30 percent over budget, didn't start making water until five years after its scheduled completion date, and took yet more years to provide the amount of drinking water it was designed to.[16]

But unlike Toowoomba, where city officials locked in on a single solution, Gill took both the aquifer and desalination seriously. He sent staff members to a German engineering firm with experience building desalination plants, then he went to Germany himself.

"I did have quite a bit of doubt about whether we could do it right," says Gill. "I went to Stuttgart and I said, Make your best case. I needed them to convince me it would work, that it would not take three to five years to come to nameplate capacity. They were very convincing."

For Gill, even years later, it wasn't a close call. "The preferred source was very clearly the Yarragadee," he says. "There was minimal environmental impact, it was less expensive, less energy was required."

But the Yarragadee opponents were loud and potent. And Gill came up with two masterstrokes that helped make desalination publicly acceptable.

First, he proposed powering the desalination plant with all-new windmills, installed at a windmill farm up the coast from Perth, so the water factory would use only renewable energy. No climate impact at all.

To tackle the problem of discharging 42 million gallons of brine a day into Cockburn Sound, Water Corporation designed a discharge pipe for the water factory that over its final six hundred feet has forty nozzles sticking up like porcupine quills, a nozzle every sixteen feet, at different angles, to widely disperse the flow of double-salty water. And to satisfy the critics that Water Corporation's design would work, Gill hired an outside consultant to assess the outfall pipe. He hired the most vocal critic of the desal plant, a prominent scientist, professor, and water expert named Jorg Imberger, who had said publicly, "Desal is like taking an aspirin for a tumor."

Imberger's study showed that the porcupine-nozzle design worked perfectly. The nozzles completely dispersed the high-salinity water from the plant, even in the enclosed bay. "It doesn't cause a problem," said Imberger.

Although Perth thinks of itself as having confronted its crisis and had a reasonable public debate about it, in the end the decision was made based on politics, rather than a technical analysis—as it was, in the end, in Toowoomba.

Geoff Gallop, the premier of Western Australia and both Jim Gill's boss and admirer, was briefed right along on the science and the politics of choosing either the aquifer or the desalination plant. He chose the A\$340 million desalination plant. "I was always a bit nervous about the politics of taking water from underground, despite the science," says Gallop. "We've got our wine industry there, all the agricultural industry down there. The science was okay. But the politics of desal proved much better for me, whatever the science."

Once Gallop chose desal, Gill didn't hesitate. He controlled the plant construction tightly. Companies bidding to build and run it, for instance, were assessed by two separate teams—one evaluating engineering and construction competence, the other evaluating financial stability. (Tampa Bay's desalination plant suffered through bankruptcies of three of its main contractors.) During construction, no tours, or reporter visits, or politician photo ops were allowed. "Managing that was very difficult," says Gallop. "If the desal construction process had become politicized or had union problems, the press would have started to write, 'The ill-fated desal plant.' They avoided that completely. On time, on budget. That was a brilliant bit of management."

The Perth water factory is the first facility of its kind in Australia—the first city-scale desalination plant being used to provide drinking water anywhere on the continent. Gill got it built in crisis mode—contracts let in July 2004, spring-pure water flowing to Perth's water mains in November 2006.

The heart of the factory is a vast building, three acres under one roof, a warehouse-like space three stories high that is stuffed with pipes, pumps, electronics, and a noise so loud that you can feel it thrumming your bones. The roar drowns out everything, and the noise never rests, it never pauses or cycles. That's what a modern reverse-osmosis facility sounds like.

Turning seawater into drinking water is a straightforward business—military and passenger ships have been doing it at sea for decades—but if you want to make city-size quantities of drinking water from the ocean, you need muscle. You need to suck in city-size quantities of water, and you need to slam that water through reverse-osmosis filter cartridges so fine that almost nothing gets through but the water itself.

Perth's desalination plant guzzles 55,000 gallons of water a minute. To get everything out of the water—to squeeze out the salt and the shark poop, the seaweed, the decaying jellyfish, the bilgewater from a passing freighter—the water factory has to crank the water up to a pressure of 940 pounds per square inch, the same pressure the ocean exerts 2,100 feet down. It is almost enough force to crush the protective pressure hull of a nuclear submarine.[17]

Inside the main building at Perth's water factory, there are thousands of white filter cartridges arrayed in racks—55,000 gallons of water a minute jetting through them, each cylinder holding fast against 940 pounds per square inch of water pressure, all that water jacked along by a line of enormous pumps right down the center of the building. That's where the noise comes from: the pumps pushing the water through the tiniest of reverse-osmosis filter pores. It is, in fact, the sound of Perth rescuing itself from water disaster.

Although it sits incongruously on a resort-worthy slice of coast, facing a vista of turquoise Indian Ocean, the desal plant is a massive industrial operation, wedged next to a power plant. One of the striking things about it is that it is immaculately clean. There is not a weed or a scrap of litter visible on the sixteen-acre site. Even the gravel is well groomed. Just outside the main reverse-osmosis building is a small blue-gray box that looks like it might hold electronic controls. In fact, chief engineer Steve Christie pops open the hatch, and there's a spigot inside: This is freshly desalinated ocean water, the water made by the water factory. Ninety minutes from Cockburn Sound to your palate. It tastes wonderful.

And Christie says the cleanliness of the facility—which is not open to the public—is no accident. "The whole system is sealed," says Christie. "Our aim is that not even an ant could get in. We want to be able to take the water and send it, clean and fresh, straight into the piping system to people's homes."

The Perth desalination plant has operated smoothly since opening— it consistently supplies 17 percent of Perth's annual 300-gigaliter water thirst. And it was a reasonable choice that solved the immediate problem, although whether it was the best choice is a different, perhaps unanswerable question.

"There's no solution to the water problem without the people being involved," says Gallop, of choosing the plant over the aquifer. "When the politics of water get hot, you need public support. I really did want the people on side with this."

And there's no question that Jim Gill, with his bar graph, rescued Perth—perhaps from the very crisis that the Australian scientist Tim Flannery had predicted in that famous newspaper headline "Perth Will Die, Says Top Scientist."

But there is a much larger question about Jim Gill's handling of Perth's nerve-rattling plunge into water scarcity: Did Jim Gill waste a good crisis?

Gill doesn't shrink from the question. He thinks, in fact, that it is the most important question.

"In some ways, we defined the problem and the solution within the existing framework of abundant water," says Gill. "And we've still got abundant water here.

"What I really believe is, we just use too much water. It's an amazing leap of arrogance that to get rid of five hundred milliliters of urine we use six liters of drinking water—twelve times the amount of the urine. That's crazy. It's just an example of how we are not serious in our life about how we manage resources.[18]

"Perth people use a lot less water than they used to"—20 percent less, per person. "And you could argue that if we hadn't pushed back the problem so effectively, if the water hadn't continued to flow from the taps, we'd be further along on the long-term goals.

"You could argue that. But we had to stave off the crisis.

"I do think we're living an unsustainable lifestyle. We need to wean ourselves from abundant water."

IN AUSTRALIA, IN UNDER TEN YEARS, one vitally important point has been made bluntly clear: Despite their utter reliability, our water systems are anything but robust. They are durable. But they are rigid, locked into their own assumptions of where the water will come from and where it will be needed. They have no flexibility, no adaptability. When the rain they rely on falls somewhere else, when the river stops flowing, when the underground aquifer we've been tapping starts to fall, we look around in astonishment and betrayal. What just happened?

Our water habits rely on those same assumptions of water availability. Abundant, flowing water is its own invitation to indulgence. That's true in nature—who can resist putting a hand in the flowing current of a creek? And it's especially true in a world where literally no signals tweak us about our water use, either as we're using it or even in our monthly or quarterly water bills. The astonishing monthly electric bill is a sharp reminder of why we use the air conditioning carefully in summer; the price of a gallon of gas encourages us to consider cars with good mileage when it comes time to buy a new one.

But we've set up a system that treats water differently. A short shower is virtuous, but hardly rewarding, whereas a long, steaming bath—using in twenty minutes the water many Australian cities hope is a full day's supply—is alluring, easy, and relaxing. Why not? You could take a bath four days a week without actually knowing how much water you were consuming; you could take a bath four days a week without noticing a change in your water bill. Indeed, despite water scarcity from one end of Australia to the other, even in cities like Perth and Toowoomba confronting crisis conditions, the typical home water bill remains less each month than the cell phone bill. Imagine how water habits would change if our bathroom and kitchen faucets simply displayed a small, digital readout showing how many gallons we were using, and had used from that faucet that day? It would be fascinating, it would be fun, it would save a lot of water.

In Perth, Jim Gill retired as CEO of Water Corporation in December 2008, and was succeeded by Sue Murphy.[19] "You cannot overstate how hard it is to follow Jim Gill," she says. "He is the water god in this state, and in some ways in Australia."

Murphy shares with Gill a career steeped in engineering and construction. She spent twenty-five years at one of Australia's leading engineering firms before Gill hired her to manage construction of that first desal plant. And she, too, knew nothing about water when she joined Water Corporation. "For me," says Murphy with a smile, "water was just a source of large construction projects—dams, reservoirs, pipelines."

But as CEO, Murphy's mission is completely different from Jim Gill's, although she feels it with at least as much urgency. She wants to confront the problem Gill didn't have time to confront: Perth's water culture. Water

Corporation is using a new slogan, "Water forever," that in just two words captures the challenge. Murphy wants to reset the assumptions about water, to dig into the sociology and the psychology of water use and permanently alter how people think about it.

"The drivers of water have been supply and demand," she says. "Our job as the Water Corporation has been to simply increase the supply. But now our job is not simply to accept the demand and meet it with supply, but to take demand and manage it down." She likes the phrase "nega-liters"— the liters of water she doesn't have to deliver because people find more efficient ways to use water. "We treat nega-liters as the cheapest water we supply." It is cheaper and easier to teach people to use the water they already have more carefully and productively than to build more dams, desal plants, and pipes.

"From my point of view, the mantra of the whole twentieth century has been 'cheap,' " Murphy says. "The mantra of the twenty-first century is going to be 'value'—what's the best value?"

It's an idea that, in fact, is almost never applied to water, at least explicitly. What's the best use of a gallon of drinking water? Flushing away some pee? Watering a backyard? Growing rice? Making a microchip?

Murphy agrees with Gill that Water Corporation didn't use the crisis either to scare or to force people into living differently with their water. "Fundamentally," she says, "we haven't changed anything." That's why she wants to ask harder, more complicated questions.

"If you look at how we set out our society here in Perth, we have quarter-acre blocks, with a nice lawn, so we can play football in the yard. In the last twenty-five to thirty years, things have changed. Mom and Dad aren't home in the afternoon that much. The kids are in [day care]. Dad doesn't really want to spend the weekend doing the yard. The backyard has become a place for entertaining, but not really a place for the kids to play football.

"That's why we need good shared spaces—that can be watered with recycled water."

This isn't water as an infrastructure problem, it's water as a sociology problem.

Because of the dramatic water scarcity, and the relentless public focus on how to find new water sources, says Murphy, "We've got some public acceptance of the idea that using less is good. But we haven't locked in new behavior. We've locked in gains like dual-flush toilets. Behavior is less tangible, and I don't think it's changed.

"It's like the difference between a crash diet and changing your eating habits forever.

"If we are going to change the way we construct society, we have to distill what we love and what makes it great," says Murphy. "That's not the mandate of the water utility. But someone's got to start that conversation."

And, of course, water has always shaped society—economically, sociologically, geographically. So the modern water supplier is not a bad place to start the conversation about "water forever." But it's also true that, crisis or no, it is in fact a lot easier to build a desal plant than to get 2 million people to build new habits.

Australia is taking its water scarcity seriously, but Australia is unusual in that, with 85 percent of its people within a thirty-one-mile strip along its coastline, almost all of its urban water problems can be solved with desalination. And desalination has what Australians have come to see as a standout advantage: It's climate independent. The oceans aren't going anywhere. If you're willing to pump in the energy, the water factories will pump out clean water. But "climate resilient water" isn't cheap.

Perth has a second desal plant under way, Brisbane and Sydney have plants built and operating, and Melbourne and Adelaide each have plants under construction. Overall, just the desalination plants that Australia's state governments have built, or committed to building, will cost A$13 billion—A$600 for every resident of Australia, just for construction.[20]

It is quite possible, in other words, to spend enough money so we don't have to do what Jim Gill insists we must: wean ourselves from abundant water. At least some of us can.

IF YOU STEP ONTO THE PORCH of Laurie and Deborah Arthur's farmhouse at five-thirty in the morning, it's easy to be surprised by the perfectly camouflaged, and momentarily motionless, visitors.

The Australian bush spreads out before you, dawn is just spilling over the flat land. There are tall trees near the house, giving way to wide vistas of scrub, crossed by a dirt road. If you step down off the porch, the ground itself suddenly seems to ripple, whirl, and race away from you, as if a wave were bunching under the brown dirt and rolling out.

It takes a minute to get your bearings, to figure it out.

It's the 'roos.

The kangaroos have quietly congregated in the Arthurs' yard overnight—perhaps fifty of them—and to the unpracticed eye, a kangaroo in the bush is invisible, even fifteen yards away, until it starts to fly.

The kangaroos have the coloration of deer—brown except for white bellies—but what is so striking is how they move. It's like no other creature—the kangaroos seem to hover, defying gravity, their legs pumping an unseen blur, their bodies angled forward like in a Road Runner cartoon, moving at an arresting speed across ground they do not seem to touch. It's hard to get your eyes to keep up with them—they are moving so fast, so silently, so effortlessly that you can lose them against the brown scrub while you are looking right at them. Their movement is unearthly—their speed, their ability to change direction without a flicker of exertion, their ability to be one place, disappear, and reappear someplace else in an eyeblink. Close to the Arthurs' farmhouse, the kangaroos are stubbornly territorial. They are reluctant to give way, they go off a bit and turn back and gaze at the house, then instantly they're racing for the horizon.

The 'roo posse is part of Laurie Arthur's unspoken covenant with his land. Lots of farmers consider the kangaroos a damaging nuisance. But as with the eagles, Arthur thinks the kangaroos are as entitled to roam the Murray Valley as a guy in a John Deere combine.

"We certainly have 250 kangaroos on the farm now," he says. "We've had four or five as pets for a while. A joey will get separated from its mum, and Deb will take it inside, make a little pouch for it, which she hangs from a doorknob, so they can climb in when they're feeling nervous."

Arthur does a similar thing when he gets cranky. He fires up his tiny red helicopter, with the cockpit that is nothing but an acrylic bubble, and he floats up to a thousand feet.

"When I get in a nasty mood, I get up here, and I just feel relaxed and happy," says Arthur. "When I'm up here, I just can't sweat it, you know?"

The chopper's engine is so loud, two passengers in the cockpit need headsets to talk, even sitting with their legs practically touching. The view is majestic—a thousand feet of altitude over the flat Murray Valley takes the horizon out a dozen miles or more in every direction.

It is late fall in Australia—May—and spread out below Arthur are thousands of acres of his land, all chocolate brown, like pans of brownies. The ground should be gleaming green with crop.

"This is my forest"—Arthur traces his finger along a wide band of dense trees, not green but khaki—"that's red gum forest. This is my forest, crying out for water."

Suddenly he smiles. "As you can see, except for the drought, we've got the world by the throat."

It may be Arthur's thirty years working the land that has given him a sense of calm in the face of a crisis no less urgent, no less potentially devastating than that Jim Gill faced. It isn't fatalism—Arthur is working as hard as the lack of rain will let him—but he knows he can't make it rain, and he knows that there is no one to rescue him.

Arthur is an unusual farmer in many ways, not least because over the last decade he has also slowly migrated into the top ranks of water policy and water politics in the Murray basin, and in Australia. He sits on the seven-member National Water Commission. He chairs the water task force for Australia's National Farmers Federation. And he sits on the board of directors of Sun Rice, Australia's nearly A\$1 billion rice growers cooperative—"although I'm not sure how long they'll let me stay, since I'm not growing any rice."

In all, Arthur's off-farm work consumes a couple days a week, and often finds him in Australia's capital, Canberra, or in Melbourne or Adelaide for meetings. Although those roles are both demanding and important, the ten thousand acres in Moulamein anchor Arthur's attention. "I'm a farmer," he says. "Full stop."

But the work in conference rooms and offices gives Arthur a helicopter perspective on water policy, to go with the everyday, dirty-boot-level perspective.

As in Perth, there are really two layers to the frustration of the irrigators. The first is, simply, no rain. There's not much to be done about that. Farming with no water is like trying to farm with no seed or no dirt.

The larger issue is how the Murray River itself is managed. "In Australia," says Arthur, "everybody wants river health, as long as the other guy does the health part."

Mike Young, one of the country's most prominent and well-regarded water economists, suggested that every Murray River irrigator give up 1 percent of his water rights a year, until farming's water stake is down to a more sustainable level. Of course, the water entitlements are property, like land or buildings. And giving up 1 percent is like giving up the right to farm 1 percent of your land each year.

"Why us?" asks Arthur. "There is a parochialism, a defensiveness, when it comes to water, no question. I can feel it slinking out of my spine sometimes." He winks. "I have to restrain myself.

"But the failures of the past are being brought to account on one group of people. Our position is, if you want our bloody water, come and buy it. If the very basis of your business is water, how can they take that away without compensating you?"

Arthur's defensiveness isn't paranoid. Wayne Meyer is a professor at the University of Adelaide, who has spent much of his professional life studying farming, irrigation, and the most effective use of water—living for years among Australia's irrigation farmers.

"The irrigators have been living in a dream world," says Meyer, "just like the cities of Perth and Adelaide. The irrigators have developed an entitlement sensibility.

"The big problem for irrigation is, it never generates enough money out of the irrigated production"—the food—"to be competitive with the productivity of water used in an urban or industrial setting.

"Irrigation," Meyer says, "has to be a societal decision. Because it doesn't win in a bid system. The Hyatt Hotel in Adelaide will always outbid the orange grove."

Arthur knows that too. He doesn't think he's entitled to a fair shake from the sky, just from his fellow Australians. "The irrigation systems were set up without an understanding of the soils or the groundwater," he says. "It's been a miserable way to manage the Murray basin. And an event like this has caught everybody short. It has brought everything to a head.

"Is there an argument that this isn't the place for flood irrigation farming? If someone wants to say this isn't the place for an irrigation scheme, an argument that the whole thing should never have built, I could listen to that." Just don't decide that without acknowledging that it's society pulling the ground out from under a whole way of life that society had previously promoted.

Meanwhile, Arthur works as hard as a fifty-six-year-old man can. In 2008, he took his combine and a crew north for seven weeks to do contract harvesting for Australia's largest wheat grower—just to make money.

The new woolshed got finished, with help from many neighbors, just ten hours before the sheep shearing started. Although the newly revived sheep flock yielded good wool, the price was low, and the total sale didn't cover the cost of the new shed, not to mention the cost of taking care of the sheep themselves.

Arthur pushed forward with his new water pipeline, to improve the efficiency of his irrigation, but just to the end of the first stage—running to one of four fields he hopes it will supply. "I'm out of money. The pipe is going to cost $450,000, and I've spent $195,000 so far."

In a twist on the classic O. Henry story, the first stage of the pipeline to improve Arthur's use of water was paid for by selling water. Arthur sold a tiny slice of his entitlement to the government—3 percent of his total water, for which he received A$195,000. The Murray River gets to keep that water. But that water is gone forever—it's the first time Arthur has ever sold water—so the new pipeline had better pay off.

Arthur's strategy, his skill, his determination, his work ethic mean nothing without water. There are, in fact, many ways to die of thirst in Australia, and Laurie Arthur's farm may be dying of thirst right in front of him.

"The big question is, is this climate variability or climate change? I have a couple million dollars' worth of gear—everything from what I'm driving, this old ute, to the new woolshed, riding on that question—$2 million worth of capital eating its head off with depreciation and opportunity cost. That's my bet."

Arthur thinks the rain—some rain—is coming back.

"If I really knew this was a step change in the climate, I'd cash out. Right now. If this is a step change, I've backed the wrong horse."

Why is he up until 2 a.m. five days a week, working on a farm that is yielding only frustration? Arthur doesn't hesitate. He breaks into a smile just like the one he has a thousand feet up in his helicopter. "Life is short," he says. "I can see the end of it from here."

Just steps off Arthur's front porch, to the left, is a patch of fenced ground tucked under the trees along the creek, easy to overlook as you come and go to the house if you don't know it's there.

It's the grave of Laurie Arthur's brother Neil, who died in April 2001.

"He was just getting divorced," says Arthur. "He died in a plane crash, piloting the plane we owned together." Neil Arthur's girlfriend was also killed when the Beechcraft Bonanza went down near the town of Goulburn. "He was flying with his new love."

The Arthur brothers farmed together—bought land, picked crops, planted and harvested, made decisions, together.

"You know, when my brother was alive, we talked about every decision. And if we agreed on something, then we did it. If one of us didn't agree, we usually didn't do it. And if we agreed, it always worked out, you know?"

In 2001, the Big Dry hadn't taken hold yet. The choices Laurie Arthur is making now are harder, with less margin for error.

"I'd really appreciate having him here now."

8

Where Water Is Worshipped,
but Gets No Respect

This is a society where people believe if they take one dip in
Mother Ganga, they are going straight to heaven. That's irrational, of course. But people are irrational. Thank goodness.

—*Ashok Jaitly,*
director, Water Resources Division,
The Energy and Resources Institute (TERI),
Delhi, India

IT IS STILL DARK at five-forty on a Thursday morning in late October, along
a quiet boulevard in a neighborhood of Delhi, India, called Vasant Kunj.
The street has a generous median in the center lined with wide-spreading
trees, and Vasant Kunj is on Delhi's south side, tucked between the international airport and Jawaharlal Nehru University.

Sixty or eighty people are gathered on the sidewalk. Some have come
on bikes, and many have brought young children. Although it is before
dawn, everyone is dressed for the day—the women and girls in bright orange and turquoise and red skirts and blouses, the men and boys in slacks
and untucked long-sleeved shirts, shorts, and T-shirts. Scattered at their
feet are hundreds of battered containers of every kind—open-topped
five-gallon buckets, two-and-a-half-gallon jugs with wide mouths and
molded-in handles, ordinary pails, old bleach bottles, huge metal cooking
pots. Every container has had a previous life—carrying paint or chemicals
or fuel. Also sitting in the dirt on the ground, or leaning upright against the
legs of men and women, are coiled hoses. The hoses, too, are a hodgepodge
of every kind and diameter—some are ordinary green garden hoses, some

are as thick as fire hoses—and many of the loops have been patched over and over with tape.

Out of the darkness roars a truck that is both imposing and rickety. Its cab has the profile of an old-fashioned milk-delivery truck. The back end is an enormous blue tank, the size of a fuel-oil delivery truck in the United States.

The tanker grinds to a stop at the curb. For the hundreds of people living in the slum area of Vasant Kunj, tucked invisibly behind a gas station, Thursday's water has arrived.

Everyone moves with speed and intensity. Two tall adolescent boys hoist themselves up onto the top of the tanker truck, and ignoring the hoses and valves along the truck's flanks, they pop open a large hatch on top. Men and women on the ground toss their hoses up to the boys. One end of the hose goes into the big hatch, and deep into the tank; the other end goes into someone's mouth, who sucks until the water starts flowing. Sometimes it's the person on the ground, sometimes it's one of the boys. Once you get the siphon going, the water pours from the hose in a forceful stream.

Within minutes, there are fifty-five hoses draped out of the top hatch, trailing down the sides and back of the truck, filling the buckets, bottles, and jugs.

The equation here is pretty basic. You bring as many containers as you have, you bring as many people as you can to help you fill and carry the containers, you fill as many containers as you can, as quickly as you can. The amount of water you get depends on all those things—and on the diameter of your hose too.

This is today's water, Thursday's water. There will be more, of course—Friday morning at about five-thirty.

The tanker comes from Delhi's municipal water utility, the Delhi Jal Board (DJB), and it carries 10,000 liters, about 2,500 gallons. The fifty-five hoses, each primed with mouth power, empty the tanker in fourteen minutes.

When the water is gone, the hoses are quickly pulled back to the ground and coiled. No one fusses about keeping them clean.

Once the hundreds of containers are full and the tanker is empty, the handling of the buckets and pails and pots becomes a matter of great delicacy. Any water splashed overboard is water you can't use to cook or drink, to wash or brush your teeth with.

People stand surrounded by their day's water—one family member stays with the containers, everyone else carts water down a path, back to their shack in the slum, walking with that distinctive counterbalanced posture that involves carrying something that is both heavy and sloshing. Even if you manage to fill six or seven five-gallon containers, maybe you've got forty gallons of water for your family for the day. A single American seven-year-old uses forty gallons of water in a single bath; if an American family were to start the day surrounded by the day's water, there would be 40 or 50 five-gallon containers, nearly 2,000 pounds of water, to be hauled.

One boy uses hooks and straps to sling water containers on either side of his bicycle—a five-gallon bucket slung on the right side, a two-and-a-half-gallon jar opposite it, another five-gallon bucket bungee-corded onto the back—105 pounds of water wobbling home on a two-wheeler.

Many girls walk off with five-gallon buckets balanced on their heads—42 pounds of water gliding off into the dawn, steadied with both hands and a well-conditioned neck.

The cleanliness of this water is dubious. Even if the water that goes into the tank is pristine, and the inside of the creaky tanker is, by some miracle, cleaned routinely, the water that people walk off with is only as clean as the containers it is siphoned into, and as clean as the siphon hoses themselves. Actually, your water is only as clean as the dirtiest hose from your neighbors that goes into the big hatch.

Getting your daily water from one of the DJB's tankers requires a scramble that wrings the dignity out of you before the sun has even come up, except that it is so routine that it becomes more chore than insult. It is how each morning begins—it is how each morning must begin, because you can't skip a day any more than you can skip a day of having water.

The ritual of the water tanker wheeling to the curb, with dozens of people waiting to fill their jugs, is repeated thousands and thousands of times a day across Delhi. The DJB runs a fleet of water trucks like this to deliver water to people in what are bureaucratically called "unauthorized settlements," but which everyone in India calls slums, ad hoc communities that have no routine water service.

Just delivering tanker water is a vast enterprise. The DJB has a fleet of a thousand trucks—if each delivers water for only two thousand people each day, that's 2 million people a day siphoning their water from a truck, then hauling it back to the shacks where they live with their families.

It is a kind of water slavery that millions—tens of millions—of poor Indians are trapped in every day, in India's fast-modernizing cities, and also in its villages. People are literally captive to the daily task of fetching water—their ability to go to work, to send their children to school, to get a full night's sleep, to be healthy, all hostage to the schedule on which the water is available, and hostage also to the quality of that water.

For India's poor the battle to get water has to be re-fought daily. But in India, in fact, everyone battles to get water.

Vikram Soni is a theoretical physicist who lives in Vasant Vihar, one of Delhi's better neighborhoods, which is just on the other side of the campus of Jawaharlal Nehru University from the Vasant Kunj slum. Vasant Vihar is home to two more universities, a sprawling, luxury indoor shopping mall, the DLF Promenade, and the embassies of Taiwan, Argentina, Iraq, South Africa, Saudi Arabia, Spain, and two dozen other countries.

Soni, who recently retired from India's National Physical Laboratory, has a special interest in the astrophysics of compact stars, and has lectured at MIT. He is as unprepossessing as you'd expect a theoretical astrophysicist to be—his three-story home is filled inside and out with well-tended plants, the furnishings are comfortable but well used, and like most professional Indians, he has servants. Soni has an intellect that is nimble and impatient, and in middle age, he has become what he calls a "half-time environmentalist," because he's worried, not to say disgusted, with how India is managing its water.

Right there in the Vikram Soni home, in one of the most upscale and politically potent neighborhoods of Delhi, the management of the household water not only isn't under control, it's a constant fight.

"The water in this house?" Soni shakes his head. "Once or twice a week, the water comes on in the pipeline. The pump goes on—the pumps are illegal, of course, but everyone puts on a suction pump, so finally everybody succumbs and gets one." That is, every couple of days, for a few hours, there is water pressure in the pipes to Soni's house. When his pumps sense the water, they immediately kick on, sucking as much as they can from the water main to fill Soni's storage tanks, which are in turn connected to the plumbing inside the house.

The normal state of affairs in Vasant Vihar, in other words, is no water service. And if you don't have a pump to pull water into your storage tanks when the water pressure does come on, because everyone else on your water main *does* have a pump, you won't get any water. Everyone else's pumps will suck it away.

In fact, in Delhi, a city of 20 million people, every household, from the most pinched shack to airy homes like Soni's, has water storage tanks of some kind. So when the water arrives briefly, whether it's at the curb or in the water mains, you can bank as much as possible. For Soni's family, an hour of water a couple times a week doesn't do it.

"We get the rest of our water from a tanker truck. The Jal Board"—*jal* is the Hindi word for "water"—"they provide the tankers.

"But there are a huge number of water [customers], and to get the tanker to come, you have to keep calling and calling. And you have to tip the driver when he arrives—300 rupees or 500 rupees.

"We get ten thousand liters, twice a week. If we call. If you don't tip the drivers, you don't get the water next time."

Soni's voice has more than an edge of exasperation. "I'm just asking for a public service, and I have to provide a five-hundred-rupee tip!"

Although Rs 500 would, in fact, buy a whole tanker of water in many parts of India, it's not the money that burrs Soni. Rs 500 is just $10. It's the ceaseless aggravation, the petty corruption, the minor powerlessness, the idea that if you don't call, if you don't pester, if you don't tip, you don't get . . . water. The water that should, in any case, be coming to you in the water main that's already connected to your house, but that is empty forty-seven hours out of forty-eight.

Just talking about the jury-rigged system, the astrophysicist gets so cranky, it's hard to imagine how he deals three or four times a week with calling the DJB, a notoriously murky, not to say stagnant, water bureaucracy.

"Oh, I don't call," Soni says, chuckling at the very thought. "My brother manages all the water. Fortunately."

India is modernizing at a furious pace—new bridges, new highways, new subways, a new space-science university that received 86,000 applications for 150 slots. At the same time, India is allowing its water system to become not just an inconvenience or an embarrassment but a crippling impediment to its own future. Just one of the thirty-five largest cities in India has twenty-four-hour-a-day water service.[1] Like Delhi, most Indian cities provide water pressure just a couple hours a day. But Indians connected to a water system, even an intermittent one, are fortunate.

Forty-five percent of Indians do not have routine access to safe drinking water—that's 540 million people who don't have reliable water, every day.[2]

"Water is crimping economic development in India, absolutely," says Soni. "It's crimping human development. It's actually impacting on people's survival—it's that bad."

Water is doing what it always does—quietly working away on both the largest scale and the smallest. In India, water is carving points off the growth rate of the economy of the whole country. Water is also killing people. Just in India, forty children an hour under five years old die from contaminated water. One Indian toddler, not even old enough for kindergarten, dies every ninety seconds from bad water, twenty-four hours a day.[3]

One of India's most prestigious scientific research centers, The Energy and Resources Institute (TERI), analyzed the state of India's environment, economy, and quality of life after fifty years of independence. The resulting report, issued in November 2009, concluded bluntly that water is "arguably the biggest crisis facing the country today."[4]

Tainted water is such a serious problem that the TERI report went to the trouble to calculate the impact of diarrhea on the Indian economy. The conclusion: Diarrhea's total cost to India is Rs 900 billion a year—US$20 billion. That's 2 percent of the country's economy, spent dealing with diarrhea caused by bad water.[5]

India can seem to have an almost unsolvable tangle of water problems—urban water systems can't keep up with growth; villages don't have basic water infrastructure; the water people get, whether in the city or the countryside, isn't clean; wastewater isn't treated; the nation's rivers are polluted almost beyond human imagining—and Indian farmers are so inefficient with irrigation that they waste roughly half the nation's water.

But, in fact, almost all of India's problems are really one problem. "Money is not the issue," says Ashok Jaitly, one of the wise men of water in India, director of TERI's Water Resources Division. "We are flush with funds. And technology is not the issue. We have the whole range of technology, from the sand filter to the reverse-osmosis plant.

"Everybody knows what the problems are. We also know what the solutions to the problems are. We don't always get down to solving them. Which is the real problem."

India, quite simply, mismanages its water. It has enough water, it has more than enough smart people and more than enough resources. What India doesn't have is a culture of taking water seriously—water, water service, and water management.

"Government officials are aware of the problems," says Jaitly. "But they are not aware as you and I are aware of it. It's not pinching someone important hard enough."

Most major Indian cities, for instance, had water service twenty-four hours a day in 1947, when India became independent from the British Empire, at least in the urban cores, and many cities still provided twenty-four-hour-a-day service in some areas right into the 1970s and 1980s. That level of service, and more important, the expectation of that level of service, has slipped away, both in the professional water community itself and among Indians.[6]

People expect water to be a struggle, they have learned to make do, and they are skeptical, not to say cynical, about the likelihood that water service will improve.

The result is that water in India is not invisible the way it is in much of the developed world—Indians who live on $1 a day and Indians who study the physics of collapsing stars both have to make securing their daily ration of water part of the texture of daily life.

The poor line up for hours at public taps—the fixed version of the water delivery trucks—where the water is turned on a few hours a day. In villages that don't have wells or taps, the poor carry water home from wells in neighboring villages, often paying for those daily buckets with the futures of their daughters. In India's 600,000 villages, it is girls who traditionally make the walk to fetch water for the family, typically twice a day. What that means is that they often don't go to school.

The well-off urban professional class does, in fact, have 24/7 water—but it is fake 24/7 water. As with astrophysicist Vikram Soni, their homes are equipped with expensive, energy-consuming pumps and tanks and sophisticated filtration systems. So when they turn on the tap in the kitchen or the bathroom, water usually comes out. But it's water that their systems have hoarded for later use—it is the illusion of 24/7 water. And even for those with money, managing the water system, thinking before they drink or cook or wash the dishes or brush their teeth, makes water a constant worry in their lives too.

The daily wrangle to secure water is corrosive not just to economic development and to health but to the spirit of the Indian people. Water problems become a creeping attack on human dignity, a kind of erosion of the spirit.

The curb where the water tanker pulls up to deliver water before dawn each morning, just outside the slum in the Vasant Kunj neighborhood—that street, and the slum area itself, are tucked amid much better-off neighborhoods, a situation that is quite typical throughout both Delhi and India.

Just a car ride of 120 seconds away—1.5 kilometers, not even a mile—is a hotel called the Grand, an imposing granite edifice hidden behind guarded gates. When you step inside, you are in a dazzling white marble lobby, as big as a ballroom, whisper quiet. The entire back wall of the lobby is a window of floor-to-ceiling glass, and through the glass you can see the hotel's garden, the central feature of which is a tiered expanse of water stretching back 160 feet, with fountains, waterfalls, and pools. The long pool is lined with palm trees and blazing pink bougainvillea. The pool is purely decorative. Birds have free access to it—flitting along the edges, drinking, bathing, playing in the shallows.[7]

The contrasts in India are perhaps a little too facile, but this one is hard to ignore. It's not just that the tourists who stay at the Grand have access to things the people in the Vasant Kunj slums don't. It is the birds. The birds who live in the garden at the Grand have access to clean water 24/7 in a way that would be a dream for the people of Vasant Kunj, just a two-minute ride away.

THE NAMES OF INDIA'S CITIES today ring with energy and possibility, even for people who've never been there: Bangalore and Hyderabad, Delhi and Mumbai. India is the place where the people are smart, disciplined, hard-working, and ambitious enough to stay up all night (their time) talking to us about our credit card bills, gratefully doing work we ourselves used to do, for pay one-fifth what we wanted to be paid. India is the place our X-rays often get read; it's the place companies are buying their legal and IT work.

India is China's partner in the twenty-first century's great leap forward—India skipped over building the laptop computers and went right to explaining to us how to use them.

In 2008, the Indian economy grew 9 percent. In 2006, 2007, and 2008—three years combined—the U.S. economy grew 8.4 percent. In the midst of the great recession of 2009, the U.S. economy creaked out 1.1 percent growth. The Indian economy slammed along at 7.4 percent.

The transformation from that growth is truly astonishing. Between 1985 and 2005, according to an analysis by McKinsey & Company, India managed to cut the number of truly poor people in the country in half—even as the population grew dramatically. And the remaking of the Indian economy and society has only just begun. McKinsey estimates that between 2005 and 2025, if India sustains its economic growth, more than seventy thousand people *a day* will move from poverty to middle class, every day for twenty years.[8]

The evidence is everywhere. In Bangalore, at one point, nine hundred new cars were being sold per day. India has 636 million cell phone subscribers—that's twice as many cell phone talkers as the United States has people.

And India's economy isn't just creating consumers who want cars, cell phones, and good jobs. India is making millionaires and billionaires. On the *Forbes* 2010 list, two of the five richest people in the world are Indians. And except for the United States, no country has more billionaires in the top fifty than India.[9]

But India is so large—nearly four times the number of people as in the United States in a country only one-third as big physically—and still so poor that twenty years of soaring development and growth have taken Indians only so far. Forty-four percent of Indian homes have no electricity. Twenty percent of Indian homes use either crop residue or cow dung as their primary cooking fuel. The Census of India measures eight kinds of "fuel used for cooking," and "cow dung cakes" is the fourth most common.[10]

And in some ways most stunning of all, 39 percent of Indian adults cannot read or write, with the burden falling most heavily on women. Fifty-two percent of India's adult women are illiterate—more than 200 million Indian women who can neither read nor write.[11]

Some of that deprivation is simply a function of the fact that India remains largely rural—three-quarters of Indians live outside urban areas. Economic modernization spreads out to the countryside much more slowly than it energizes the cities.

The contrast between the ambitious, media-savvy, high-tech India and the impoverished India is nowhere more dramatic than in the world of water.

In September 2009, in a discovery that changed forever the way humans will look at the Moon, NASA announced that its scientists had discovered water on the Moon's surface—a thousand pounds of dusty Moon rock would yield two eight-ounce glasses of water.

In March 2010, NASA announced an even bigger discovery of water on the Moon: Instruments scanning forty craters at the Moon's north pole had found ice, lots of it. Each crater contains iceberg-size quantities of ice—660 million tons in just the craters examined.[12]

It was, in fact, India's space program that made possible the discovery of huge quantities of water on the Moon. The NASA instruments that found the water rode to the Moon on Chandrayaan-1, India's first Moon-bound spacecraft, designed, built, and launched by the Indian Space Research Organization (ISRO). Although India's instruments didn't discover the water, the Indian press overflowed with pride:

The *Times of India*: "One Big Step for India, Giant Leap for Mankind."

The *Hindustan Times*: "Water on Moon Is India's Discovery, Says ISRO Chief."[13]

Nothing quite captures the dichotomy of modern Indian society like India's twenty-first-century spacecraft finding water on the Moon. The Indian scientists and engineers who created Chandrayaan, at ISRO head-quarters in Bangalore, don't themselves have running water at home. Officially, municipal water service is provided in Bangalore just 4.5 hours a day.

None of India's global brand-name cities—Mumbai or Hyderabad or Delhi—does any better. Quite the contrary, the water service in India's great cities is nothing short of primitive, and getting worse.

In Delhi, most homes receive water just an hour or ninety minutes a day.

In Mumbai, that's the goal—an hour or ninety minutes a day. Many people receive water only every other day.

In Hyderabad, some areas have water four hours a day, some two hours, and some people get just ninety minutes every other day.[14]

Thames Water, the water utility for metropolitan London, is one of the companies that have outsourced many critical IT functions to India. Thames Water uses Wipro, the renowned Indian technology provider based in Bangalore, with 100,000 employees. So the biggest water utility in Britain, where uninterrupted water service is the unquestioned basic, gets its IT from people in India who don't themselves have water service most of the day.[15]

Indians have adapted to human-created water scarcity, but that doesn't mean they've lost sight of either the frustrations or the ironies. No less a body than the Indian Supreme Court, in an April 2009 order, demanded that the Indian central government immediately tackle the nation's water problem "on a war footing."

The justices, despairing of what they described as "serpentine queues of exhausted housewives waiting for hours to fill their buckets of water," ordered the immediate establishment of a scientific panel to, within months, come up with solutions to a whole range of Indian water problems that took half a century to develop.

The Supreme Court's order was more an expression of official vexation than it was a practical effort to provide water. "It is indeed sad," the Court wrote, "that a country like India, which scientifically solved the problem of town planning . . . during the Indus Valley Civilization and which discovered the decimal system in mathematics and plastic surgery in medicine in ancient times, and is largely managing the Silicon Valley in the USA, has been unable to solve the problem of water shortage till now." [16]

Ashok Jaitly, of TERI, is on the panel advising the Supreme Court's water reform effort. "We were ordered by the Court to find a way of providing everyone in India with safe drinking water in three months.

"Three months! Could anything be more ludicrous?

"I'd be satisfied if we could do it in three hundred years. Well, okay. If we could do it in thirty years."

FOR WESTERNERS, India combines the familiar and the exotic in a way that camouflages, for a while, how differently Indians think about some things. The newest urban landscapes, for instance, seem on the surface no different from what you'd find in fast-growing parts of the United States. Many of the main highways around Delhi are new and have four or five lanes in each direction, vast flyovers are under construction everywhere, and one of Delhi's booming suburbs, Gurgaon, is home to large, glassy office buildings with logos from PriceWaterhouseCoopers, Convergys, Ericsson, and GE. IBM, Google, and Microsoft all have offices in Gurgaon, which looks like Tysons Corner, Virginia (Gurgaon has the shopping too—it is considered India's "mall capital"). The zippy pace of urban life has a distinctly hectic vibe. One of India's cell phone companies uses this tagline in its ads: "Impatience is the new life." Indians are friendly, many people speak English, and with the infusion of Indian immigrants into U.S. society, India seems closer than ever. [17]

But Indians think differently about matters both minor and profound. The idea of driving between painted lines in a somewhat predictable fashion is as new as the highways themselves, and there are regular road signs exhorting, "Lane driving is safe driving." India's roads are a cacophonous experience that verges on insanity. Drivers talk to each other in a ceaseless conversation, conducted exclusively with the car horn, and every truck in the country has a colorfully painted sign on the back reading in bold letters

"HORN PLEASE!" Indians will often shake their heads side to side—the gesture Americans learn from infancy means no—to mean just the opposite. *Yes! I agree with you! That's why I'm shaking my head back and forth!*[18]

America seems to have lost its sense of smell, compared with India. Indians still live amid the aromas of daily life. You can often smell meals cooking, in both cities and villages, and walking around urban areas, you often catch a whiff of what the Indians call "drains"—the open sewers where almost all waste ends up. The animals that Indians live with, in the villages and in the cities, smell like the livestock they are. And in the villages, neat round cakes of cow dung—the size and shape of Frisbees—are everywhere drying in the sun and perfuming the country air. Many places, including taxis, have plug-in air fresheners to muffle the ambient odor. The big luxury hotels actually perfume the air in their lobbies (and burn incense in their public restrooms).

Water is as complicated a cultural element in India as any, and it's easy to leap to what seem like logical conclusions about Indians and their water, which turn out to be utterly wrongheaded. Several of India's major rivers are considered holy—they are, in fact, considered to be goddesses, and their water is thought to have great powers of cleansing and redemption. The Ganges (or Ganga), which flows through what many regard as the most sacred city in Hinduism, Varanasi, and the Yamuna, which flows through Delhi, are regarded as the holiest of India's rivers. They are also among the most polluted rivers in the world. The water in the Ganges and the Yamuna is so fouled with routine city pollution and industrial waste that it is too dirty to be run through American sewage treatment plants—it would have to be cleaned up before the sewage treatment plants could take it. In many places, the rivers are chocolate brown or charcoal black.[19]

And yet both rivers attract millions of pilgrims who believe that the goddess rivers can, with a simple dip, wash away a lifetime's sins. Funerals by the thousands are conducted on the banks of the Ganges and the Yamuna, with the remains of the deceased burned and committed to the water.[20]

How is it possible to worship a river that you also treat as an open sewer? For Westerners, that contradiction is so vivid, it seems unresolvable. A fundamental hypocrisy.

For Indians, the condition of the rivers is terrible, but it isn't a contradiction.

Praveen Aggrawal is a general manager at Coca-Cola for public affairs and sustainability in India and South Asia. Criticism of Coke's use of local water supplies in India helped trigger what has become a sustained water consciousness and conservation effort for Coke worldwide. Aggrawal, based in Coke offices in Gurgaon, works every day with water issues. Coke, of course, needs water to make its drinks, and Aggrawal is also part of a group of companies trying to change how Indian businesses approach water issues.

"For us," Aggrawal says, "water is a bounty given by the gods. You can't commoditize it. This is a deeply religious and spiritual subject for our population.

"For most people, water is sacred. In most rural homes, in the area where you store the water, you go barefoot—that kind of respect."

Why wouldn't that lead to the cleanest rivers in the world, instead of the most toxic?

"Most Indians, the inside of their houses are really clean," Aggrawal says. "They leave the muck just outside. We have an expression, 'Shit in public, eat in private.' Which means, basically, My house is clean—the outside is someone else's responsibility."

And so for Indians water that isn't in their immediate control also doesn't feel like their responsibility. "How can a factory manager who regards the Yamuna [River] as sacred pipe toxic waste into it? Well, the water leaving the factory with acids and toxins is going to the gods. They'll be responsible for it. There is no emotional connection there."

It is a leap that's hard for Westerners—but for Indians, the very fact that the rivers are goddesses means you can't hurt them by polluting them. The power of the Yamuna and the Ganges transcends anything a factory's wastewater pipe can do.

Ashok Jaitly of TERI says that you can't tackle the problems of water in India without taking account of water's cultural significance.

"It goes very deep in India, our attitudes about water," Jaitly says. "It's so complex. It's so fascinating. Water and religion, water and music, water and literature, water and spirituality—water is integral to all of those. It has all those connotations, and you have to deal with that as part of the politics of water in India.

"This is a society where people believe if they take one dip in Mother Ganga, they are going straight to heaven. That's irrational, of course. But people are irrational. Thank goodness."

The irrationality is layered through how Indians use water, as well as how they treat it.

Farmers in India use 80 percent of the water consumed in the country, but because both water and the electricity necessary to pump the water into farm fields are basically free, farmers have no incentive to be careful about how they use water. Irrigation efficiency—the amount of irrigation water that actually helps grow crops—is between 25 and 35 percent in India. Which means that roughly 70 percent of the water Indian farmers use is wasted. And since farmers account for 80 percent of water use, 56 percent of the water available to the country is wasted.[21]

India's big cities—the focus of so much of the country's entrepreneurial energy and growth—don't supply water to the people they've got now, and millions of new residents each year are pouring into those urban areas. Indians talk more seriously about Delhi running out of water than they do about Delhi getting water service twenty-four hours a day.

Millions of Indians in villages of a thousand or two thousand people—where 70 percent of the country's people live—rely on water-gathering systems that would have been regarded as primitive in the developed world a hundred years ago. Yet getting water to the villages is harder than it seems, because even small-scale solutions, like compact, solar-powered water purification systems, require the kind of continuous support from someone with technical skill that has proved hard to sustain.

Finally, there is the complexity of Indians' sometimes conflicted attitudes about water service itself.

One of the interesting things about water is that it is one of those rare areas where the gold standard of service and the basic level of service are the same thing: Water should be provided twenty-four hours a day, seven days a week, in pipes that keep it clean and safe. In many parts of the world, 24/7 water is *the* fundamental municipal service.

And many places with more modest resources, and equal challenges, have decided that continuous water service is simply part of being a twenty-first-century city, and have done the work to put 24/7 water in place. Hanoi and Ho Chi Minh City, in Vietnam, both have water service 24/7. Phnom Penh, the capital of Cambodia, and Kampala, the capital of Uganda, also both have continuous water service.[22]

In India, which takes such pride in its economic and technological achievements, there is virtually no political support for the idea of 24/7 water. "In fact, when someone says water 24X7," wrote a senior official of the Delhi Jal Board in 2006, "we laugh it off as an absurdity."[23]

Puzzling as it may seem, there is resistance to the very idea of 24/7 water. Wealthy and middle-class people have created a world of artificial 24/7 water that meets their needs; however inconvenient, they are all too familiar with the Indian bureaucracy, and skeptical of what 24/7 water might require. Poor people already spend a huge portion of their time and limited income to get water. For them, promises of 24/7 water simply ring hollow; why spend more for something that seems unlikely to materialize?

The slow deterioration of India's municipal water systems owes something to an attitude that is a parallel to the attitude about the sacred rivers. The water infrastructure was a given. It was installed under the British Raj, it was the job of the government to provide water, the water itself needed to remain cheap, and the pipes, pumps, and treatment plants would sustain themselves—just as the sacred Ganges and Yamuna absorb their pollution.

As a result, Indian water utilities charge so little for water, and let so much water leak away unbilled (as much as half in Delhi), that they don't come close to covering the cost of their basic operations—payroll and energy—with revenue from water bills. Customer payments cover only 60 percent of the Delhi Jal Board's operating costs. With huge operating deficits (covered by their local governments), it's not surprising that expensive maintenance, upgrading of the systems, and expansion aren't high priorities for Indian water utilities. India's cities have grown far faster than their ability to keep up. Streets, highways, and traffic are a nightmarish tangle, and that's something that everyone suffers through every day. The water infrastructure gets far less attention than the roads under which it is buried.[24]

As for actually restoring 24/7 water service, at this point, no major city could simply turn the water on for everyone twenty-four hours a day. In fact, all Indian cities are providing water twenty-four hours a day to someone. Ninety percent of the city will be valved off at any given time, and 10 percent of water customers will be receiving their two hours of water. To provide twenty-four-hour pressure to a whole system again requires, first, that all major pipes be able to hold pressure for twenty-four hours; leaks aren't that important when the water is on for only ninety minutes a day, but they are devastating when the water is always on. Every customer needs a meter, and many Indians with a connection have no water meter. Most utilities would need to change the infrastructure—the valves and control systems—of a water system long adapted to intermittent supply, and many would have to dramatically increase their pumping capacity to be able to provide citywide pressure twenty-four hours a day. Customers would have to adapt their in-home systems, disconnecting automatic pumps and giving up the use of storage systems.

One thing that probably wouldn't be necessary is more water, except for the initial startup. It turns out that in the few cases where Indian cities have tested 24/7 water, they discover that people who can get water anytime they need it actually use less than people who are hoarding water against the next time the pressure comes on.

As it is, the neglect of India's water system leaves Indians to make their own ad hoc arrangements. That forced, and inconvenient, self-reliance breeds skepticism and cynicism about official vows to make things better. And India's growth, rather than being harnessed to fix water problems, ends up making them worse.

But India's drift into a self-created water crisis—the result of poor management and of a failure of Indians themselves to understand the support that good water service requires—is what makes water in India more than simply an Indian problem. The invisibility of the water system in the developed world breeds its own kind of indifference to the needs of the water infrastructure. India is simply an extreme version of the kind of benign neglect of water systems found in many other places.

It's the kind of attitude that allowed Atlanta to slide into a water crisis, it's what makes us think Las Vegas (and Los Angeles) can sprawl without regard to their water supplies, it's the kind of attitude that undermines a century-old consensus that tap water is fundamental, and worth paying for.

In the United States in 2009, water projects necessary to keep systems operating through 2014 totaled $255 billion in cost—which comes to $170 for every person in the United States every year, just to keep the pipes well maintained. As the Indians have learned, water problems get worse, more complicated, and more expensive if you push them off into the future.[25]

India's problems are about human nature, and about how hard it is for communities to plan for the future. India's water problems are already becoming our water problems—not just the way we take water for granted but the fact that we fail to appreciate two things: the value of easy, reliable water service in our daily lives and our economy, and the level of investment that kind of water service requires.

Maintaining water systems is like saving for retirement. It not only provides no immediate satisfaction, it actually reduces your ability to enjoy yourself right now. But if you don't save slowly and steadily for retirement—just like if you don't steadily maintain and improve the water systems you've got—at some point it becomes too late. You can't possibly save enough to retire if you don't start until you're sixty—and if you haven't started, you're in trouble. Likewise, at some point it becomes almost impossible to rescue water systems that fall too far behind, as the people in Delhi and Bangalore know all too well.

V. S. Chary is an Indian expert on water systems and water attitudes, based at the Administrative Staff College of India. Chary has made bringing back 24/7 water service a mission, and he also wants to restore the Indian public's expectation that 24/7 water service is the basic they should get.

"It's easy to slip into interim water supply," says Chary. That is, it's easy to go from twenty-four hours of water a day, to nineteen, to fourteen, to four hours in the morning and four hours in the evening, to three hours whenever the utility can manage it. "It's easy to slip into it, but it's very difficult to recover back to twenty-four hours a day. Physically and psychologically."

Chary knows something else, which drives him to insist on nothing less than 24/7 water for the cities of a country audacious enough to find water on the Moon. "It's not rocket science," he says. "It's plumbing."

~~~

DEFENCE COLONY, which started as a residential area for officers in India's military, is one of the nicer neighborhoods in Delhi, a calm oasis in a city so big, so busy, and so complicated that it can be overwhelming. Defence Colony is a patch of land not even a mile on a side, with about 25,000 residents. The quiet streets are lined with multistory homes and small apartments, all set behind walls and gates. The parks are well tended and crowded with joggers, walkers, kids playing soccer, and seniors doing yoga. Defence Colony has a market, with good restaurants, stands specializing in fresh fruit and flowers, along with a couple outposts of the global economy, including a Citibank with a 24-hour ATM and a Baskin-Robbins.

Early each morning, if you walk the streets, you'll find that almost every car is in the process of being washed. Virtually every family has at least one car, and a lot of those who have a car have a driver; cars in Delhi get dirty quickly, and water problems notwithstanding, many drivers in Defence Colony spend the first hour of the day washing the grime off their employers' cars. It's not typically a running-hose-style wash; it's more bucket-of-water-and-a-cloth. Still, it's striking: almost every car washed every day.

For about a week while I was visiting India, I stayed with friends in Defence Colony who had an apartment on the first floor of a three-story house. Defence Colony is one of those spots on the globe where it is possible to experience firsthand the idea that upper-income people around the world—in Boston or Beijing, Managua or Damascus or Delhi—can seem to have more in common with each other than they do with their fellow citizens from lower income brackets. It wouldn't take a professional from Denver more than a few hours to get her bearings in Defence Colony.

Except, perhaps, for the water.

Because in Defence Colony, as in Vikram Soni's diplomatic neighborhood across town, the water supply is both spotty in frequency and sketchy in cleanliness. The house where my friends live has the requisite pumps and tanks. The Jal Board opens the valves to Defence Colony's water mains each day for a little more than an hour, and the pumps come to life automatically with a muffled roar. When you turn on the faucet in the kitchen or flush the toilet, there's usually water, piped from tanks. The quality of the water is something else again.

The kitchen has a reverse-osmosis unit to clean the water from the house plumbing—it is literally a tiny countertop version of the same technology being used in the vast seaside water factory to save the city of Perth, Australia. That RO water is the water to cook with, it's the water stored in the big water bottle in the refrigerator, it's the water for making a cup of tea and for filling the water bottles propped on the bathroom sinks.

You don't want to rinse your mouth—or, more difficult to remember, your toothbrush—with the faucet in the bathroom. So, in fact, there are two water systems inside the house: the regular municipal water, and the thin stream of water produced by the RO appliance. The house water is assumed to be non-potable—not just unpalatable, but dangerous. The two systems are supplemented with cases of bottled water. Not bottled water for convenience, but bottled water because the RO appliance needs electricity, and the power goes out a couple times a week.

People do shower in the tap water. My first couple showers, I thought there was a vague, odd smell in the bathroom, but I couldn't quite place where it was coming from, or what it was. During my third shower, I figured it out. Or rather, it hit me—the smell and the source—like a bucket of water in the face. The odor was the very faint tang of raw sewage. And it was coming from the shower spray.

When I realized this, I was facing the showerhead, rinsing the shampoo suds out of my hair and off my face. I immediately closed my eyes and my mouth. The water was unquestionably dirty, in ways that had been described vividly.

The cross-contamination of water supply pipes by sewage pipes isn't just a function of the deterioration of India's water system—it's actually a debilitating side effect of not maintaining 24/7 water service.

In Indian cities—and in other cities around the world—water supply pipes and sewage pipes sometimes run alongside or on top of each other, having been laid together in the same trench for convenience. When poorly maintained sewer pipes leak, it is possible for the liquid they are carrying to seep into water supply pipes. Possible, but in most instances, not likely. The water supply pipes are protected by the pressure necessary to move the water along—the pressure means clean water leaks out of cracks or sloppy joints, but it is hard for contaminants to leak in against a well-pressurized system.

In Indian cities, though, the water supply pipes don't have any pressure in them most of the time, so the pressure that would otherwise keep leaking sewage from seeping in doesn't exist, leaving the supply pipes vulnerable.

And the practice of every home, apartment building, and business in India's cities being equipped with its own suction pumps makes the problem much worse. Instead of the water being pushed steadily through the water mains by the Jal Board's pumps, the battalions of individual pumps suck the water along through the pipes, creating negative pressure in the water mains. If there are any breaks or cracks in the mains, they will suck in water and contaminants, aspirating any leaking raw sewage right into the clean water mains.

That's why there was no question what the funky smell coming from the shower water was. The only question then is, How much faith do you have in your soap?

The cross-contamination of supply pipes with sewer pipes that results directly from not having 24/7 water service is neither unusual nor trivial. In May 2009, in Hyderabad, the problem killed nine people, including at least three children, sent two hundred people to the hospital, and resulted in angry protests over the poor quality of municipal water—at the hospitals where the sick people were taken and at the headquarters of Hyderabad's water board. Hyderabad officials were so grimly certain of what had happened that the day after the sicknesses started, when only four people had died, they had already announced they would pay compensation of Rs 200,000 (US$4,000) to the survivors of each victim who died. Within three days, Hyderabad's Institute of Preventive Medicine had found E. coli—human intestinal bacteria—contaminating the drinking water from the neighborhood where the deaths occurred.[26]

In Defense Colony, in homes that look no different from those in nice neighborhoods in Coral Gables or London, providing yourself and your family with safe water is merely an inconvenience. But it is an inconvenience that is never far from your mind—you have to think about what water you are using each time you use it. You have to worry about whether the pump is working and whether the tanks have been cleaned, and you have to pay enough attention to the water-main service to know if your tanks actually got filled. If water service doesn't come for a couple days for some reason, you need to summon a private water tanker.

Comparatively, of course, the residents of Defence Colony have water service that doesn't typically limit their lives. But the population of Delhi is 20 million—there are almost as many people in Delhi as in the entire nation of Australia. One detailed analysis of Delhi's residents and their access to water concluded that no more than half the city's residents lived in buildings that had a water connection.[27]

The Jal Board reports that it has at least 11,500 public "standposts," that is, neighborhood water spigots, for those who live in buildings without water service. If only a hundred people line up for water each day at each of those public spigots (a very conservative estimate), and if each of those hundred people is only collecting water for a total of five people (also conservative), that's 6 million Delhi residents who stand in line each day, surrounded by jugs, buckets, and containers of every kind, waiting to collect water, when the pressure is turned on at the public spigot. The spigots don't get better service than the homes—they, too, only get an hour or two of water each day, and someone from the family has to be standing ready with the containers when the water is turned on, whether it's four in the morning or one in the afternoon, whether it prevents kids from going to school or prevents adults from holding down a job with fixed daytime hours.

So between the 2 million people relying on water delivery trucks and the 6 million using standposts, at least 8 million Delhi residents a day rely on water carried home by a family member—which means that more than a million Delhi residents devote hours every day to waiting for water, which is itself of dubious quality, right in India's capital.

THE VILLAGE OF JARGALI is a strange place in the twenty-first century, if not in India itself. Jargali is just fifty miles south of Delhi. You drive past the high-tech office parks of Gurgaon to get to it, but beyond the office towers, you are quickly in rural India. Jargali is a media-dark zone—no one listens to the radio here, no one watches TV, there are no newspapers, no Internet. Homes have the flag of India painted on their outside walls— three crisp stripes, saffron on top, white, then green. In the white stripe is writing that indicates what kind of public assistance that home is entitled to. The people of Jargali do have water from a well, but it isn't potable.

So twice a day, the women and girls of Jargali follow the dirt road out of town, headed for a well near the village of Molhaka. As water walks go, in both India and the wider world of people with limited access to water, the women and girls of Jargali are pretty lucky. The round-trip takes just an hour, so a trip each morning and each afternoon frames each day, but doesn't consume it.

The Census of India has an unself-conscious approach to cataloging a country that encompasses both the Grand hotel and Jargali. To record where drinking water for home use comes from, the census has three broad categories—"within premises," "near premises," and "away" ("away" is at least one kilometer, round-trip).

In India, 32 million households get their water from "away"—17 percent of the country. About 170 million people drink water every day that has been carried home by foot, one out of six people in a country of 1 billion.[28] That's the number of people in the United States who live east of the Mississippi River. It's as if everyone from Maine to Key West, from New York to Chicago, from Memphis to Atlanta, relied on water that someone had walked to collect every day.[29]

The number of people worldwide who have to walk to get water is one of the few estimates related to water that no one seems to have made. But clearly, in the twenty-first century, perhaps 100 million people are making the water walk every day, with hundreds of millions depending on water that has been carried, almost always on the head of a woman or girl, from a well or a spigot back to home.

I joined the women and girls of Jargali for their water walk one sunny afternoon in October. I was given two pieces of equipment: an open-topped metal pail, and something called an *eendhi*—pronounced like "India" without the final "a"—a tightly woven ring, about the size and thickness of a bagel. The *eendhi* was the first surprise—water is darned heavy, the top of people's heads is a bit touchy and not all that flat, and the *eendhi* is a way of establishing both a cushion for your head and a reasonable surface on which to settle a heavy, awkward container of water.

We set off walking east, along the road out of town, and then branched onto a well-trod dirt path that threaded along the edge of fields planted with millet and mustard. Ten girls and two women made the walk that afternoon. Anjana was one of the girls who walked with me. She said she was eleven, but might have been twelve or thirteen (Indian villagers tend not to keep close track of birth years). Anjana wore loose, fire-engine-red pants, an elaborately embroidered black top, and purple flip-flops, and she carried two jars on her head, a fat orange plastic one that held 3.5 gallons and a second, smaller stainless-steel one that held another gallon or so. Balance was key for Anjana—her slim arms could only just reach the top of the first jar she was carrying once both jars were on her head.

Basdevi was one of the adults making the walk that afternoon—she was tall, a little severe and skeptical. She carried a single, large yellow bucket with a snap-on lid, the kind of five-gallon container sold at home improvement stores in the West.

Almost everyone balanced two empty containers on her head—the five-gallon bucket, then a smaller, narrow-necked metal jar or pot on top of that. My own wide-topped metal pail held about 2.5 gallons of water—and there weren't any experienced water carriers using something with a wide-open top, for good reason, as soon became clear.

Because a foreign man was on the walk, we were swarmed by curious, highly amused boys, ranging in age from six to eighteen. Not one male but me carried a water container.

Headed to the well, the girls effortlessly wore the empty water containers on their heads, walking along briskly in bright clothes and tired flip-flops, using just a single hand to steady their buckets. I was the last to arrive at the well, twenty-two minutes after I set out.

The well itself was wide-mouthed, dark, vertiginous. It was a classic well, in the sense that it was a deep, round hole, lined with carefully placed stones all the way around, and as far down as you could see.

It was, however, completely unmarked—simply a hole in the ground, eight feet across, in the middle of a farm field. If not for the girls surrounding it and already hauling up their water, I would have walked right by. There was no ledge or rail, just a flat circle of stones around the lip. The top of the water was twenty feet down.

Hauling water up out of a well is, simply, hard. Water weighs 8.3 pounds per gallon, so even 2.5 gallons of water means hauling twenty-one pounds of water up from twenty feet down.

You tie your water container carefully to a cloth rope, so as not to lose it, you pitch it down toward the center of the well, you let it sink well below the surface, that is, you "drown the bucket" to get both a fresh slug of water (not filling over the edges from the surface—that might be dirty) and a full container.

The hauling up is not just hard, you have to lean out over the well, while simultaneously reeling in your bucket hand-over-hand; you need to keep your balance, and you need to keep your twenty-one-pound (or, in the case of the girls, forty-two-pound) bucket from banging the sides of the well. Otherwise you scrape dirt and debris into the water, and you splash out the water you're trying to bring home. The girls had no leverage to help with the hauling. I could easily imagine losing my balance and falling in. The stonework lining the well was so smooth that I wasn't sure how I would get out if I did fall in.

Hauled up to ground level, the water itself was crystal-clear and cool. How clean it really was—from a shallow, open well in the middle of farm fields—was questionable. But it looked good.

Walking back home, we fell into single file, using both hands to steady the buckets. The girls walked from the waist down, assuming the stately, hip-rolling walk that, it turns out, you need to keep forty or fifty or sixty pounds of moving water well behaved on your head. You use your hips as a kind of gyroscope—the bottom half of your body, ideally, copes with the changing terrain; the top half of your body stands tall, keeping the load of water on an even keel.

My own return trip was a bit of a mess. Keeping both hands above your head while walking two or three kilometers takes conditioning. My arms tired quickly, and even with the *eendhi* as cushioning, the metal bucket with twenty-one pounds of water in it seemed to grind into the top of my head. You feel the weight of the water in your neck, your shoulders. The soft, fine, sandy soil becomes a trial to walk through, even without slippery, thin flip-flops. And there is a reason the regulars don't use open-topped vessels. A quarter of my water sloshed out, splashing across my face and shirt. Any water you spill is water you don't get to use or drink. By the time I was back in Jargali, my shirt was soaked, my neck and shoulders ached. And I was thirsty.

I carried half the water, and half the weight, of the typical twelve-year-old girl walking ahead of me. And I stopped twice to give the crown of my head a break.

As I wobbled back into Jargali with the pail of water on my head, laughing villagers shouted at me in the local dialect as I walked by. Translated, the gentle heckling included:

"You are learning that life in India is hard."

"I would like a drink of your water."

"You look beautiful carrying the water!"

The whole trip took fifty minutes. Thirteen of us carried water. If the average haul was 5 gallons (19 liters), we managed to bring back 65 gallons (227 liters) total. Global health experts agree that the generally accepted minimum of water per person per day for all routine activities is 13 gallons (50 liters).[30]

So, technically, thirteen of us went on a water walk and managed to carry enough water for only five of us for the day. A grown woman, experienced at carrying water, can handle perhaps 8 gallons (30 liters) herself—that's walking four kilometers round-trip, the second half of the trip carrying 67 pounds of water on her head. And in that single trip, the woman can't even carry enough water to meet the world health minimums for herself for a day.[31] Which means, of course, that the 32 million families getting their water from "away" in India—and the half-billion people around the world who rely on foot-supplied water each day—don't get anything like the 50 liters of water per person per day that is considered the minimum.

If the typical American had to walk with the hundred gallons of water she herself uses every day—not to mention the hundred gallons that her husband and her children each use—that hundred gallons of water would require twelve round-trips to the well (thirty miles of walking in Jargali), each time carrying sixty-seven pounds of water on her head.

Walking is simply not a great way to move water.[32]

Back in Jargali, I talked briefly to Anjana and Basdevi. Anjana had carried twenty liters of water that afternoon—forty-four pounds, for a girl who was lucky to weigh twice that. Anjana does two walks a day, taking the morning walk early enough so she can be in school by eight. She has never lived in a house with running water—in fact, she said she didn't know that some houses had water piped directly to the inside of the house. Anjana said, "The men and boys customarily do not fetch water." I asked, "Do they get to drink water, then?" When that was translated, she burst out laughing, as did the crowd of forty villagers who had gathered to listen to the conversation.

Basdevi has lived in Jargali for fifteen years, and has eight children. She makes the water walk twice each day, trying to bring home thirty liters a trip—for a total of sixty liters of water.

Basdevi thinks that walking for water is ridiculous. "Water should be delivered to every individual's doorstep," she says. And she knows. She grew up in a home with tap water.

MEHMOOD KHAN KNOWS THESE VILLAGES, because he grew up in Nai Nangla, the village next to Jargali. "The gap between this kind of village, this kind of area, and the developed parts of India is even wider than fifty years ago," says Khan, who is fifty-five years of age. "Water was not an issue at all in my childhood."

If there is an Indian version of the Horatio Alger story, Khan qualifies. When he was growing up in Nai Nangla, one of five children, the village had 300 people, perhaps 60 families. His father was a farmer, and head of the village council for three decades, and dug two wells that still exist, each just a short walk from the heart of Nai Nangla.

Khan walked to school each day, a 6-kilometer round-trip. There was no electricity or refrigeration. "I did my homework under a kerosene lamp." He left Nai Nangla at age sixteen, in 1970, went to college, then got an MBA from one of India's most distinguished business schools. At age twenty-eight he went to work for Unilever. "I started out procuring buffaloes for them, then I was a trader, trading tea, coffee, black pepper." Khan went on to help open Eastern Europe for Unilever, and then to run its Asian business, based in Vietnam. He has traveled the world many times. He's been to seventy countries, including North Korea, which he's visited six times. Unilever is a global consumer giant that is equally comfortable selling laundry detergent in Jargali and Nai Nangla—in small packets like those for ketchup, at 1 rupee (2 cents) each—and Ben & Jerry's ice cream in Whole Foods markets in the United States.

Khan's last posting was at Unilever headquarters in London, where for eleven years he was head of the corporation's global innovation process. All that time, Khan had been visiting Nai Nangla every year, and in 2003 he started returning several times a year to try to find ways to improve life in the village, which has about 1,200 people, and in the surrounding region of Mewat.

"I got involved with people doing things with education," he says. By 2003, the literacy rate for girls had fallen to 2 percent; overall in Mewat it was just 23 percent. And although Nai Nangla had a school, which it did not when he was growing up, "these kids, even the fourth and fifth graders, they could not read. They could not write. I was better off forty years ago than these kids were."

In Nai Nangla today, there is still no refrigeration, most people have no electricity, the store is a single, small room with products in dusty jars and cans on the dirt floor, and each day the villagers collect the dung from their cows and water buffalo by hand, form it into dinner-plate-size disks, and set it out in the sun to dry as a source of fuel. Although there is a pond on one side of the village, its water is unfit even to wash clothes in, though the women of Nai Nangla do anyhow. The village has no other water supply.

It is a little arresting to imagine that a senior corporate manager—for global innovation, no less—at one of the world's largest companies started life here. (There was something in Khan's family, and perhaps in the Nai Nangla water then as well: Khan's brother is a cardiologist in Florida.)

The Mewat region has 550 similar villages, each with about 2,000 people. Khan can tick off the cascade of problems: poor schools or no schools, low literacy, no jobs, primitive agricultural practices leading to low food production, a quicksand poverty that means people struggle to survive.

"They are living on today," says Khan. "Not even for tomorrow. Just today. If I talk about 'one year from now' or 'two years from now,' people say, 'who knows?' That's the kind of pathetic condition that the community has gotten into."

When Mehmood Khan started to dig in, to hold community forums, to peel back each problem, at the bottom of almost every problem he found: water.

Girls don't go to school because they have to fetch water.

Girls drop out of school because the schools have no working bathrooms.

Rainfall is less than it used to be; groundwater levels and well levels have fallen; the water used on fields is brackish, producing less food and slowly poisoning the soil with salt.

Villagers' health is poor because they don't get enough food or water, because they live amid not just the waste of their animals, but their own waste as well. Children who get diarrhea are in serious trouble. The scarcity of water means it isn't often used for hand washing.

"Of all the issues in Mewat," says Khan, "the most hopeless of all is water."

That is the first lesson of water poverty—in rural India, but anywhere else as well. Water poverty doesn't just mean your hands are dirty, or you can't wash your clothes, or you are often thirsty. Water poverty may mean you never learn to read, it means you get sick more often than you should, it means you and your children are hungry. Water poverty traps you in a primitive day-to-day struggle. Water poverty is, quite literally, de-civilizing.

In August 2009, Khan and his wife decided they couldn't keep visiting Nai Nangla's problems. He resigned from Unilever to work in Mewat full time. "I decided to bring my day job back to the village," says Khan. "We ditched our flat in London to return to Nai Nangla." They keep an apartment in Gurgaon, and Khan has turned his parents' old white stucco house, with its open courtyard, into a charitable foundation that includes a computer training school and a sewing school, along with a spartan bedroom for Khan.

Khan has used his charitable foundation to unleash venture-capital-style entrepreneurial philanthropy. The computer school is linked to a bank and an insurance company—students who do well take their skills right to jobs. The sewing school is linked to a clothing manufacturer in Gurgaon who is buying the work the girls make.

Khan persuaded India's largest milk processor, Mother Dairy, to set up a milk-buying storefront right in Nai Nangla. Villagers used to sell their milk to itinerant brokers, at whatever price was offered, typically Rs 12 a liter. The Mother Dairy outlet, locally staffed, has a generator, a refrigerated holding tank, and computerized equipment so that villagers can instantly see the quality, quantity, and price of the milk they bring in daily. The dairy outlet pays about Rs 20 per liter; in a year this one room has raised the total income of those living in the village by 40 percent.

Khan has started an orchard of fruit trees and perennials, and underwritten construction and start-up of a dairy farm. "People have gotten out of the habit of thinking long term," says Khan. "The orchard is something you have to cultivate and take care of, that comes back year after year. Patience is required. The initial work is more, but it pays off."

Khan, who is white-haired at fifty-five, typically dresses all in white and wears open-toed sandals, giving him an air of ascetic focus. He doesn't do the work himself: He enlists young people, he gives them responsibility and financial incentives, and then he coaches, cajoles, runs interference with bureaucrats, and takes great joy in watching things move along. Although the fact that Khan grew up in Mewat gives him credibility, "I'm seen as a kind of alien here," he says. Villagers say, "Which planet does he come from?"

Water is constantly on Khan's agenda. Following in the footsteps of his father, Khan is bringing water back to Nai Nangla, or at least much closer. Khan enlisted a local college student named Ronak to find a new supply of "sweet water" nearby, to irrigate the farm fields Khan's family still owns, to revive the two wells nearest Nai Nangla, which had long since become dirty and stagnant, and to sell irrigation water to other farmers.

Near Molhaka, the same village the women of Jargali walk to, test wells found a great spot for a new tube well. "Ronak found the sweet water. He negotiated to buy the piece of land, he bought the pipe, he got the rights to lay pipe along the farm fields, he got the machine to dig the trenches, he laid the pipe!," says Khan. Ronak laid 5-inch pipe to irrigate the farm fields, and 1-inch feeder lines that will flow constantly into the two old wells on either side of Nai Nangla, cutting the routine walk for water from a 4-kilometer, half-hour round-trip to 5 minutes. Water for the villagers is free, but the land and equipment cost Khan's foundation Rs 1.5 million (US $30,000).

That's another lesson of water poverty in Nai Nangla: It is possible for one relentlessly determined person to start unraveling the problems. But Khan brings a combination of qualities that illustrate not the possibility for success, but why failure has so much inertia. Khan is simultaneously an outsider and a local. He has operated at the top of the economic pyramid—in London. He knows not only how things should work, he knows how to make them work. But Khan is a product of the village culture he is trying to fix; he knows what daily life here is like, he knows how to win political support, he knows how to goad the bureaucrats who often mummify progress in India. And Khan has financial resources and a sense of the connection between investment and return. (Ask him why he likes dairy cows, and he says, "If you feed a cow 50 rupees of fodder, she gives you 150 rupees of milk—300 percent ROI [return on investment] per day.")

It would be hard for anyone who had never left Mewat—no matter how energetic or talented—to see what is so clear to Khan. It would be hard for anyone from outside Mewat to have the political subtlety to do what Khan is doing, no matter how brilliant their ideas. Even in places like Nai Nangla or Jargali that seem simple, water problems are never simple, certainly not as simple as they seem. Each problem can be solved, but it often turns out to require solving two underlying problems first.

Coming back from visiting Ronak's new well, Khan slips into Nai Nangla's primary school. Education is the thing Khan is most proud of making progress on in the last six years, especially for girls—he talks often of "unlocking woman power." He greets by name the women and girls washing clothes along the edge of the foul pond, then says later, "You can see it in the eyes of girls at the wells washing clothes—the strength. They will not be BS'd by their husbands."

In the schools across Mewat today, says Khan, "among girls age six to fourteen, 86 percent are going to school," a remarkable leap. But 75 percent of them still drop out around fifth grade because of lack of toilet facilities as they reach puberty, get their periods, and value privacy.

Nai Nangla's school has stout cinderblock buildings arranged around a dirt courtyard, but as a launching pad for children into the global economy it is otherwise flimsy. None of the teachers has a desk, and in a school of 320 elementary-age children, not one student has a chair. In fact, there is not a single chair in the entire school. In every classroom, immaculately dressed children sit on the floor facing their teachers, who teach from blackboards at the front. The school has no water—teachers ask the children to bring them water from home—but it is supposed to have about five pit toilets. This is what Khan is curious about.

He sweeps in like the prime minister of Nai Nangla, requests to see the headmaster, then asks the wide-eyed man to be taken to the latrines. Two are padlocked. Several have an unappealing odor even from outside the door. None is functional.

Khan is furious. He upbraids the headmaster, because he knows that without working toilets, as soon as girls reach adolescence they will drop out. "I will pay to fix these toilets," Khan says. "We will do it this week. But you must take responsibility for making sure they work.

"And I will tell you how we will judge whether you are keeping them working: We will ask the girls whether they are willing to use the toilets!"

The headmaster nods with fearful reserve. It is quite easy to imagine Khan coming back in ten days, gathering a group of thirty girls, and asking whether the toilets are clean enough for them to use.

Khan sweeps out, his anger fresh. The children—320 children—simply use the field adjacent to the school when they need to go to the bathroom. "It is a crime," says Khan.

In some ways Khan's school-toilet tantrum is classic India. Why did the headmaster even listen to Khan, who holds no official position and has no direct power over the school? "Because he knows who I am!" says Khan, in a rare burst of realpolitik. "I met with the education minister just last week, and he knows that." More puzzling still, why would the teachers and the headmaster tolerate working in a school with five nonworking latrines? Khan shakes his head. "The sense of initiative you want in teachers is often totally missing," he says.

Other progress notwithstanding, it is the kind of moment in which Khan knows that he did the right thing in leaving Unilever to return to the villages, and also in which he is discouraged about his chances of changing this most basic element of Indian life.

"I will beat the system on a lot of things," he says. "But I'm not sure I can beat it on water."

EVERYBODY IN INDIA HAS A WATER STORY, and not just a water story but a story about water management and water mismanagement. Each story is a kind of Indian water allegory.

In Hyderabad, the high-tech metropolis eight hundred miles south of Delhi, Hanumantha Rao determined to provide 24/7 water to one of the three sections of the city whose water service he supervises. Hyderabad is a city of 4 million, and Rao is a general manager for the city's water utility. He embarked on the project in 2006. The neighborhood that was most easily isolated for the 24/7 experiment had five thousand water connections, and 55,000 people, just over 1 percent of Hyderabad's residents. It had been receiving water ninety minutes, every other day, and to turn the intermittent water service on and off to that one neighborhood, workers had to open and close sixty-seven valves by hand. Before moving to 24/7 service, every customer in the test section had to have a water meter installed, and significant leaks in the mains had to be fixed to hold 24/7 pressure. The preparation took nine months, with twenty-four people from Rao's staff working on it full-time. Pipes were fixed, meters were installed, and in late 2006, the neighborhood of Adikmet got 24/7 water. "The customers were happy," says Rao. Some customers did get huge initial bills, because they had leaks, or because servants let faucets run for hours. The big surprise was about water consumption.

"They were not using any more water in twenty-four hours a day than in ninety minutes every other day," says Rao. "A person getting 24/7 water isn't going to take ten times the water they used to get." In fact, people quickly adjust, and use only the water they need. All went well for six months. "My personal opinion was, it was successful," says Rao. "Senior management decided it was a 'costly experiment.'" And it ended. Adikmet gets five hours a day of water now; Rao was ordered to cut that to two hours. But his own testing has shown that homes take 10 percent *more* water when they get only two hours of supply—a hoarding effect. "I am not reducing it to two hours," Rao says. Given his success, is 24/7 water coming to Hyderabad anytime soon? Rao squints silently as if the meaning of his experiment might not be clear after all. "It is a dream," he says, meaning, It is a fantasy.

Pentair is a $3.3 billion global water infrastructure company, based in Minneapolis, with factories and operations worldwide, including in the Indian seaside city of Goa and in Delhi. Pentair makes components critical to the huge desalination plants being built around the world. It also produces a line of freestanding reverse-osmosis units, ranging from sleek countertop models that residents of Defence Colony might buy, to closet- or room-size units to supply clean water for a school or an office building. Those mid-size units are perfect for Indian villages—installed in a storefront, they can become a kind of high-tech well, taking whatever water supply is available and turning it into safe drinking water, typically for a charge of a few rupees.

Pentair supplies these units at just the cost to make them—the company thinks it is the largest supplier of village RO units in India, through a partnership with an NGO. Pentair wanted to show off how this changes village life, and made arrangements for me to tour the unit installed in a village about forty-five minutes east of Delhi.

"We supplied that one for free, to a third party," says Mukund Vasudevan, who heads Pentair's India operations and has an MBA from the University of Chicago. "It was a couple years ago. It took us a while to track it down and check on it. It turns out the person who was supplying service to it stopped. And the RO unit itself is apparently gone." Pentair doesn't quite know what happened.

It's one of the frustrations for Pentair, whose staff in India thinks of it as an Indian water technology company with the potential to help solve an urgent problem. The challenge is that the very villages that could benefit the most from Pentair products don't have the wherewithal—the human infrastructure—to use those products. Councils of village elders don't have the kind of taxing authority or financial resources that even small U.S. towns have. They don't have employees or buildings or a budget. Many villages don't have electricity. So while Pentair may have the "solution" to their water problems, there is no one to buy the solution or take possession of it if it is donated, no one to keep an RO unit connected to a source of power, to operate it and maintain it and fix it when something goes wrong. Vasudevan sums up the odd situation with sad precision: "There is no customer in the villages for Pentair."

Jyoti Sharma has some of the most hopeful stories about water in India—in her stories, Indians don't assert control over water, but they do assert control over their water supply, over the fate of their water.

Sharma, thirty-nine years old, left school with an MBA from the Indian Institute of Management Bangalore, one of the country's most demanding business schools, and started out at one of India's biggest ad agencies. But after she took time off to have kids, the world of high-dollar corporate marketing lost its appeal.

"I thought, You only have one life. Is that how you use your good brain? I wanted to do something where I could make a change.

"Water appealed to me because water is a problem which *has* a solution."

For Sharma, water has become too much of a big-picture, big-project issue—the problems are big, the solutions must be big, everyone sits around with his arms crossed waiting for someone with big resources to do the big fix. Sharma thinks in the smallest possible units of solution.

She also thinks, as she puts it, "We spend too much time blaming each other for the water problems. In the process of blaming, we don't actually do what we can do. My personality is, no blame games. If the politician is corrupt"—she smiles lightly, this is India, of course—"oh, let him be."

Sharma founded a group called FORCE (Forum for Organised Resource Conservation and Enhancement), and after five years, she has a vast tapestry of connections. She knows the water-deprived and water-privileged communities in Delhi, she can make a call and get senior officials from Coca-Cola on the phone, and just as easily reach senior engineers from the Delhi Jal Board. It is, in fact, her ability to connect to everyone—

to treat water as both an open-source community and susceptible to open-source solutions—that gives Sharma and FORCE their power.

FORCE has driven a rainwater harvesting effort in Delhi, to capture monsoon rains and channel them into recharging the city's failing aquifers. Delhi has gone from having two rainwater harvesting structures to having more than three hundred, and the city now requires all new buildings to collect, clean, and send into the aquifer any rainwater that falls on the footprint of the building.

Sharma is a master of patiently watching for the simplest solution, for the solution that clicks perfectly with real human behavior about water.

She thinks the daily scramble by the most impoverished Indians to the Jal Board's trucks—with the hoses trailing from the tankers, "like the head of Medusa"—is absurd. "A lot of fighting happens," she says, "a lot of wastage happens, and those hoses lowered into the water, they dirty the whole ten thousand liters." At the Shivanand settlement in West Delhi, the tankers come each day around 8 a.m., but there is no scramble by the three hundred families who live in the slum. The tankers fill three black, plastic water storage tanks that hold about twenty thousand liters of water. The tanks are clean. Clean water goes directly from the tankers into the storage tanks. Someone from the family can come anytime and claim the sixty-liter-per-day allotment. The water is waiting for you, rather than the other way around. Instead of daily water worry, says Sharma, "there is water security." How hard is that? Sharma helped broker the arrangement—even helping persuade the Jal Board to lay the concrete pad for the tanks and install them. "It does require a more human connection," she says.

One of the things that's clear is that water problems across India—magnified by population growth, by economic growth, and by severe drought—have grown so chronic, and yet are so untended, that rather than making people discouraged, it has energized them. Many communities have simply stopped waiting for the government, for "someone else," to fix their water problems.

In a much larger Delhi slum cluster, Rangpuri Pahadi, with six hundred homes and 3,500 people, residents got tired of standing in line at hand pumps to collect the day's water. The Jal Board will not provide "permanent" water connections for unauthorized settlements. But the people of

Rangpuri Pahadi were gradually becoming more prosperous. "They work as domestic help in the better neighborhoods, or in construction," says Sharma. They wanted water delivered to each of their shacks, as many of them experienced water delivered to the homes where they worked. "They wanted a distribution system. We said, 'This is how much it will cost.'" The residents collected the money to have better wells drilled, then bought a pump and a two-thousand-liter storage tank. Then the residents laid small water pipes to every individual shack that wanted to "subscribe" to water service—five hundred of the six hundred homes. "They did all the work themselves," says Sharma.

What they created, in fact, is a miniature Jal Board. Each dwelling gets water through a stainless-steel conduit laid in the dirt—half an hour a day for those who pay Rs 125 a month, an hour a day for Rs 250 a month. The water comes on a set schedule. The tariff—$2.50 or $5 a month—pays for electricity and the salary of an employee who is constantly opening and closing valves for customers. Rangpuri Pahadi's residents might earn $90 or $100 a month, so the water is costing a whole day's wages or more (the equivalent of an American spending $150 a month on water). But the water often makes the wages possible.

"This saves standing in line at the hand pump, which was at least an hour a day," says Bhawan Devi, a fifty-year-old woman who is on the five-member water board that oversees the system. "That meant you often couldn't work a job with regular hours. That's why people are willing to pay to get the water."

Sainik Farms is a Delhi neighborhood that seems to be from a different universe from Rangpuri Pahadi—the homes are large, airy, with acres of wide lawn, styled like haciendas or villas, set behind large walls. Sainik Farms is one of the more well-to-do communities in Delhi—home to retired cabinet ministers, prominent business and entertainment figures, lawyers, and government officials.

But it shares some important qualities with Rangpuri Pahadi. Sainik Farms, too, is an illegal settlement, outside the jurisdiction of the city of Delhi, built up over the last two decades on former farmsteads. It has been overtaken by metropolitan Delhi, but it remains "unrecognized," receiving no city services. Residents work together to provide their own electricity, garbage collection, security, lighting, and roads.

Sainik Farms residents handle their own water needs, as if each were a solo homestead, drawing their water from wells.

The problem is, Sainik Farms' fifteen thousand residents are taking so much water from the aquifer that the wells keep running dry. "In 1985," says Pradeep Bhagat, sixty-four, who has lived in Sainik Farms for twenty-four years, "we had wells going down forty feet." He has had four wells drilled at his house, and three of them are dry. The one that works, he says, "is at two hundred and twenty feet, and it's almost over. It's erratic."

Over the last thirty years, every Sainik Farms home has simply drilled wells deeper and deeper—Bhagat says some are now four hundred feet deep, and there may be ten thousand wells in this one four-thousand-acre neighborhood. The wells are unmetered, unbilled, unmonitored, and people take as much water as they need, without regard for anyone else or for the health of the aquifer.

Bhagat is a retired executive for a global paint and chemical company who runs an organization that helps educate children in slum neighborhoods. He knew the water situation in Sainik Farms was a slowly evolving disaster. Today, in fact, 60 percent of the private bore wells in Sainik Farms are dry—and more than half the residents now have their water delivered in tankers. "Yes," says Bhagat, "we are trucking in water today, just like the slums."

Dried-out wells, right there in your own yard, are hard to overlook. They are as potent, and as scary, a symbol of mismanaged water as mains that are dry 23½ hours a day.

Bhagat shakes his head. "Something had to be done. Together."

FORCE has partnered with the Sainik Farms residents association to educate residents about the value and simplicity of rainwater harvesting—taking the almost unmanageable floods of monsoon rainwater that fall in the neighborhood and returning them to the failing aquifer. As it happens, the best way to turn rainwater into groundwater is to use the dry bore wells in reverse, to send harvested rainwater down after a basic filtration.

Rainwater harvesting has caught on in Sainik Farms. But there has been a much more tectonic shift in thinking. The residents have decided to give up their thousands of individual wells, and have asked for both approval and assistance in moving to a system of twenty community wells, with distribution pipes and monthly fees—much like the slum colony of Rangpuri Pahadi.

"It would be a little water utility," says Bhagat. "We've said we will design the system, pay for the wells, pay for creating the distribution system." The Jal Board seems positive, but isn't sure about legal issues.

"The attitude in Sainik Farms had been, we don't need to create our own water security," says Jyoti Sharma. "We can just buy our way through. It has taken years to get to this point. And some hardship."

It is a huge leap—the leap from selfishness to community, the leap from thinking of yourself as an individual with entitlements to thinking of yourself as a member of a group whose behavior affects other members of the group. Water issues, in particular, are often made worse when everyone operates independently—all those pumps sucking water from mains in Delhi and Hyderabad and Bangalore make everyone's water dirty. The collective solution is usually cheaper, more efficient, less wasteful, and better for the fate of the water itself. Money and technology are often not the best solutions to water issues—rainwater harvesting is simple, low-tech, and it's a lot easier and less expensive than finding new sources of water.

Bhagat isn't quite as sanguine as Sharma about the motivations of his Sainik Farms neighbors. "Not everyone is with us, even now," he says. "Only the people who are suffering are behind it. People don't act until the time they can't get their own supply. They don't want to see beyond their own nose. That is the unfortunate part of our character."

JUST NORTH OF THE CITY OF DELHI, in the neighboring state of Haryana, the Yamuna River is wide and flat and peaceful, winding through undeveloped land, with a scattering of people net-casting from boats or from shore. The river isn't a pristine mountain stream, but it has an appealing naturalness. It is hard to believe that the northern edge of the fifth-largest urban metropolis in the world is just around a bend and under a bridge.[33]

At the border between Delhi and Haryana, there is a barrage, a low dam across the Yamuna, and from behind the dam, Delhi has a water in-take to supply drinking water to the thirsty city of 20 million people. Al-most no water passes beyond the dam—the Yamuna is reduced to a trickle, not even 10 percent of the width of the river above the dam.

For perhaps 150 feet, the mighty and revered Yamuna River is little more than a creek. Then, all at once, the river's flow is restored from the west bank—a vast cataract of water comes rushing in from the mouth of a tributary that is 60 feet across, and if you close your eyes, the volume and the sound of crashing water are such that you can imagine a wilderness river. The surface churns with turbulent rapids.

But the smell and the color of this tributary are astonishing, arresting. The water is India-ink black. The smell is barnyard-organic fermented with chemical-plant acrid—manure and methane. At the point where the black tributary joins the Yamuna's riverbed, the smell is almost too strong to bear.

This is one of the "drains" that collect the wastewater from Delhi's residents, its hospitals, factories, and businesses, and pour it back into the Yamuna, almost a billion gallons a day. It is, in fact, a black river of raw urban sewage—this one drain puts out 1 million gallons of wastewater ev-ery four minutes.[34] Out in the middle of the restored flow floats the carcass of a dead water buffalo; along the banks you can easily spot every sort of debris—flowers, clothing, take-out food containers, a Bacardi rum bottle, a hypodermic needle.

In Delhi, if you are washing your clothes or brushing your teeth, clean-ing out an industrial mixing tank or flushing a toilet—it's all going right into the Yamuna.

Just beyond where that first canal of sewage empties into the riverbed, there is an informal beach, where people come to make sacred offerings. You can rent the services of a boat and a pole man and float out onto the black river. The smell eventually dulls the nose. The surface of the Ya-muna is odd—it's flat and shiny black, but pocked everywhere with tiny bubbles fizzing up, like the surface of a dark ale. Methane is bubbling up out of the fermenting river.

People at the beach wash clothes in this water; people come to take a dip in the Yamuna's sacred flow, to wash away their sins, as if anything they have done could be cleansed by water this dirty. In fact, it's not wise to even put your hand in the water. It is unimaginably, almost immeasurably, filthy.

In a 2009 study, India's Centre for Science and Environment reported that as the Yamuna flows through Delhi, the pollution isn't just unsafe for swimming. India's standard for safe swimming is 500 bacteria per 100 milliliters of water—500 bacteria in about half a cup of water. (The U.S. EPA standard is half that.) Halfway along its trip through Delhi, the coliform count in the Yamuna was typically *10 million* bacteria per 100 milliliters, and was often 100 million. One eyedropper of Yamuna River water is enough to make six bathtubs full of water unsafe to sit in. Says the CSE report, "The river is unfit for any human purpose."[35]

The Yamuna flows for fourteen miles along the eastern side of Delhi, mostly hidden from easy public access, or even routine public view, by Delhi's ring road, an eight-lane highway that circles the city. By the time the river leaves the city limits, twenty-two drains have emptied into it. The Yamuna just gets dirtier as it goes. It is a real-life river Styx. The venerated Ganges River is, if conceivable, dirtier. And the Yamuna is the main tributary of the Ganges, contributing 60 percent of its flow.

It is easy to get discouraged about water in India—and the state of the Yamuna is a perfect example. It's hard to imagine, in a country that can compete in brain jobs with the United States, that can send science missions to the Moon and put six billionaires among the fifty richest people in the world, it's hard to believe that it treats its great rivers this way.

But two things are worth remembering. First, India is capable of dramatic, even inspirational, change. In 1998, responding to pollution that was turning Delhi's air into a toxic soup, the country's Supreme Court ordered all public transit vehicles in Delhi, all taxis, and the entire fleet of small, ubiquitous, three-wheeled vehicles called auto rickshaws, converted to low-emission compressed-natural-gas fuel. Today, all ten thousand of Delhi's public buses run on CNG, as do all five thousand taxicabs, along with every one of the signature green-and-yellow auto rickshaws. Conversion of the privately owned auto-rickshaw fleet was painful, but also impressive: There are at least 53,000 of the auto rickshaws, barreling with suicidal speed and heedlessness along every street and through every intersection in Delhi. (New York City has thirteen thousand yellow cabs.) Delhi's air is still often gray and smoky, but it would be dramatically worse without the shift to CNG fuel of the last decade.[36]

Second, while India seems to have accommodated itself to some truly intolerable water circumstances for a modern nation—no major cities with a basic, always-on water service, one in six people relying on water that is transported by foot—we all get used to appalling circumstances, some just as stunning to outsiders as India's water compromises. The public schools in many of the great cities of America—Philadelphia, Detroit, Los Angeles—aren't just bad, they have become nonfunctional: ineffective at teaching, often not even safe. The schools in the nation's capital are, quite simply, a national disgrace. That's a failure at least equal to intermittent water service in Bangalore.

In both cases, change is hard at least in part because the rich have opted out of the public system, and the people who are left often have dramatically less money and political clout. Well-off Indians create their own twenty-four-hour water systems; well-off urban Americans send their children to private schools. Poor Indians don't have the power to demand improved water service, and fear they might not have the money to pay for it; low-income Americans don't have the time and influence to insist that public schools do a proper job of educating their children.

As with schools, there is no quick fix to water problems.

India is in the middle of a water crisis, but it is a slow-motion crisis, which is why it is important not just in India but for everyone beyond India. India's water problems can seem as hard to clean up as the Yamuna River—so messy that they are beyond solution. Where do you start?

But most of India's water problems aren't really water problems, they are people problems—problems with how they think about water and how they manage it. That is the most important lesson for Indians and the most important warning for the rest of us. It means the problems are, in fact, solvable—because we let them happen in the first place. You can have a perfectly good water system—as much of India did—that you let slip away because you don't take it seriously.

That's a lesson that Americans, and those in much of the rest of the world, are in danger of unlearning. We upgrade our running shoes every year, we upgrade our cell phones every eighteen months, we upgrade our laptops every three years. But our water systems have become like a feature of the natural world, a kind of man-made geography that we have come to believe needs no more attention than a waterfall or a mountain.

It is just the opposite. Water systems fall behind fast, and catch up slowly, and only with grinding effort. There is no leapfrogging over an aging water system, the way, for instance, cell phone service or satellite TV service or wireless Internet service allow quick leaps forward. You can't beam water through the air.

ONE BIG CITY IN INDIA has brought back always-on water service—with both remarkable effort and remarkable results.

Navi Mumbai is a gawky adolescent of a municipality on India's west coast, whose many personalities can be hard to reconcile. Navi Mumbai's main thoroughfare, a modern, high-speed, four-lane boulevard that runs along the Arabian Sea, is called Palm Beach Marg (Palm Beach Highway). Palm trees line the median, and there are mangroves in places along the seashore. Navi Mumbai has dozens of apartment blocks, but many are somber, sterile, Soviet-style concrete towers, un-Palm-Beach-like, in fact, utterly untropical. Navi Mumbai's commercial zone stretches for ten miles, and includes gleaming glass office campuses, like those in Gurgaon. Wipro is here, and so is Tata. Right alongside the office parks are chemical factories and oil refineries. The city's old central business district is a few uninspired blocks of tired concrete buildings, but its parkways are home to elegant hotels and a half-dozen flashy shopping malls. Navi Mumbai's taxes on new cars are so low that many well-off Indians come here to purchase luxury vehicles—in 2009, 996 luxury cars were sold in Navi Mumbai.[37]

But a better indicator of Navi Mumbai's economic well-being is a fat green water main that runs right alongside Palm Beach Marg for many kilometers. Like the factories alongside the office parks, the prominently visible water main isn't attractive. For much of its length it is two meters in diameter, a single pipe big enough for anyone under six feet tall to stand inside, and it is painted the uninspired, flat green of bridge girders and boiler rooms. But the water main is the spine of a system that, by the end of 2010, supplied 24/7 water to 65 percent of Navi Mumbai's residents—continuous water, anytime they turned on their taps, to 800,000 citizens, including all slum dwellers. That water main would make always-on water service available to every resident in 2011.

Vijay Nahata is Navi Mumbai's municipal commissioner, the equivalent of its mayor. With a grin, he says, "For water, we decided to have the ideal system."

Navi Mumbai was conceived around 1970 as a planned community designed to take pressure off Mumbai, which even then was bursting with unmanageable growth. So unlike many Indian cities that are hundreds of years old, Navi Mumbai is only forty years old, laid out with care—and it has attracted residents exhausted by Mumbai's crowding and cost. The city has 1.2 million people—double what it had just ten years ago. It is large and fast-growing, but relatively small by Indian standards—about twenty-eighth in size.

The effort to make Navi Mumbai the first modern Indian city with continuous water started in 2002, and came from the opposite problem: chronic water shortages. The city was buying its water from two state agencies that couldn't keep up with Navi Mumbai's demand. The city was routinely receiving one-third less water than it needed. Searching around for its own water supply, Navi Mumbai ended up buying a dam from another government agency that couldn't afford to finish it, and once that dam was finished, it more than tripled the city's supply.

"When we got our own dedicated source of water," says Sanjay Desai, executive engineer for water in Navi Mumbai, "the idea of 24/7 water came up."

"We liked the idea," says city engineer Mohan Dagaonkar. "We knew it can be done. How it can be done is another question."

"There were hundreds of hurdles to making this happen," says Desai.

The idea of Navi Mumbai being a "planned" city seems to have seeped thoroughly into its culture. The city counts, and reports, on almost everything. At the end of 2009, for instance, city land had 157,283 trees, and the city operated 320 public toilet facilities, with "total seats of 3,626." Navi Mumbai has become obsessive about good sewers, good landfills, and recycling. Precisely to avoid the cross-contamination between water mains and sewer pipes that plagues the rest of urban India, it is standard practice in Navi Mumbai for the water main to run alongside the road, buried one meter down, and the sewer line to run in the center of the road, buried two meters down. No possibility of confusion, and almost no chance that a sewer leak will find the water main.

City officials take a sly pride in all this, because they know how unusual it is. Says city engineer Dagaonkar, "We do know where our pipes are. Here everything is on the network—it's on a geographic information system"—plotted down to the inch with GPS coordinates.

It has taken six years of planning, politicking, and construction to move the city's government and residents to 24/7 water. (Hoover Dam took five years.)

The big green water main had to be laid twenty-one miles from the reservoir to the city. Every home and business got a new, tamperproof water meter, and the city's 220 miles of water mains were inspected, repaired, or replaced. Every dwelling needed a supply pipe, including those in slum clusters and villages, which had traditionally been supplied with water tankers. And there were lots of meetings to explain that water bills would go up—with 24/7 service, they double, from Rs 65 to Rs 130 a month, that is, from about $1.50 a month to $3. "People were opposed to the increase," says municipal commissioner Nahata. "But it was small relative to income."

City engineer Dagaonkar says public support took time to muster. "Everyone wanted 24/7 water—they love the idea of 24/7," he says. "But they didn't want meters. They didn't want increased rates. They wanted us to do it for free. It took five or six years to convince them."

And one thing was evident—with roads being dug up for water mains, with the huge new green water supply pipe moving into town in kilometer-long leaps, with every home being visited to receive a new water meter, it was clear where the money would be going.

Navi Mumbai has discovered another technocratic technique that is not common in Indian governments: outsourcing. Navi Mumbai government is relatively lean, and many functions are outsourced with contracts, which include tight performance standards, and then managed. Garbage is outsourced, sewage treatment is outsourced, and the operation of the water supply system is outsourced. City officials say they find it easier to insist on performance from their contractors than they would from an army of employees. "We find that once you have a permanent employee in a job, they become lazy," says Nahata. "That's why we've outsourced."

The rollout of continuous water appears to have actually reduced water use per person in the city—from seventy-five gallons a day to sixty gallons a day, in part because of the metering, in part because people don't have to hoard water just in case. And Navi Mumbai has gone to the trouble and expense of installing something not of much use in most Indian cities: fire hydrants. Now that the city's water mains will have water pressure all the time, it will be possible to roll up to a fire hydrant and find water in an emergency.

Navi Mumbai appears to be the very leading edge of a critical shift in attitude. Just ten years ago, the idea of taking an Indian city from intermittent to continuous water wasn't even discussed, according to V. S. Chary, from Hyderabad's Administrative Staff College, the country's leading proponent of bringing Indian cities back to 24/7 water. "When I would talk to people about it here in India, they would look at me puzzled. They think I'm a fool."

Chary now has a roster of Indian cities that have committed to moving to 24/7 water and are receiving grant money to help them. It's not four cities, it's not a dozen, or three dozen. The list of cities publicly committed to restoring 24/7 water service is forty-five.

"It's not a tipping point," he says. "That is yet to come. But it is huge. It is a delight."

In Navi Mumbai, many who have gotten 24/7 water—especially the poorest residents—have a different life as a result. Sharda Sonawane lives in the slum area called Shivaji Nagar, in a small apartment with her husband and her in-laws. They got 24/7 water at the end of 2008—a sink with a tap, on the wall in the main room. Alongside the sink sits a small, portable washing machine for clothes.

The Sonawanes used to stand in line, along with everyone else in Shivaji Nagar, to use a public tap, an hour or more of waiting a day. "I worked for a cosmetics company then," says Sharda, "and I was often late for work because I was standing in line. My wages would be cut as a result."

Her mother-in-law, Vandana, a former municipal councilor, says her home was occasionally robbed while she was standing in line for water. "While we stood in line, they would steal things. And because there was so little water, we would use the water in a miserly way. It was not very hygienic—we would use less water for bathing."

Everyone in the house has regained the precious hours each week they used to stand in line for water.

"We bathe more," says Vandana, "and the water is clean."

"We wash clothes every day," says her daughter-in-law.

The cost of their water has doubled. "Yes, we are paying more," says Vandana. "But we are getting more water. And it is better water. So we are happy."

The Shivaji Nagar neighborhood is in some ways representative of Navi Mumbai's best qualities. Although it is poor, the public pathways are spotless, and there is no stream of raw sewage running along the streets.

Bhimrao Rethod is an auto mechanic in his mid-thirties who has lived in Shivaji Nagar for thirty years, since he was a little boy. He shares an apartment with a total of eight people—his parents, his brother, his wife, and their three kids. As a boy, he says, "I used to stand in the queue, sometimes all night, I would skip school to stand in line. And then, we wouldn't get that much water."

How is the tap inside his apartment better? Rethod laughs.

"First, you save time. Second, you don't have to store the water. I can go to work on time. Illnesses and sicknesses are reduced. It's a hassle-free life! My children have 24/7 water!" Rethod has been speaking the local language through a translator. But his enthusiasm causes him to switch into English.

"Even rich people in Bangalore and Delhi don't have 24/7 water. I'm a lucky man, and I am richer than those people!"

9

It's Water. Of Course It's Free

You can "own" a glass of water, but only until you drink it and pee. Once you pee, you don't own that water anymore.

—Mike Young,
water economist,
University of Adelaide,
Australia

WHEN YOU CHECK INTO one of the Starwood chain's Four Points hotels in the United States and Canada (there are 109), you'll find a couple bottles of water in your room. Nothing unusual about finding bottled water in a hotel room—it has become standard, as if the hotels didn't have much faith in their own tap water, and the glasses and ice bucket they provide.[1] What is unusual is that the bottled water is a gift from Four Points—no extra charge for the water, any more than there's a charge for using the bottle of hand cream by the bathroom sink. Four Points puts a little tag around the neck of the bottles:

"It's water. Of course it's free."

The water is a treat, and the tag is just the right touch of hospitality. Pause a moment and appreciate that Four Points has gotten the customer service right, and it has picked a way of presenting that service with an edge of irony, guaranteed to make its guests smile.

"It's water. Of course it's free."

But why is the tag funny, exactly?

First, of course, it's funny because in many hotels almost nothing these days is free—from the WiFi to the gym. And of course, bottled water itself is almost never free. But it's really funny because everywhere else in life water is in fact free, or essentially free. Can you believe the nerve of all those other hotels, charging you for something as elemental as water?

Four Points underscores the idea on its Web site, where it has a page dedicated to the free bottled water it offers. Underneath the slogan "It's water. Of course it's free" is the line "What's next? Paying for air?" The page is illustrated with a picture of a crystalline waterfall with a sign propped in the current reading "$3.99."[2]

The silliness is self-evident. Who can put a price on water? Who would put a price on water?

In fact, we have even gotten used to paying for air—at least in one circumstance. Most gas stations now charge 50 cents to use the air hose to fill your tires; if you can find the water spigot at a gas station, though, you can use it without charge.

Four Points hotels is tapping something primal in our relationship to water—we don't think it should cost us anything. The water bills most people in the developed world get each month from their local utility are nominal—in the United States, it's $1 or $1.50 a day for always-on, never-fail, unlimited water service at home. At that rate, you have to flush the toilet a hundred times before you've spent a dollar on water. London's water utility, Thames Water, provides clean water to one of seven people in the United Kingdom, and brags that its unmetered customers pay 88 pence per day (about US$1.30) and that its customers with water meters pay 73 pence (US$1.08).[3]

In some places, water service is almost literally free. Traverse City, Michigan, the charming tourist town along Lake Michigan's eastern shore, charges $10 for the first 4,500 gallons of water per month. The next 25,400 gallons also cost $10—total. In Traverse City, you can use 30,000 gallons of water a month—1,000 gallons a day, enough to take twenty-five baths—and pay $20. Forget flushing the toilet—in Traverse City, you can fill a swimming pool for less than the price of a bottle of Merlot.

In most communities, in fact, the water bill isn't for water at all—it is typically just for the cost of getting the water to us, the pumps, the electricity, the staff to monitor water pressure and water safety and to be on standby for water-main breaks. The water itself costs nothing.

But if you had to pick a single problem with water, if you had to pick a single reason that our relationship to water is so out of whack, it is captured in that perfectly turned slogan: "It's water. Of course it's free."

Although we don't often realize it, free isn't that great. The lack of a price, on water or on any other resource, leads to all kinds of inequities and inefficiencies. Water may be the most vital substance in every aspect of human endeavor, but the economics of water is a mash-up of tradition, wishful thinking, and poor planning.

Las Vegas—where the water supply is in such desperate shape that homeowners are paid $40,000 per acre to rip up their lawns and water police officers enforce the water rules—has among the lowest residential water rates in the country. A typical family's bill there is $23.62 a month. In Atlanta, the same amount of water would cost you $50. In suburban Philadelphia, an area with ample rain, where water scarcity isn't an issue at all, it would bring a bill of $80, three times the cost in Las Vegas.[4]

In Napoleon, Ohio, the Campbell Soup factory, the largest soup factory in the world, simply puts its huge water intakes into the Maumee River and takes all the water it needs. No charge.

Out in California's Imperial Valley, the vast agricultural basin from which much of our carrots and lettuce and broccoli come, the water is imported from Lake Mead, on the Colorado River, delivered through a vast network of canals. The water is essential to farmers in the Imperial Valley, which, with average rainfall of three inches a year, qualifies as a desert. Every square foot of cultivated dirt in the Imperial Valley has to be irrigated with six feet of water. The Imperial Valley is the eleventh most productive agricultural county in the country; it is also the nation's largest irrigation district.[5]

The water is essential, it is imported, it is sucked out of a river that is itself increasingly dehydrated. And the water is cheap. The price for farmers is a flat rate: $19 per acre-foot. That's $19 for enough water to cover a one-acre field in one foot of water—$19 for 325,851 gallons of water.

The average home in the United States pays $3.24 for 1,000 gallons of water.

The average home in Las Vegas pays $2.71 for 1,000 gallons of water.

A farmer in the Imperial Valley pays six pennies for 1,000 gallons of water.

And here's the really astonishing thing: The farmer in the Imperial Valley is using *exactly the same water* as the mother giving her daughter a bath in Las Vegas. The mother is just paying forty-five times as much as the farmer.

A carrot farmer in the Imperial Valley can get about thirty thousand pounds of carrots from an acre of land. Those carrots will use about $114 worth of water to grow. What that means in the grocery store is that the big, three-pound bag of carrots from California required an astonishing 217 gallons of water to grow—and that water cost 1 penny.[6]

We need carrots, of course—we need affordable carrots. We need farmers, and farmers need huge quantities of water. But why does the water that costs a California farmer $1 cost a Las Vegas homeowner $45?

Las Vegas is often held up as the perfect example of squandering water—not just all those lawns and golf courses, but the casinos with their fountains and canals, all at water rates less than Atlanta or New York City. Without the Colorado River, without Hoover Dam and Lake Mead, Las Vegas really would be a fantasy.

But the Imperial Valley, which depends as surely as Las Vegas on the same imported water, actually uses five times as much water as Las Vegas. And because Clark County, home to Las Vegas, has twelve times more land than the Imperial Valley, the farmer is using sixty times as much water per acre as the casino mogul.[7]

And why are we growing carrots in the Imperial Valley of California? Just one reason: cheap water available for the farmers. We could grow carrots somewhere else. We've decided to have farms in the Imperial Valley as surely as we've decided to have casinos in Las Vegas—by providing water to both places, from the very same overtapped Colorado River.

On the surface, Las Vegas somehow seems "unnatural," a frivolous use of precious water; the farmland, on the other hand, seems smart, a judicious use of the precious water.

Both the Imperial Valley and Las Vegas have huge economic impacts. The water siphoned off to the Imperial Valley generates crops and cattle worth $1.5 billion a year. The water piped into Las Vegas generates gambling revenue alone—not accounting for all the related economic activity, from grocery stores to movie theaters—of $8.8 billion a year. In terms of bang for the bucket, the water we're using in Las Vegas is having a geometrically larger impact than the water in the Imperial Valley—per gallon, it's generating twenty-four times more economic activity.[8]

Once you start to unpack the economics of the way water is used in Las Vegas and in the Imperial Valley—what the water costs and what it helps produce—Las Vegas doesn't look quite so sinful.

But the really important thing to understand is that Las Vegas casinos and Imperial Valley carrot fields both represent judgments—choices about economic development, water, and water economics. It's just that we don't typically acknowledge that they are choices, or that there is an economics of the water required to create both.

There is nothing wrong, in fact, with deciding that casinos can afford more expensive water than carrot farmers—as long as we realize that's a decision we've made.

If we doubled the cost of water to the Imperial Valley farmers—which would surely occasion outrage—the amount of water in the three-pound bag of carrots would be just over two pennies. And the farmers would still be getting water at one-twentieth the cost to the casinos.

Here's the real reason the economics of water is skewed: The prices the farmer and the casino owner pay for water aren't prices in the way we think of them. Water prices aren't fixed in the market—by matching people's demand for water against its supply. The price of water for both farmers and Las Vegas is simply the cost to deliver the water to each, and nothing more.

So it's not quite fair to use the term "inequities" for the odd price differentials in the world of water—it is rare to be able to compare such dramatically different water uses and water costs as growing carrots and running a roulette wheel within the same water system. But if the differentials are not necessarily water inequities, they are often water absurdities.

The dramatic price differences exist, in fact, for a quite noble reason: We want to make sure people get the water they need, so we've developed a practice of not charging at all for the water itself, we just charge for the water service, the delivery. And in a century of water that was abundant, safe, and free, we've haven't had to ask how much water people used, how much water they should use, how efficient they should be with it—or what they should pay for that water, and what we should pay for their products as a result.

No less a chronicler of capitalism than Adam Smith captured the polarity of water's value and its price in *The Wealth of Nations*, in a chapter called "Origin and Use of Money." Smith compares water and diamonds:

> The word VALUE, it is to be observed, has two different meanings, and sometimes expresses the utility of some particular object, and sometimes the power of purchasing other goods which the possession of that object conveys. The one may be called "value in use"; the other, "value in exchange." The things which have the greatest value in use have frequently little or no value in exchange; and on the contrary, those which have the greatest value in exchange have frequently little or no value in use. Nothing is more useful than water: but it will purchase scarce any thing; scarce any thing can be had in exchange for it. A diamond, on the contrary, has scarce any value in use; but a very great quantity of other goods may frequently be had in exchange for it.[9]

Smith published *The Wealth of Nations* in 1776, when getting water required a lot more effort than it does today. So his insight is truer today than it was then. Gather the half-finished bottles of water from the cup holders and the floor of your minivan and see what you can get for them. Forget "value in exchange"—not even their original drinkers will finish off "old" bottled water. Their best use turns out to be filling the dog bowl.

The irony, of course, is that bottled water is the only water we seem happy to pay for in the first place.

NOT MANY PEOPLE DREAM of owning their own water utility, and Bob Eckelbarger certainly didn't. The water utility in his tiny hometown of Emlenton, Pennsylvania, was such an antiquated, leaky mess, though, that with a stubborn sense of civic duty, he insisted on buying it in order to save it.

Eckelbarger was hired as the lone employee of the Emlenton Water Company when he was twenty-two years old. Two years later, tired of the

unwillingness of the five elderly owners to invest their modest profits in up-
dating the crumbling system, Eckelbarger told them he was either quitting
or buying the company.

And so, in 1962, Eckelbarger and his wife, Beverly, came to own
their own water utility—the water treatment plant, the storage tank, the
pumps, the pipes in the ground, and the water meters. They ran the water
company like they would have run a small-town doctor's practice: always
on call. They had four hundred customers—all the homes and businesses
in Emlenton, serving about twelve hundred people. Bob did the pump and
pipe work. Bev did the bills.

Bob was a young man when they bought Emlenton Water Company,
and he was nearing retirement age when they sold it thirty-six years later.
"Sixteen hours was the longest we ever had an outage," he says. "We never
had a customer out of water overnight."

When they took over in 1962, the typical customer's monthly bill was
$2.40. "We had total revenue of $960 a month," says Eckelbarger.

When he bought the company, Eckelbarger got all the records, going
back to its 1876 founding, so he knew that in 1912, the year the little utility
first installed a water filtration plant, the typical bill was $1.97 a month.

In the half-century between 1912 and 1962, the price of water service
in Emlenton had gone up 43 cents, not even a penny a year. In fact, back
in 1912, $1.97 was a substantial sum. As much as such comparisons are
meaningful, $1.97 in 1912 is equivalent to more than $43 now—the water
bill from 1912 would be 25 percent higher than the typical water bill today.

It was a typhoid outbreak in Emlenton in 1912 traced to the water supply that motivated construction of the filtration plant. For water that didn't
make you sick, $1.97 a month might have seemed a bargain.

The filtration plant that Eckelbarger got with the company in 1962
"was pretty much like it was when they built the plant in 1912." The town's
pipes were leaking 80 percent of the clean water being pumped, and the
pumps, Eckelbarger says, "were not steam pumps, but just one generation
beyond steam, belts and jackshafts and pulleys and so on."

In the first few years, just to raise the money to buy new pumps and
pipes, Eckelbarger persuaded his parents to mortgage their farm three
times.

He also went to the Pennsylvania Public Utilities Commission for approval for a rate increase. "I wrote a letter to the PUC in Harrisburg," says Eckelbarger. "I had some numbers, our expenses, our income. And I told them we needed more money. After we wrote the letter, they said they would require us to meet with them, so I packed stuff in a briefcase and I went down there"—220 miles east—"and I sat and talked to the man who was head of the water rates for the state. I showed him what we had done by way of improvements, and what we had to do.

"I was asking for $3.20 a month"—80 cents more. "Well, by the time he looked through all the numbers, he said, 'You need a nickel more. You won't make it at $3.20.' He approved a rate higher than we asked for."

And so in 1963, the rate for water in Emlenton went, in a single leap, from $2.40 a month to $3.25 a month, a 35 percent increase.

Even at $3.25 a month, the water was still just half the relative cost it had been fifty years earlier. In 1963, it was also comparatively cheap. Americans in those days were spending $7.50 a month on alcohol. Any family with a telephone in Emlenton in 1963 was paying $5.65 per month. So the 80 cents was a big one-time jump, but $3.25 a month for water wasn't a burden, it was a bargain.[10]

That wasn't how the people of Emlenton received it, though. "I nearly started a revolution," says Eckelbarger. "People weren't very happy with me."

Even in 1963, even in a small town with its own water company, the treatment plant right downtown on River Street, two things were true: People didn't know the condition of their water system or the money required to sustain it, and any increase in their monthly water bill was met with annoyance, if not resentment.

Recently, water rates in Indianapolis, which had been frozen for five years, needed to be raised on an emergency basis in the middle of the recession because of rising interest rates on the utility's debt. During months of hearings and debate in 2009 and 2010, residents expressed their unhappiness.

"We paid our bills in good faith and get nothing for it," said one woman, apparently discounting the actual water in her anger at the proposed 17.5 percent rate increase.

"We already pay hellacious water rates," said another woman, who owned two apartment buildings.

Indianapolis's "hellacious" water rates, before the increase that was ultimately approved, were $25.50 for a home using seven thousand gallons a month. In the end, Indiana's regulators approved a 12 percent increase—$3 a month, bringing Indianapolis's rate to $28.50, still 20 percent below the national average.[11]

Out in El Dorado County, California, in 2010, the water district was proposing to raise water rates for the first time in ten years—a hike of 35 percent, from $23.75 to $32.07 a month, on average, for a stunning ten thousand gallons of water per home. The El Dorado Irrigation District has 42,000 routine water customers; it received 4,500 letters protesting the rate increase.

Over several months, the water district decided to cut the rate increase to 18 percent—a rise of about $4.25 a month instead of $8.30. The president of El Dorado's board wrote a piece for the *Sacramento Bee,* explaining that he and his colleagues were simply trying to make up for years of neglect of the district's pipes and pump stations by past boards, but apologized for "the upheaval caused by the proposed rate increases."[12]

In Washington, DC, in February 2010, the city council approved new water rates that would help dramatically accelerate the replacement of the city's water mains. The 17 percent rate increase will allow the nation's capital to replace its water mains over a hundred years. At the previous rates, replacing the water mains was scheduled to take three hundred years.[13]

The attitude about water and money has a universality that crosses cultures and crises. In Australia, in the midst of a grinding, relentless, potentially permanent shift in rainfall that left the country without enough water to function, elected officials lost their jobs just for talking about raising water prices, or because they did raise them.

Jim Gill, who took Perth and its Water Corporation through the water crisis as CEO, says, "A water utility like ours produces water so cheaply, people expect us to continue that. We are just too good at what we do."

His successor as CEO of Water Corporation, Sue Murphy, articulates the practice of water managers worldwide: "We don't charge for the water at all. We charge for the pumping, for the storage, for the convenience."

And like her colleagues, she shares the view that water is what's called in economic terms "price inelastic": If you change the price, you won't change the behavior of people using the water.

"If you change the cost of water by a factor of ten," says Murphy, whose customers have water bills similar to U.S. bills, about $40 a month—"well, the rich people would still have gardens, and the poor people would live in dust bowls."

No one, of course, is talking about changing water prices by even a factor of two—and well-off people would, in fact, change their behavior if the routine water bill went from $35 a month to $350 a month.

It is, in fact, people from outside the water utility fraternity who have begun to question the idea that the price of water doesn't matter anyway, so you might as well keep it cheap. Robyn McLeod, commissioner for water security for South Australia, the vast state that includes part of the shriveled Murray-Darling Basin, started life as a history teacher, and came to water through jobs involving the environment and the private sector. Observing the conflict in Australia over scarce water between cities and farmers, between commercial plant growers and suburbanites with brown lawns, she says, "Water is about the economy. And water is far too cheap."

People have no incentive, in other words, to pay the slightest attention to what water they use or how they use it—because whether they are farmers flood-irrigating fields or wealthy people filling swimming pools, the price is zero, or close to it, so it doesn't matter how much they use or what the productivity of the water itself is.

The culture of universally cheap water also means that water systems worldwide rarely charge enough to sustain themselves—whether in Delhi or Washington, DC. They perform worse and worse over time, which not only erodes confidence in the system but also creates resistance to the very price hikes that would help solve the problem.

That, in fact, may be the ultimate irony: Keeping water at, or near, free eventually ends up depriving people of the water that the free water is supposed to make available to them. Free water means there isn't enough revenue to keep the water flowing.

The very organizations we expect to do a good job of providing water to everyone—including people of limited means—end up starved of resources, whether in the developed world or the developing world.

Free water has a cost, and not a trivial cost.

WHY ARE WE SO STUBBORN about the cost of water coming out of our taps?

Well, as a species, as a culture, we are used to finding the water we need and using it. Adam Smith wrote about the gap between the value of water and its price in 1776. But in his footnotes, Adam Smith cites Plato, in his dialogue "Euthydemus," in which Plato gives Socrates this line: "For only what is rare is valuable; and 'water,' which, as Pindar says, is the 'best of all things,' is also the cheapest." As an epigrammatic summation of water, Plato's is hard to beat: the very best of all things, and the cheapest. Plato wrote "Euthydemus" 380 years before the birth of Christ (and 2,156 years before Smith wrote *The Wealth of Nations*).[14]

But it's not just that we're in the habit of free water. We certainly got into the habit of free TV—a much shorter habit than free water—and now 90 percent of U.S. homes pay for TV, either cable TV or satellite TV.[15]

No, our attitude about water and money is more fundamental. It's not so much that we're price-sensitive—it's almost the inverse. We think pricing water would be wrong. Water is one of the very few things that we don't think should be distributed based on your ability to pay. If you put a price on water, by implication, there will be someone who can't afford the price, and so won't get water. Clearly, that's inhuman—in Indianapolis or India.

Actually there's nothing unethical about managing water demand with price, and there's nothing immoral about allowing the market to help allocate water—just so long as we solve what might be called the first-glass problem, so long as everyone has access to water for their basic needs at the lowest possible cost. Beyond that, a little application of the market might help us use water more wisely, more equitably, keep water cleaner, and leave some for Nature herself.

First, let's clear up a couple things that may be puzzling or contradictory about water and money.

Although the general view in the world of water utilities is that water is "price inelastic"—that people don't change their water use based on how much it costs—that's a piece of accepted wisdom that simply isn't true, even in the confines of reasonable water charges.[16] If water coming into our homes cost twice what it costs now, if the monthly water bill were suddenly $60 or $70 instead of $30 or $40, even that would cause people to be more careful with their tooth-brushing and, especially, their lawn watering. If people were receiving water bills of $80 or $100 a month—right where the typical cell phone bill is—they'd start thinking about how full the dishwasher and the clothes washer were before running them, and even whether to buy the new models of those appliances that use much less water.

The problem isn't that water consumption is price inelastic, it's that water bills are so low, it doesn't pay to be sensitive to the price. Water bills may rise by large percentages, but a 30 percent increase in a bill that's $25 means $7.50 a month. People don't bother remaking their lives to save 25 cents a day. We're not behaving with disregard to price, we're doing just the opposite: instinctively making the judgment that the effort required to change water consumption won't save us enough to be worth it.

Large consumers of water—factories, farmers—are very price-sensitive. That, in fact, is the point: If water is almost free—$19 for enough to cover an acre of land one foot deep, as in the Imperial Valley—or if it is literally free—as for Campbell Soup in Napoleon, Ohio—there's no incentive to spend money on expensive, sophisticated systems that use less water but get the same crop or soup production. If the water gets more expensive, then it pays to retrofit your irrigation system, or your soup-cooking system, with equipment that costs money to install, but reduces the amount of water you use.

The water bill that the typical home or business gets simply covers the costs of delivering the water. It costs a huge amount of money to build and maintain reservoirs, pipes, pumps, treatment plants, monitoring systems; to put staff in trucks, in water plants, in testing labs. And water is heavy stuff. Moving it, pumping it through treatment membranes and filters in plants, maintaining water pressure in water mains—all requires huge quantities of energy. Water utilities use 3 percent of the electricity generated in the United States, equal to the output of 162 power plants, making water utilities the largest single industrial users of electricity in the country. In California, 20 percent of the electricity in the state is used to move or treat water. When the cost of gas goes from $2 a gallon to $3.50 a gallon, the cost of moving your water goes up as well.[17] But there is no charge for the water itself—the water is "free" to the customers, as it is typically free to the utilities.

Prices vary wildly for water, in the United States and around the world, because while the water itself is free, the cost of acquiring it and delivering it varies wildly. Traverse City, Michigan, which will happily deliver you a thousand gallons of water a day for $20 a month, sits on the shores of Lake Michigan, one of the largest bodies of clean, fresh water in the world. The city doesn't have to do much work to get the water it is distributing to its residents' faucets.

Water is a good deal harder to come by in Santa Fe, New Mexico, which also has among the highest rates for water customers in the country. A family of four using just two hundred gallons of water a day would have a bill of more than $40 a month.

In Las Vegas, or the Imperial Valley, the "price" of the water reflects the cost of supplying the water, but it doesn't reflect the demand for it. Traverse City could supply water to a lot more people without raising the price much. Las Vegas paid people to take out their lawns to avoid the cost of having to find some new supply of water to keep those lawns—and future lawns—green.

Price differences have to do with the age of the local infrastructure, variations in the cost of energy and labor, and even with the sophistication of utility managers in handling issues like debt and politics.

Some water utilities have begun to price water use in tiers—it's not just that you pay more if you use more, but there are steps in the prices, and when you jump up above a basic level of consumption, the price per thousand gallons jumps as well. That's a modest effort to "price" water in relationship to demand—or more precisely, to price it according to how much it will cost to find more water. Bigger quantities of water require more energy and more infrastructure to deliver. And it is much more expensive to acquire the next gallon of water supply, and to deliver it, than to distribute the water a utility already has access to. It's expensive to plan and build new dams, new treatment plants, new wells, new canals, or, as in the case of Australia, to build new factories that turn seawater into drinking water.

To be effective, such "block" pricing has to be dramatic. In Santa Fe, the city charges $4.43 per thousand gallons, up to seven thousand gallons a month, a reasonable supply for a family of four. For every thousand gallons you use over seven thousand, the rate jumps to $15.84 per thousand gallons, almost four times the base price.[18]

FOR A WATER ECONOMIST, Mike Young lives every day with the absurdly perfect example of water and price. Young, one of the most thoughtful and influential voices in shaping Australia's water policy, is a professor of water economics at the University of Adelaide, and director of its Environment Institute.

He lives in a charming neighborhood of quiet, tree-shaded streets with cafés and well-kept yards, just north of downtown Adelaide.

In Young's apartment building, there are no individual water meters, just a flat fee for water. The water economist could turn on a faucet on June 1 and let it run until July 1, and not pay a cent extra.

And Adelaide, remember, is the capital of South Australia—the driest capital city, in the driest state, on the driest continent in the world.

As Young puts it, "The marginal cost of running the water twenty-four hours a day, or running it for just a drop, is the same: zero dollars. There is no price signal at all."

Mike Young, fifty-seven, is an unusual economist in a number of ways. First, of course, he's a *water* economist, which makes him something of a pioneer. He's also an economist who puts the environment first in line for water before even human water needs, and who has a sharp economic rationale to justify putting nature first. And Young is an admirer of the Medicis of Renaissance Florence, but not, like most of us, because of their art patronage, but because the Medicis invented double-entry bookkeeping—debits and credits.

"If you run a really good water accounting system," says Young, using a phrase rarely heard in either the world of accounting or the world of water, "if you give someone more water, you have to give someone less water.

Governments have always been keen to find a way to give more water to somebody, but they forgot to work out who is going to have less water."

Coming out of a century of unthinking water abundance, the situation of Young's own apartment building, where no one can even measure the residents' use, let alone price it rationally, is surprisingly common, a symbol of water so abundant we didn't need economics to help manage it.

The largest apartment complex in New York City is Peter Cooper Village and Stuyvesant Town, a parklike urban enclave on eighty acres along Manhattan's East River. Peter Cooper Village and Stuyvesant Town together are a single development that encompasses fifty-six apartment buildings, spanning nine New York City blocks, with 11,232 apartment units and 25,000 residents. And no water meters. Like Mike Young, on the other side of the globe, residents pay a flat fee for utilities, and they can use as much or as little water as they want. Not only no charge—no one even knows how much water any apartment is using.

In London, the situation is even more striking. Just 22 percent of the customers served by the water utility, Thames Water, even have water meters. Three-quarters of Londoners can use as much water as they want for a flat rate. You can't effectively charge people for the water they use if you can't measure the amount they use, and you can't give those people a sustainable water system if you can't charge what it costs to secure and deliver their water. Thames Water's current goal: to have half the homes in London on water meters by 2015.[19]

What the century-long golden age of water has done, says Young, is create an incredibly sophisticated system for gathering water and distributing it—an engineering system. But the engineering system assumes one thing: plenty of water. What we don't have—in Australia, in the United States, in most of the world—is an equally sophisticated system for figuring out how to allocate the water, particularly when there isn't enough.

"I think petroleum is better managed than water," says Young. "With all its ugliness, at least it works."

That's where economics comes in. Economics is a way of managing scarcity, and market economics—pricing—is a way of letting the people who want something that's scarce participate in deciding who gets it. We get to vote with our money on how important something is to us—whether we're consumers lining up for the latest iPhone, or real estate developers lining up for a block of midtown Manhattan land, or drivers balancing the price of July gasoline against the drive to the beach.

Young is also refreshingly realistic. Ask what seems a really basic question about water economics—Who owns water?—and he says, "Nobody does. You can 'own' a glass of water, but only until you drink it and pee. Once you pee, you don't own that water anymore.

"The debate over ownership of water is misinformed. The question is, Who is allocated the opportunity to use the water?"

Young's laboratory for thinking through his ideas about how to allocate water—how to decide who gets it—is Australia's Murray River, a sadly perfect case study of the need for an allocation system to match the engineering system. Australia's rainfall shifted so dramatically in the last decade that from 2006 through 2010, the Murray literally ceased to flow. The result has been a painful mess—economically, politically, socially.

The lack of water in the Murray River, where both farmers and communities are desperate for their share, for their entitlement, illustrates one of Young's principles: A water market—trading in water, paying for scarce water—is often not a bad solution, or an immoral solution, but just the opposite. A market can be the best, smartest, quickest, even fairest, way to distribute water, especially in the middle of a crisis.

"If you're running an irrigation system as big as the river Murray," says Young, "and you have a very dry winter, and you can only give out 30 percent of the water you normally give out instead of 100 percent—how do you organize the political process in a way to have the decisions sorted out in six weeks, before the growing season ends?" No government bureaucracy moves that swiftly, with results that are also effective and widely seen to be fair. But farmers—who know their own situation, their own finances, the state of their land, and the market for their crops—can decide quickly how much water they might buy, or how much they might sell, depending on the price.

A market in water only works, though, if a community, a region, a watershed, has done two things first: set up the water allocation system in advance of the crisis, and set up the water system in a way that acknowledges that water is a different kind of resource. It's not a new iPhone or a block of valuable city land, precisely because of the first-glass issues.

Some water needs to be secure and guaranteed for everyone, at the lowest possible cost, outside the market system, and some water needs to be unleashed in such a way that the market helps distribute and manage it much more effectively than it does now.

To explain, Young does something that we almost never do with our water systems. Sitting at a café table, he moves aside the coffee cups, takes a clean sheet of paper, and starts to sketch out a water system from scratch.

It's a water system in the shape of, well, basically a water glass.[20]

At the bottom of the glass, the first layer of water is for the environment. It's the water necessary to keep the natural water system itself—the river, the aquifer—alive, stable, and healthy.

Young, who is more economist than poet at heart, calls this "maintenance water." "That's the water necessary to maintain the environment, to maintain the system."

If that seems like a forehead-smackingly obvious way to start, well, in truth it's a radical concept. Around the world there are rivers that are so overused and so stressed by climate change that they often no longer flow to their own mouths—starting, in fact, with the Murray River, which needs a dam to hold back the sea, and including the Colorado River, the Rio Grande, and the merged Tigris and Euphrates rivers in Iraq, the Shatt al Arab, which no longer holds back the Persian Gulf, so that salt water is now flushed a hundred miles inland.[21]

"Without water for the system," says Young, "there is no water for anyone." That is, a good system starts out safeguarding the very renewable resource you're trying to allocate.

The next layer of water in the glass has almost the same standing as the "maintenance" layer—it's what Young calls "critical human needs." That's just what you'd guess: "Water to flush toilets, to take a shower, drinking water and water needed by the city and by industries, all operating on very tight water-use restrictions."

The "critical human needs" layer of water isn't water to run the hose while washing your car, or take a twenty-minute shower, or run decorative fountains in the plaza of your shopping center. It's the basic water necessary to keep people and the economy alive—and it's priced much the way water today is, at cost.

Together, the environmental and the critical human needs might take 20 percent or so of the water from a system—although that percentage depends on the system, how it renews itself, and who depends on it. That's the solution to the first-glass problem. The first water goes to sustain the ecosystem we all depend on, and to sustain us.

Those first two layers of water in the glass are both simple and self-evident, but in practice, in the way we've come to manage water around the world, they are rarely specified and protected with such clarity.

Of course, the volume required for each will not be easy to discover, and certainly won't be immediately easy to agree on. Upstream users will have every reason to argue that a river needs minimal flows, to keep more water for themselves; urban users will have every reason to argue that decorative outdoor landscaping is a vital economic interest that needs to be protected and watered. Indeed, what qualifies as "critical human needs" beyond sanitation and drinking water will require debate and setting priorities—hospitals are a critical human need, and water parks are not, but what about the auto plant, the steel mill, the plant nursery?

Bringing this kind of economic planning and principles to water will not take the politics out of water, any more than sound economic principles take the politics out of zoning decisions or tax policy. But agreeing on those two numbers in advance—regardless of how fractious the conversation necessary—gives everyone a measure of security, of calm, especially in a crisis.

Above the environmental and critical human needs layers of water, there's a fat layer of water that Young calls the "sharing" layer. This is all the rest of the water that powers our economy beyond the "critical human needs" allocation. It is everything from swimming pools and factories and luxury hotels to flood-irrigated rice fields. Young's "sharing regime" is an effort to bring clarity, order, and market pricing to bear on the big bulge of water that, in flush times, has been used with wasteful disregard, and that in times of scarcity, is the source of bitter dispute.

The shared water is divided into two categories—"high security" water and "low security," or "general security," water.

The idea is almost transparently simple: People who have high-security water get their water first, and they pay a premium for that guarantee. People who have general-security water pay much less, but they get their water second.

You pay the price you are willing to in order to secure the supply you need. If you're running a microchip factory or a vineyard of expensive, drought-sensitive grapevines, you might think it's worth it to invest in high-security water. If you're running a rice farm—water intensive, but not perennial, and perhaps not profitable enough to justify expensive water—you buy low-security water. If you're running a city, on top of your critical human needs allocation, you might buy both general-security water and high-security water, to go with whatever other sources you might have—your own wells, reservoirs, or desalination plants.

"If you've got high-security water and low-security water, everybody can determine how much risk they want to take," says Young, "user by user." They can determine how much risk they can afford to take—economically.

What percentage of the total available water is high-security and what percentage is general-security—as with the slices that go to the environment and critical human needs—depends on the dynamics of each water system, how much water is available, who uses it, how variable the flows are, and how a community is growing economically. The point is to understand the water system and to set the rules for how to use it in advance.

In Young's ideal scheme, the high-security pool of water is 30 percent of the ten-year moving average of available water. If a river typically has an annual flow of 10,000 gigaliters (half the historic flow of the Murray River), then 3,000 gigaliters—30 percent—would be designated high-security. That leaves 50 percent as general-security water (20 percent has gone to the "first glass" allocation).

If the river, in this example, does flow at 10,000 gigaliters in a particular year, you take off 20 percent—2,000 gigaliters—for the environment and the critical human needs, and 3,000 gigaliters for the high-security users, and the remaining 5,000 gigaliters goes to the low-security users.

But if there is a dry period, and the river only flows at, say, half its normal rate—if there's a year where it's just 5,000 gigaliters—the low-security users get no water. The high-security users get all their water first.

And even the high-security users have some risk: If the flow falls even lower, the high-security users only get what's left, on a proportional basis, after the environment gets its water and the critical human needs get their water.

Young wants the rules set up in advance. Again, it seems obvious, but most watersheds—the interlocking and interdependent systems of sources of water, reservoirs, lakes, rivers, and the users of water, cities, industries, and farms—exist in a kind of patchwork quilt of agreements, jurisdictions, assumptions, overlapping law, and practice.

"The beauty of this system," says Young, "is that you don't have to know in advance what's going to happen with the future of water. Everybody knows the water they're going to get"—or, if there's an actual trading market each year, the water they can buy—"and they can focus on, What's the best outcome I can get with that water?"

People know what proportion of water they're going to get each year, they know their own circumstances, and instead of fighting, they can make decisions.

For those who might find the layers-of-water-in-the-glass explanation a little too abstract, Young has a metaphor that seems like a wild leap of imagination, but turns out to be very useful.

"A really good comparison," he says, "is the way corporations are defined. A corporation, a company, is a legal artifact." That is, corporations aren't in any sense natural—we invented them. "Corporations have really exciting features, which apply to water as well."

"Exciting" may be an overstatement—but interesting, certainly.

"If you own a corporation," Young says, "you have a 'unit share' in it—the stock. That's each owner's share of an uncertain future.

"The rules about how the corporation operates are all specified in advance.

"If there's a profit, you get paid according to your share of ownership.

"If there's no profit, then there's no dividend, and even though you own stock, you get nothing that year.

"The law around corporations is designed so that they must reveal when something is wrong—with tremendous drama sometimes, but also clearly, transparently.

"Corporations provide the opportunity to take risk.

"And the governance structure is functional. The people who run the company day-to-day aren't the shareholders, they are the managers. They need a structure that allows them to get quick decisions, decisions as fast as the world is changing.

"Similarly, if we're running out of water this week, we need decisions this week, not in six months, after everyone with a 'stake' is consulted."

If the comparison between corporate management and water management isn't, in fact, exciting, it is in fact quite brilliant.

Corporate law doesn't care what you're doing with your company—opening a convenience store or writing software—and the rules of the game are set in advance, so you start a company, you go to work for a company, you buy shares in a company, with clarity about the rules and the consequences. You decide what risk you're comfortable with.

In the water scheme that Young can draw on a single sheet of paper, it likewise doesn't matter what you're doing with the water, and the rules are specified in advance. If you're in a highly water-sensitive business, you buy high-security water, and shoulder the cost. In most years, the low-security users will also get water, but not in every year.

It's like the difference between preferred shares of stock and common stock.

And like publicly traded corporations, the water system would require openness—about the state of the water, the volume, the availability, the management.

And in a real crisis, the rules are well settled about water for the system itself, and water for the basics of human civilization.

Young thinks this kind of system—which is relatively simple to describe but would require a fair amount of negotiating and politicking to implement in any given community—has something else in common with corporations: It is a robust system. Robust, in his terms, is something very specific: "A robust system for water can evolve and change as the conditions evolve and change."

If the climate changes, and the water that's available changes, the buckets of water adapt. If a factory owner, or a farmer, figures out how to improve water productivity, he can sell some of his water to someone who needs it more. The very price of the water itself encourages efficiency and investment—a city facing increasing population may raise rates, or require low-flow plumbing fixtures and appliances, to avoid having to either buy more water or build new treatment plants.

We don't often think of water itself changing, or of our relationship to water changing. But it does change, just the way our relationship to electricity or transportation changes.

"What looks like the right way to use water this year," says Mike Young, "won't look like the right way to use water ten years from now. A robust water system continues to work in harmony with itself, with the environmental conditions, with what people need.

"With a robust water system, you look back and you don't regret the past, and you look forward and you don't fear the future."

Most of us, in fact, do not live in a water system where we can look back without regret or look to the future without fear.

There are two final twists to Young's model framework.

First, Young sets aside a slice of both the low-security pool and the high-security pool of water for the environment—for the river, the mangroves, the marshes, the lakes, the aquifers—on top of the initial environmental allocation. His point is that the environment has typically been a victim of water neglect and water starvation, when times are flush and when there is no water at all. Humans always put themselves first.

"This way, if anybody gets water, the environment gets water too. That protects the environment from the politics of water."

A second twist has to do with the actual market for water. Water is a funny commodity—it's not traded the way copper is traded, or oil, coffee or wheat, or even something intangible like carbon credits. It's not traded, because in most places, you can't move the actual water and deliver it the way you can the wheat or the carbon credits. It's nice that someone in Orlando has a lot of water, but even if someone in a drought in North Carolina is willing to spend money for that water, there's simply no way of getting the water itself from Orlando to North Carolina.

Water is rarely priced, and rarely priced smartly, and some of that has to do with our attitude about it—but some of it also has to do with the character of water itself. Water and water systems are fundamentally different because the water itself is so independent. Even if you have a willing seller and a willing buyer, the water can be utterly unwilling to be traded.

But there are watersheds where water could be traded. The basic criterion is very simple: If I sell you my water, you have to be able to get the very water that I would otherwise use. A river or an aquifer is a good example—one farmer or factory or community doesn't take the water it is entitled to, and the buyer of the water takes exactly that amount of water.

Indeed, the Murray River has just such a nascent water-trading market—which has helped, even with the tiny allocations in the Big Dry, as some farmers sell what little water they have and others buy enough water to keep some crops in the ground.

What's happened in Australia is a reminder of how spoiled we've become. In the developed world, water has been so readily available, we've completely lost sight of the fact that it is a vital economic resource—one, in fact, that we have no well-constructed economic system to give out. As Young puts it, "Allocating the opportunities to use water gives us the quality of life we have."

Sometimes the economics of water is as clear as water itself.

Before the start of the 2009–2010 NBA basketball season, the Cleveland Cavaliers quietly removed all eighteen water fountains from their home stadium, the Quicken Loans Arena in Cleveland, known as the Q, which seats 20,500 people.

Where the fountains had been, the Q eventually taped up paper signs that said "For your convenience, complimentary cups of water are available at all concession stands throughout the Q."

For more than three months, through dozens of events and hundreds of thousands of visitors, there were no water fountains in the Q. If you wanted a drink of water, you had to stand in line at the concession stand, and when it was your turn, you could receive a free nine-ounce cup of water or you could buy a bottle of chilled Aquafina for $4.

On February 8, 2010, Cleveland's daily newspaper, the *Plain Dealer,* ran a page-one story about the then five-month-old removal of the water fountains headlined "At the Q, No More H_2O, at Least not From Fountains."

Team spokesman Tad Carper said the Cavaliers had removed the water fountains because they posed a risk of spreading germs and illness during the winter flu season, and he told the paper the Cavaliers had followed the recommendations of the NBA and the International Association of Assembly Managers, the trade association for big arenas.

The *Plain Dealer* reporter called both the NBA and the IAAM, and neither had ever recommended removing water fountains to fight the spread of disease. Carper then said the Cavaliers had simply wanted "to provide the healthiest environment for our fans."

As to the possibility that the Cavs and the Q were trying to drive bottled water and food sales by sending thirsty fans to the concession stands, Carper said, "That's simply absurd. That never crossed our minds."

Cleveland city councilman Anthony Brancatelli wasn't buying it. "It's clearly an opportunity to sell more drinks," he said. "If there were health reasons, we'd be taking [water] fountains out of every school and institution."

Two days later, the *Plain Dealer* raised an issue it hadn't in the first story: The Cavaliers' removal of the water fountains was illegal on at least two counts—the arena hadn't gotten a permit to remove the water fountains, and Ohio's building code requires an arena of the Q's size and age to have eighteen water fountains.

Cavaliers spokesman Carper backpedaled, telling the paper the Cavaliers were considering reinstalling the fountains now that flu season had peaked.

The day the second story was published, the Cavaliers posted an announcement on the NBA Web site saying that the removal of the water fountains was always intended to be "temporary"—something Carper hadn't previously mentioned—and promising that the fountains would be reinstalled "as soon as possible." In an attempt to mollify the anger caused by the water fountain removal, the Cavaliers set up temporary water stations around the arena for the games scheduled before new water fountains were installed, so thirsty fans wouldn't have to stand in line at the concession stand.

During the time the fountains were absent from the Q, the Cavaliers alone hosted 29 sold-out games at the arena. If just 10 percent of fans bought a bottle of water they wouldn't have otherwise, the Cavaliers sold sixty thousand extra bottles of water, bringing in nearly $10,000 in additional concession sales per game.[22]

One of the few real gifts the bottled-water industry has given us, in fact, is some important insight into the economics of water. Two things are clear: We are willing to pay for water, even to pay what amount to ridiculous sums, more than gasoline costs, and we intuitively understand that different waters, in different settings, have different values and therefore different prices.

And, of course, as the Cleveland Cavaliers discovered, like hundreds of city councils before them, we also think that basic gulp of water should be free.

Why, after ten thousand years of organized human civilization, do we suddenly need an economics of water?

As we've said, the last hundred years of human society have been an unusual period with regard to people and water: If you lived in the developed world, not having to think at all about where your water would come from was a whole new human experience.

But that flush, unthinking period is over. Many areas that have never experienced water scarcity are being hit with dramatic reductions in natural water availability. Growth in population, millions of people moving into the middle class around the world, and the spreading of factories to developing countries—those three trends put additional demands on water supplies. And population growth carries a dramatic hidden water tax— remember that even in the developed world, where our daily water use is indulgent, we require far more water to make our electricity and our food than we use for drinking and sanitation. The typical American uses about a hundred gallons of real water a day; the food the typical American consumes requires five hundred gallons a day to produce.[23] As more people rise into the middle class, their unseen, but very real, water needs increase in the same way.

A second problem is that the hundred-year golden age of water has caused us to think that water delivery is a kind of natural phenomenon—

you turn on the faucet, the water comes out. It's like opening a window and having a cool breeze come in. Except, of course, it's not. Vast national water circulation systems that were installed a century ago are crumbling and need to be upgraded. The cost to refurbish them is modest compared with their value, it's modest even compared with what we spend on bottled water, but it's still tens of billions of dollars a year for many years. An explicit, well-thought-out economics of water would help us with both problems.

One of the most striking changes in our relationship to water in the next hundred years will likely be that we will start using the right water for the right purpose. We won't use purified drinking water to flush our toilets and water our lawns. We won't hesitate to tap the most readily available source of water for most cities—our own wastewater. And that layering of water uses dovetails perfectly with Mike Young's economic framework for water: different waters, different prices.

The most basic ration of water for all of us will be low-price. Beyond that, there will be tiers of water with all kinds of qualities—the cleanliness of the water, the quantity available, the reliability or security of the supply in times of scarcity—and those qualities will all come with prices.

The result should be a richer appreciation of the value, the uses, and the costs of water. It should also mean a water system that is more transparent, that makes more sense, and that generates the money necessary to sustain and improve itself.

None of that flies in the face of the idea of water as a basic human right—pricing water doesn't require further squeezing people in developing nations who don't have good water access now. In fact, poor people around the world pay a terrible daily price for their water today, a far higher price than the $34 a month our water costs us. That price is the billions of people in the world's developing nations who give up education, or the possibility of employment, just to stand in line for hours a week, or walk for hours a day, to get their water; it's poor people whose children are dying at a rate of a hundred kids an hour from diseases they catch from tainted water.

Indeed, as the residents of the Delhi slum Rangpuri Pahadi remind us, poor people, too, are willing to pay for water when it makes economic sense. Rangpuri Pahadi is the Delhi neighborhood that organized its own miniature water utility, installing storage tanks, laying water pipes, and charging each family that connects a modest monthly fee that pays the costs of the system. Those families, in fact, are much closer to their water supply than the rest of us. They can see the pipes along the side of the dirt paths, they know how their monthly cost relates to the staff and electricity costs, they remember what it was like just six or seven years ago, when they had to stand in line for water of lesser quality. That water was "free" only in the most pinched sense; in fact, water you have to walk for, water you have to stand in line for, is the opposite of free—it's a kind of water bondage that desiccates the whole rest of your life.

An economics of water should be liberating—both for people and for water itself. It doesn't mean turning water supplies or water infrastructure over to remote, self-interested, profit-driven corporations. It means putting not just a price, but a value, on the most important substance in our daily lives, and putting a price, and a value, on the work necessary to make sure that substance is available in the quantity and quality that sustains the kind of communities we want to have.

Price is incredibly potent. Indeed, if you had to pick one thing to fix about water, one thing that would help you fix everything else—scarcity, unequal distribution, misuse, waste, skewed priorities, resistance to reuse, shortsighted exploitation of natural water resources—that one thing is price. The right price changes how we see everything else about water.

Mike Young thinks this moment in our relationship with water is as rich with energy, invention, and possibility as the water breakthroughs of a hundred years ago. "I think we're at an exciting threshold," he says. "In twenty or thirty or forty years, water management won't be the exciting, intellectually challenging stuff it is now. We'll have water largely sorted out, in both the developed and the undeveloped world. We'll be using water to be prosperous, and water management will be autonomous. And boring."

A century ago, our forefathers had to create a vast engineering system for delivering water, and then get it funded and constructed. Our task is far easier. All we have to do is change how we value that system and the water it delivers. All we have to change is how we think about water, so that a hundred years from now, a bottle of water with a tag that says "It's water. Of course it's free," won't be charming, it will be absurd.

10

The Fate of Water

The ultimate test for one of our fountains is, if you walk by it, do you say, I've just gotta watch this for a moment. We don't do babbling brooks. We try to actively engage the conscious part of your mind.

—*Mark Fuller,*
CEO, WET Design,
a fountain design company

IN THE MODERN AND AIRY TERMINAL A at Detroit's Metro Airport sits a smooth black slab of granite. It looks something like a black river rock, except for its size. Positioned on the floor, the disk is an oval, the height of a low table, and about forty feet across.

The top of the disk is covered with an inch of water that spills smoothly off the edges and down the sides, which bulge outward, softening the granite, making it seem more like a cushion than a rock. Since there is no lip, the black skirt of water slips with perfect ease over the edges, following the bulging stone like liquid silk, disappearing beneath the disk. It's as if a spring has bubbled up in the hectic travel terminal.

The disk is never at rest. Jets of water shoot up from the surface and land at random spots across the black oval. Everything about the streams of water is a little curious. They don't fire off in any pattern. Some streams are continuous for long moments, like the arc from a hose. Some are clipped bursts, sending up discrete slugs of water. Sometimes just a couple arcs of water are firing, sometimes twenty or thirty crisscross the disk at once.

The jets have a playful quality—sometimes firing off together, creating great splashing clusters of landing water, sometimes firing off separately with a mischievous call-and-response quality. The fountain and its water have a personality. And there is no railing or barrier. You can walk right up and trail your hands through the water.

What is most intriguing about the fountain is the water itself. The arcing water is clear, ripple-free, and bubble-free. The water in the continuous arcs is so smooth that even though you know it's flowing, you can't see any motion in the stream. The water has an unearthly quality; it looks heavy, like a breeze wouldn't bother it.

The behavior of water is so familiar to us that we sense any oddity quickly, even if we don't quite get it. The arcs of water shooting over the black disk seem to have their own relationship with gravity. Indeed, the fountain as a whole bends the space around it—people hurrying distractedly along the concourse spot it and alter their trajectory to circle in, often coming to a complete halt, head tilted in alert wonder.

The Detroit Terminal A fountain is the creation of a group of people who think about water all the time, about how water behaves, about its texture, its sounds, about how water reflects light, about how water moves, and about how water moves us. They work for WET Design, a company based in Los Angeles that specializes in fountains, but not of the decorative carved-fish-spouting-streams-of-water variety.

"The ultimate test for one of our fountains is, if you walk by it, do you say, I've just gotta watch this for a moment," says Mark Fuller, CEO and cofounder of WET. "Then we've succeeded. We don't do babbling brooks. We try to actively engage the conscious part of your mind."

Today, there are lots of people who think about water all the time—oceanographers and engineers designing offshore oil rigs, people who manage water treatment plants on land and who make ice sculptures on cruise ships; businesspeople who sell water filters and project managers who specialize in constructing city-size desalination plants; plumbers, politicians, and people who lay water mains; farmers who get too little water or too much, meteorologists who predict water's behavior in the atmosphere, and planetary scientists who study water's behavior on the Moon, on Mars, and on Enceladus, a tiny moon of Saturn that geysers water vapor and ice hundreds of miles straight up off its surface into space.[1]

But there are few places where people consider water from the perspective that Mark Fuller and his team of engineers, architects, designers, and computer programmers do. WET now employs two hundred people, and they think of water as a means of artistic expression. For them water is both an essential collaborator and an untamable force.

"The medium we've chosen to work with is much more independent than any other sculptural material, Michelangelo and his limestone included," says Fuller. "What we do at WET, from a scientific standpoint, is we play with the unnatural state of water. We try to control water. We go to absurd lengths to organize how we start with water—sweeping arcs of jets, grids that are strongly geometric—but the second the water comes out of a nozzle, nature takes over. The hand of God is present. It's the wonderful juxtaposition of the control of man, and absolutely relinquishing control."

That's actually a bit modest. The fountain at the Bellagio hotel in Las Vegas, which is an attraction all its own on the Strip, knits elaborate tapestries of water in the air, choreographed to music. When it was turned on in 1988, the Bellagio fountain was unlike any that had come before, in scale, or in the ways its water performed. It was WET's first truly grand-scale fountain. More than a decade later, the fountain still attracts thousands of spectators every day, and the hotel charges $50 a night extra for rooms that overlook it.

The Bellagio fountain was until December 2009 the largest in the world. That's when WET debuted a fountain in Dubai, at the base of the tallest building in the world. WET's Dubai Fountain is set in a thirty-acre lake (four times the space of the Bellagio fountain) and sends water five hundred feet in the air, across a span of three football fields. The tallest building in the world, the Burj Khalifa—taller than the Empire State Building stacked on top of the Sears Tower—is now fronted by the largest fountain in the world. At any given moment, WET's Dubai Fountain has 22,000 gallons of water dancing in the air. The fountain alone cost $218 million.[2]

The Bellagio and the Dubai, and the dozens of other signature water features WET has created—from the newly restored fountain in front of

New York's Lincoln Center, which forms patterns and moods with nothing but dozens of perfectly vertical jets of water, to the fountains in Branson, Missouri, which shoot water and balls of fire simultaneously, timed to music—all rely on technology that Fuller and his colleagues created, technology that has revolutionized what water can do.[3] WET fountains make use of an underwater mount that Fuller named an "oarsman," because it has pivots—gimbals and armatures—that allow it to aim water through three dimensions, with a range of motion and a nimbleness greater than the human wrist. The jets of water come from shooters powered not by pumps but by compressed air, giving WET's water choreographers more control, more precision, and more power when unleashing their water, and giving the jets themselves a surprising oomph and crispness.[4]

The designers at WET use the technology (which is always invisible to the viewer) to coax remarkable performances from water. They can make a circle of oarsmen spin and spray water to create a ring of whirling dancers in flaring skirts. They release water's innate sense of humor—two fans of water that seem to dance cheek-to-cheek, while in the background Frank Sinatra's "Dancing Cheek to Cheek" plays. WET fountains evoke the texture of water, the sensuousness, the sounds.

Mark Fuller has, in fact, taught water new tricks. Growing up in Salt Lake City, in the winter he would create ice dams in the gutters, "to see how far I could flood the water out into the street." Sitting in the back of a fluid mechanics class in college one day, "we were watching this 16 mm film, and it included a picture of this little stream of water that was so clear, it didn't look like water. If you turned the faucet off slowly, slowly, slowly, you got water that looked almost like a liquid icicle. Like the stem on a martini glass."

That's called laminar flow—water that is moving utterly without turbulence. No bubbles, no eddies, all the molecules of water in the stream moving at the same speed, in the same direction. "It's analogous to a laser," says Fuller.

Inspired by the 16 mm film, Fuller and some classmates built a laminar-flow nozzle as an undergraduate thesis project—a nozzle that creates that crystalline flow, but with force and authority instead of the thinnest little stream. Fuller, who was an Imagineer for Disney before founding WET, took the idea to Epcot and used it to create the leaping fountain outside Epcot's Imagination Pavilion in 1982, where the solid, clear beams of water still chase each other through planters in the park. "The Detroit [Airport] fountain is based exactly on that thesis project," says Fuller.

There is nothing "natural" about the presentation of water in the black disk—there is no place where we encounter jets of water with laminar flow, that's why the fountain's arcs of water are so arresting. The Detroit airport fountain is unpredictable—you never know where the water is going to come from, where it's going to land, how long it's going to last, what other streams one arc might trigger.

The fountain's leaping arcs of water aren't quite random. They are meant to evoke flying from one place to another. The Detroit fountain looks like nothing so much as an abstract, idealized version of the route maps in the back of the airline magazines. The whole creation has what Fuller calls a "fugitive nature. It has the allure of a sunset. The forms we weave in the air are fleeting. They are there for a split second and gone."

The fountain is so appealing to travelers that there are seventy videos of it on YouTube—including one person who has filmed it four separate times—and 46,000 people have watched them (not including the 25,000 people who have watched WET's own video of the fountain). You can watch the Terminal A fountain online for ninety minutes.

WET creates fountains that irresistibly change your mood, refresh both your sense of wonder and your sense of balance.

Far from being decorative or frivolous, in fact, a particularly brilliant fountain can restore, however briefly, your full appreciation for water, can instill a sense of respect and humility, along with a smile, for the lubricant of our lives.

The designers at WET think about water differently. They try to understand not just water's qualities and power but the emotions those qualities evoke, and why. They treat water as a sculptural medium, but also as something truly natural, something wild, independent, ultimately untamable.

The modern world allows us to approach water in strictly utilitarian terms—it is literally at hand when we need it. Our domesticated water requires zero work, zero thought, and comes with zero risk (and zero magic).

As for water's power, its wildness, we do get the occasional reminder of that—in a particularly dramatic downpour, or a hurricane, or when we see news reports of flooding. In fact, part of the appeal of water parks is that they unleash water's wildness in ways that are thrilling and uncommon. Because unless we're sightseeing somewhere like Niagara Falls, we've mostly insulated ourselves from water's power.

What the people at WET understand about water is precisely what is absent from our whole approach—whether in suburban Las Vegas or urban Delhi. There isn't really a word for exactly what's missing in our relationship to water, but it's an ingrained sense of cherishment, almost of reverence. Why wouldn't we revere water, of all the things we could revere?

In April 2009, Senator Arlen Specter hosted a town hall at Muhlenberg College, in Allentown, Pennsylvania. Specter—since defeated in a Democratic primary for reelection—wasn't just Pennsylvania's senior senator in April 2009, he was the longest-serving U.S. senator in Pennsylvania history, elected to five terms starting in 1980.

The Muhlenberg appearance was utterly routine, and Specter was relaxed and candid. Students asked questions about the impact that the drilling for natural gas that is growing dramatically in Pennsylvania will have on the safety of water supplies in the state.

"Which side are you on, the natural gas companies, or the people who have to drink poisoned water?" one student asked.

"That's certainly not a loaded question," Specter said, to chuckles. "I have never taken a position in favor of drinking poisonous water, and I don't intend to this afternoon."

The question-and-answers moved in the direction of drinking water, and Specter said, "I don't like drinking tap water because I don't trust tap water—if I have an opportunity to have bottled water."

A few moments later the senator said, "On a very serious level, I want to have clean drinking water, and I've supported legislation to help communities have clean drinking water. But I think there is a natural inclination"—and here the senator shrugged his shoulders to say he understood the inclination—"for people to want to be a little extra-sure on the water. Where I can have access to a bottle of water, I'm going to use it."[5]

Arlen Specter, the person, is entitled to any view he wants to hold. But for one of the most senior members of the U.S. Senate—a man who then sat on the judiciary and appropriations committees, and also on the Senate's environment and public works committee—for Arlen Specter to say flatly, "I don't like drinking tap water because I don't trust tap water," is astonishing, even outrageous.[6] The United States has among the safest, most closely monitored water systems in the world, a tap water system that is responsible in part for the extraordinary leaps in life span in the United States in the last hundred years.

It would be like Specter standing up before a group of college students, shrugging, and saying that he avoids bridges because he "doesn't trust" the bridge building system in the United States, or that he always drives because he "doesn't trust" U.S. air traffic controllers. If there's something so dangerous about U.S. tap water that a senior U.S. senator takes pains to avoid it, he should tell us what the danger is, and he should be leading the fight, loudly, every day to rehabilitate the system. Otherwise, Specter's comment is both corrosive and irresponsible. If he doesn't drink the tap water, where does that leave the people he represents?

The absurdity is only magnified by Specter's saying that because he wants to be "a little extra-sure," he drinks bottled water when he can. Bottled water isn't regulated with anything like the scrutiny and care that tap water is. The chance that there's something hinky about your drink of water is actually greater with a commercially packaged bottle of water than with a glass of tap water.[7]

At the opposite end of the silliness spectrum, Apple's iPhone App Store offers an application called "Water Your Body," which both recommends how much water you need each day and offers to keep track of your water consumption. "Water Your Body" even gives you "a daily and overall average grade on how well you are keeping up" on your water drinking. Because of "revolutionary research," the app's makers say, "we have learned just how critical it is, for your health, to maximize your individual water intake."

Of course, "Water Your Body" is absolutely absurd. Leave aside for the moment the question of what kind of person can manage the intricacies of an iPhone but has trouble remembering to drink water. No one who is not recovering in a hospital needs to track her daily water consumption. There is no proven need to "maximize your individual water intake." In a healthy person, you simply can't.

The finely tuned chemistry of your body quickly turns "extra" water you drink into pee—and your body regards extra as 1 percent excess water. "Water Your Body" simply maximizes your time in the bathroom, and takes 99 cents from you. Your body already has an exquisite, built-in, water-tracking app: thirst.[8]

The senator and the iPhone app both trade on our water illiteracy. The cost of "Water Your Body" is just 99 cents, the cost of having leaders who don't understand our tap water is much greater.

The world of water is changing dramatically, as we've seen. There is increasing talk from activists and NGOs, from companies like GE and Monsanto, and from the occasional forward-thinking elected official, of a "global water crisis." We hear routinely, from those trying to rouse the world to action, and from nonprofit groups trying to help, that on this very day, more than 1 billion people in the world don't have access to clean drinking water, and that before the calendar turns on another day, five thousand children will die because they lack clean water, or from an illness they got from tainted water.

And we hear that over the next forty years, the problem is going to get catastrophically worse: Between 2010 and 2050, the world will add 2.4 billion new people, equal to the populations of China and India combined.[9] It's possible that 1 billion of those unborn people will also lack clean water.

The very repetition of those statistics, though, isn't activating—it's numbing. Why will it matter more to us if there are 2 billion people without clean water in 2050—what, that extra billion people is going to suddenly wake the rest of us up? The billion people who don't have water today aren't enough of a catastrophe—but we'll certainly dig some fresh wells and lay some new water pipes once the number hits 2 billion?

We do have a crisis of water availability around the world—virtually every country has water scarcity problems, every country has water infrastructure gaps.

But talking about a "global water crisis" is absolutely the wrong approach. It's the wrong perspective, with all due respect to the water activists who devote themselves both to trying to get water to people, and trying to get the rest of us to pay attention.

We already have a "global energy crisis" and a "global economic crisis," a "global environmental crisis" and a "global health crisis." And, of course, perhaps most pressing of all, a "global climate crisis."

Except for the economic crisis, we haven't confronted any of those with either urgency or imagination; the "global water crisis" might well turn out to be one global crisis too many for most people.

Depending on which scenario turns out to be correct, tackling the global climate crisis may be the most important thing humans have ever done—our very survival may depend on it—or it may simply be more important than any other policy issue right now. But tackling global climate change also requires, literally, remaking the way the industrial economies of the world operate, every minute of every day. It's not actually something you're going to fix by swapping out your lightbulbs.

Water is completely different.

There is no global water crisis, because all water problems are local, or regional, and their solutions must be local and regional. There is no global water crisis, there are a thousand water crises, each distinct.

It turns out that if a resident of Portland, Oregon, defaults on her mortgage, that financial failure ripples throughout the interconnected financial system of the world.

It really matters what kind of new power plant a small, regional utility in China installs, because the emissions from that plant will continue every hour for three decades, and whatever that plant sends into the atmosphere quickly reaches California, and beyond.

A family in Durham, North Carolina, that switches from two cars to one, that starts walking and riding bikes to commute and do errands, has a tiny but significant impact on the entire energy system of the world.

It's an odd paradox: Our behavior in the face of truly global crises—the ones that seem most out of reach of individual action—can actually make those crises better, or worse.

But the water problems of Barcelona cannot be solved by conservation in New Orleans or Bangalore. Unlike your mortgage payments or your electricity use or your driving habits, how you shower or water your lawn has no impact on the water availability of people an ocean away, and may well not have any impact on people a single time zone away.

That doesn't relieve us of responsibility for water behavior and water habits—precisely the opposite. It means we must take responsibility for our own water issues, because no one else on the globe will. No one else can.

That's actually the good news about water: The problem is right with us, but so is the solution. Part of what is so frustrating about genuinely global problems like finance and climate is that even if you and your community, even if your country, behave with thoughtfulness, discipline, and foresight, if your neighbors don't, all the good work you do can be instantly undone. That isn't true of water. If Perth, Australia, remakes its water economy in an intelligent, sustainable fashion, that effort can't be undone by water carelessness in Melbourne, Australia—or in Istanbul.

Most water problems are, in fact, solvable. When people talk about "fixing" the health care crisis in the United States, even the smartest, most determined, most insightful experts on medical care and the economy of medicine will start by telling you that the problems are complicated, layered, and won't be quickly or easily disentangled.

Water problems involve competing interests too—farmers, industries, individuals, people with big yards versus people who live in apartments, the needs of rivers and lakes versus the needs of human civilization. But we have the technology to clean water to any level we want and within watersheds to deliver that water where it needs to be. Until 2009, even the astronauts living in the International Space Station were having all their water delivered—by rocket ship, at a cost of $41,650 per gallon. If we can get drinking water into orbit, no ordinary human settlement is out of reach.

(The space station now has an on-board recycling station that turns urine, and even sweat, back into drinking water.)[10]

We have the intellectual technology to resolve the conflicts over water, whether in the Colorado River basin or among the nations of the Middle East. That we don't manage those conflicts well, that the states of Florida, Georgia, and Alabama can grind their water dispute through U.S. courts for twenty years without resolution, often has nothing to do with the water itself. There's enough water in the southeast corner of the United States for everyone. The failure to resolve the disagreement over twenty years—while many of the core problems, like Atlanta's dramatic growth, got worse—is a failure of political leadership, a failure of water management, not a problem with the water.

In fact, our approach to water management in the golden age of water has been so cavalier, so profligate, that it's a kind of good news. We can do better quickly. We don't use nearly the amount of water we go through— we waste an incredible amount. In the United Kingdom, 19 percent of water pumped leaks away. In Italy, it's 29 percent. London loses a day's water from its mains every four days. In New York, whose water from upstate is so clean it gets no filtration, the city loses a whole day's drinking water every week to leakage, more than a billion gallons of losses a week. In the United States overall, we lose 7 billion gallons of water from leaking water mains—a day. That's more fresh water than thirty of the fifty states in the United States use each day.[11] The bad news is that there aren't a hundred big leaks in London or New York that can be quickly fixed to dramatically reduce the leaking; there are ten thousand leaks. But as water systems that are a hundred years old get replaced, water efficiency should steadily improve; big cities have a literal reservoir of clean water right in their own water mains, and smart utilities that are replacing old, leaking pipes install self-monitoring systems that make spotting future leaks simpler, quicker, and less disruptive.

Agriculture is even less water-smart; poor farming practices squander nation-size quantities of water. Worldwide, farmers use twice as much water as all other uses combined—67 percent of world water use goes to agriculture. And as much as half the water used for irrigation is wasted.[12] Agriculture needs a blue revolution to follow its green revolution—farmers need to be shown how to make better use of rainwater, how to use irrigation water more effectively, with more precision, and the financial incentives need to be flopped so that instead of water and electricity being so cheap that wasting makes sense, farmers are rewarded financially for learning to grow the same amount of food with less water.

That's obviously easier to prescribe than to accomplish, but it doesn't require any technological leaps. The efficient farming practices exist.

In the United States, withdrawals of water for irrigation peaked in 1980. Adjusted for inflation, the United States produces 60 percent more agricultural products today than in 1980. And U.S. farmers use 15 percent less water, as a group, than in 1980. The water productivity of U.S. farmers has increased 90 percent—adjusted for inflation, the water that in 1980 produced $100 worth of products, today produces $190 worth of products. That kind of leap is good news not just for areas that have water scarcity and water management problems; it means we can, in fact, produce a lot more food while using less water.[13] Along with crops, there are huge quantities of water to be harvested from the world's farm fields.

We all need to update our idea of what "clean water" means. The blossoming of micropollutants in water supplies around the world should have exactly the same energizing impact as the discovery of bacteriology at the turn of the nineteenth century. That insight led to water filtration and distribution systems that inaugurated the century-long golden age of water we've just lived through. The micropollutants are a similar red flag—the world's natural water circulation system is huge beyond human comprehension, but not, as it turns out, huge beyond human influence.

We don't understand the impact of the micropollutants—the traces of medicine, the residues from chemicals and plastics we use in our everyday lives—on either human beings or the natural systems we depend on. We don't swim all day long in our drinking water, so it's possible that minuscule quantities of chemicals won't affect our health. At the same time, we can hardly be comfortable when our "clean" wastewater is altering the physiology of whole lakes and rivers full of fish. Now is the moment to figure out the impact of what we've been unintentionally putting into water, and develop inexpensive techniques for removing the micropollutants before we return the water to its source or reuse it ourselves. Micropollutants, like most water problems, will only get worse, more difficult and more expensive to deal with, as time goes on.

But it's also true that as individuals, we often get wildly exercised over potential tap water problems, misjudging the relatively small risk from water, and ignoring the much more damaging problems closer to home. If you walk into a mall in the United States and take a stroll, the health problems of the people around you have nothing to do with what's in their tap water. In the United States, we have a long way to go in terms of improving routine health and fitness before the quality of our tap water is a central issue—especially because tap water overall is so safe. In fact, most American adults and children could do with a little more tap water, a little more of the activity that makes you thirsty, and a little less soda.[14]

It's equally important to reimagine our idea of "clean water" at the lower end, to stop using water cleaned to drinking-water quality standards for things like flushing toilets and watering lawns. In the United States, half the water supplied to homes is used outside—for lawn watering and car washing. In many communities, as an artifact of an older era, it is still against the law to use gray water, or treated wastewater, for things like lawn watering, maintaining athletic fields, or flushing toilets—or the system to provide such water isn't available. That's simply silly.[15]

In Orlando, Florida, and surrounding Orange County, recycled wastewater has been used without fanfare since 1986—going on twenty-five years—for exactly those kinds of outdoor uses. A purple-pipe lawn watering or irrigation system isn't just permitted for new homes, apartments, and developments, in Orange County it's required by local law. In fact, if the purple-pipe reclaimed water service is available in a neighborhood, residents and businesses are forbidden to water their lawns with drinking water.

That's a mind-flip, and one that has had dramatic results.

Orange County, one of the fastest-growing counties in the nation, now distributes almost as much treated wastewater as it does potable water. Orange County residents use 57 million gallons of drinking water a day, and they use 51 million gallons a day of treated wastewater on their lawns and landscaping.[16]

More to the point, Orange County has closed the loop—most of its drinking water comes back to it and is used a second time. Orange County has tapped one of the most readily available water sources that most communities overlook: the water its customers have already used, that the county already has in its pipes and treatment tanks.

The psychological change is just as dramatic: After two decades, residents in central Florida *expect* athletic fields and wide expanses of green space, golf courses and backyards, to be watered with recycled water. It's hard to retrofit communities with a second, inbound water system, but it's not particularly challenging to lay purple-pipe systems in new construction—any more than it is difficult to provide new homes with electricity or cable TV, with natural gas service or high-speed Internet lines.

Consider an interesting thought experiment. In the United States in the ten years from 2000 to 2009, more than 9 million new single-family homes were built and sold.[17] Perhaps not all those new homes could have easily been equipped with a purple-pipe system—new construction is typically concentrated in subdivisions, but not just plumbing is required; the local water utility has to be set up to treat wastewater and return it to customers. Still, if even a quarter of those homes, 2 million, had been plumbed to use recycled water in yards and toilets, that would be 2 million homes (and 5 million people) that would be using half the potable water of similar homes built before 2000. That saves enough drinking water to supply 1 million more homes—or 2 million homes with purple pipes themselves. Even if we can't instantly and affordably change the world we live in, we can change the new world, the world we're building every day. Patricia Mulroy reduced water use per person by a hundred gallons a day in Las Vegas—but it took her twenty years.

The water habits we've become accustomed to are just that—habits, customs. And that, too, is good news. Most people now would literally recoil if the passenger sitting next to them on an airplane lit a cigarette, but most people over thirty may well remember a time when smoking on airplanes was routine. It wasn't finally banned until April 1988.[18]

If we change our behavior, if we are forced to or wooed to, we quickly change our expectations, and a virtuous cycle often begins.

Changing behavior can start in the simplest ways, as many U.S. companies are discovering. Measuring the amount of water you use, and how you use it, quickly reshapes your appreciation of water. Campbell Soup, in its 2010 corporate sustainability report, started providing a measure that it has come up with for its own use: cubic meters of water used per metric ton of food produced. In 2008, Campbell used 10.33 cubic meters of water per metric ton of food. In 2009, it was down to 9.35. In human terms, a can of classic Campbell's condensed tomato soup (10¾ ounces) required 107 ounces of water to make in 2008, and 96 ounces of water in 2009. Campbell has set a public goal of getting that down to 53 ounces of water per can of tomato soup by 2020—two cans of soup for the water it uses now to make one.[19]

That kind of water reporting—along with reporting of the details about energy consumption, pollutants, and greenhouse gas emissions—should be required of all public companies. It should be as routinely available as sales data and earnings per share data and the details of executive compensation. Most companies have that data right at hand, because they pay for their water and their energy already. Looking at it, and letting everyone else look at it, would change how we think about water, and the world.

Water problems aren't simple or trivial. People are dying every hour because water is badly managed. But water problems are eminently solvable, in an era when so many problems seem insurmountable.

This hasn't been a book about how cities can more smartly manage municipal water systems or how farmers can improve their irrigation efficiency. This book isn't a polemic on behalf of urgent water activism, or a sober warning about the future of water supplies. It is not a book about California's collapsing water system, or how to rescue our crumbling water infrastructure, or the issues around water pollution. There are already many invaluable books about all those topics.

That is one of the challenges of grappling with water—any piece of the water story deserves a book of its own. There is, in fact, an encyclopedia of water, first published in 1972, called *Water: A Comprehensive Treatise*. It was conceived and edited by Felix Franks, a British scientist and academic, and a review of the first three volumes in the journal *Science* in April 1974 called it "easily the most ambitious and extended treatment of this ubiquitous liquid." Franks's encyclopedia eventually ran to seven volumes and 3,896 pages. And it deals exclusively with the physical and chemical properties of H_2O.[20] There's no way of producing a book about water that is exhaustive without also producing a book that is exhausting.

Everything about water is about to change—how we use water, how we share it, how we think about it. That makes it a vital moment to understand where we've been, and to imagine how our water future might be different, and why. That's why this book has tried to be about just one theme: our relationship to water.

Our water problems are real. Our approach to water must change, and we'll be happier if we realize that, and handle the change with creativity and forethought rather than confront it as a crisis. It's water we're talking about, so there will be no avoiding the change. What we can choose is the time and the approach and the level of panic.

This book is an effort to rescue water not so much from ignorance as from being ignored. We think about and talk about our relationship to technology and to our pets, to our stepchildren, to our weight, our food, our medicines, and our diseases. But we almost never talk about water, about the significance of water to us, about its meaning and its value. Even the shelves full of books about water tend not to be about water itself. Water is intentionally the central character of this water book—it deserves to be.

We desperately need a fresh way of thinking about water. Or, more realistically, a starting point for thinking about water. Most of us haven't ever thought about it very much. But we'll need a foundation for understanding water as water issues become more urgent. We'll need a framework for thinking about the fate of water, because the fate of water is our fate.

We are at the start of a new era of water, an era when we'll need to use water much more smartly. Most important, we should be at the start of an era of much greater water equality—an era when no one dies simply because they can't get water, of all deprivations. There's no excuse for public officials not making clean water for their citizens a top priority now. But the era of smart water should make drinking water easier and cheaper to provide even in challenging circumstances. There will be less excuse than ever for not guaranteeing basic access to water.

For the parts of the world that have luxuriated in the golden age of water, the era of smart water need not be the start of an era in which water scarcity or water limits desiccate the way we live. Quite the contrary: Using the right water for the right purpose may well open our eyes to the kinds of untapped sources of water, starting with our own wastewater, that we routinely overlook now.

The biggest water problems of all are water illiteracy and water mythology. It is our understanding of water—whether we live in Toowoomba, Australia, or the slum of Vasant Kunj, India—that will ultimately determine whether we solve our water problems. With very few exceptions, nothing stands in our way except our own attitude.

That's why the comments of Senator Arlen Specter that he simply doesn't trust U.S. tap water are so unsettling, why the iPhone app offering to permanently archive your personal water consumption glass by glass is so silly. If we're going to have an era of smart water, we're going to have to leave the era of water mythology behind.

Our attitudes are already holding us back in ways we don't ever get to see. The very same hotel executives at MGM Resorts who took a chance on a custom-designed showerhead for their luxury Las Vegas hotel, the Aria, had a second decision to make about the hotel's plumbing: what kind of toilets to install in the 4,004 rooms. Senior vice president Cindy Ortega argued for installing dual-flush toilets, with one button for flushing liquids and a second button for flushing solids. The toilets are common across Europe, Asia, and Australia, and the flush button is elegantly designed and easy to understand—a small button for the small flush (1 gallon), a larger button for the large flush (1.6 gallons). Once you've used a dual-flush toilet, standard toilets seem primitive and pointlessly wasteful.

Dual-flush toilets are so rare in the United States, however, that they are a mere novelty. Despite the urgent water scarcity issues in Las Vegas, not a single major hotel there uses them. Ortega was all for them.

The conversation was as serious as the showerhead conversation, she says. "We installed dual-flush toilets as tests for some of the executives who have bathrooms in their offices." There was a lot of discussion of whether the toilets were harder to clean, says Ortega, and about whether guests "would be ready to accept the fact that there wasn't a trade-off in having one, in a luxury hotel." There was also some question about whether guests would be able to figure out which button to push. "Would people push the wrong button? Would they push the button more than once?"

In the end, they went with a single-flush, low-flow toilet instead, 1.5 gallons per flush. "Whether it's true that the guests would not accept it, we'll never know," says Ortega. "It was 2005 when we had to decide."

Is it really possible that guests spending $400 a night for a room with a TV remote with twenty buttons, along with a second, touch-screen remote that controls the curtains and lighting for the whole room—not to mention the iPhone or BlackBerry that most visitors carry—is it really possible those guests wouldn't be able to figure out which button to push to flush the toilet?

IT IS 63 DEGREES, the sky is clear blue and bright, the air has the freshness of spring in the mountains, when it still gets cool each night and each day has to warm up from scratch. I can hear the waterfalls roaring faintly in the distance.

A couple hundred steps through the woods is a small but dramatic canyon cut by Little Bushkill Creek, in the Pocono Mountains, a hundred miles north of Philadelphia, where I live, and a hundred miles west of New York City. The creek itself is no more than five or six steps wide, flowing fast, and as it comes down the flank of the mountain, it flumes over five waterfalls in a space that you can hike in just a few minutes. A leaf that hits the current just right can enter the Upper Canyon Falls and come washing out Lower Gorge Falls four or five minutes later.

The creek is surrounded by wooden boardwalks that run right alongside the falls themselves. In a few places, there are bridges spanning the creek, and you can stand two feet over the stream and look down over the very top of three different waterfalls, an incredible vantage.

The creek is shadowed and dappled, the water is the color of dark honey. Looking downstream toward the main falls, the water betrays no hint of the precipice to come right up to the last moment. The water folds over the edge and shatters into a white cascade of foam and mist a hundred feet tall, crashing into the pool below and quietly reassembling itself into Little Bushkill Creek.

This is a place where you can get right up close to the water, to the creek, to the waterfalls themselves.

At the top of the series of falls, standing on a footbridge over a short, six-foot waterfall, you can smell the water below—woodsy—and your legs are brushed by a light, insistent breeze. That's the force of the stream and the waterfall pulling the air along with it. I take in a deep breath, and it's like breathing in a bit of the mountain and the sunshine and the creek itself.

It's hard to be in a bad mood around beautiful flowing water. Whatever cares you have are lightened when you spend some time with water. The presence of a brisk, bright mountain stream makes you smile, it makes you feel better, whether you're already feeling good or you're low. Being in water is almost always refreshing—whether a bath, or a swimming pool, or the surf at the beach—but simply being near water is refreshing.

Standing alongside Little Bushkill Creek as it breaks over the falls, I suddenly solve one of the trivial if nagging water mysteries of the modern world: why we leave the water running while we brush our teeth. The answer is simplicity itself. We like running water. Even if it's just coming out of the bathroom faucet and splashing into a white porcelain sink.

A short hike above the falls, the creek widens out, and it's clear and shallow enough that you can see the bottom. There's a spot where a huge, flat rock sits in the midday sun in the middle of the stream, a rock the size of a queen-size bed. I rock-hop out to the boulder, lie on my back, and close my eyes. Dry and warm, I am instantly immersed in water. I am surrounded by the stream, by the colorful chatter of flowing water: bubbling, sliding, slapping, gurgling. We have more words for the sounds water makes than for water itself.

In many places around Bushkill Falls, there is not the slightest hint of the world beyond the forest and the water. There is no cell phone service, there is almost no litter, no power lines or buildings. Just the path, and the sounds of the forest and the creek. It strikes me that, even for solitary pioneers crossing the continent, it would be hard to feel completely alone with such lively flowing water at hand.[21]

Water is a pleasure. It is fun. Our sense of water, our connection to water, is primal. Anyone who has ever given a bath to a nine-month-old baby—and received a soaking in return—knows that the sheer exuberance of creating splashing cascades of water is born with us. We don't have to be taught to enjoy water.

We may not know the details about how it gets to us, how much it costs, or what's necessary to protect it, but we like water. Each of us individually, and all of us together, have a huge reservoir of goodwill about water, even a sense of proprietorship. The water that comes out of the kitchen faucet—that's *my* water.

When it comes to thinking differently about water, when it comes to actually appreciating it and doing the work required to reimagine how we use it every day, that affection, that sense of protectiveness, is one of water's lucky virtues.

Climate change may or may not be caused by human activity, it may or may not be remediated with a cap-and-trade system, but in terms of the public conversation, it's hard to muster much affection for the atmosphere, or for polar ice caps. That's one reason polar bears are so often part of the conversation.

The financial system might well be in dire need of structural reform, and the stability of the global economy may indeed depend on getting that right—not to mention the stability of each of our jobs—but the banking system isn't fun, and few of us have any sense of protectiveness about it.

We all know what it feels like to be thirsty, and what it feels like to be refreshed with a glass of water. We know what a dried-up lawn looks like, what it feels like, and we know what a plush, well-watered lawn feels like. We know what water that's been sitting in the bottom of the canoe all summer looks like, and what it's like to stand at the base of a waterfall and feel the power of water, the spray and the spirit. We do have a big thirst—physically, societally, and also emotionally.

When you think about the qualities of water that are so appealing—the energy, the playfulness, the adaptability, the variety of mood, the artistry, and also the sheer everyday usefulness—what's striking is how much the personality of water mirrors our own personality as people. In the best sense, the spirit of water and our own spirit are the same.

Many civilizations have been crippled or destroyed by an inability to understand water or manage it. We have a huge advantage over the generations of people who have come before us, because we can understand water and we can use it smartly. Everything about water is about to change—except, of course, for water itself. It is our fate that hangs on how we approach water—the quality of our lives, the variety and resilience of our society, the character of our humanity. Water itself will be fine. Water will remain exuberantly wet.

NOTES

A NOTE ON SOURCES

One of the thrilling and humbling aspects of writing about water is the vast range of people who devote their careers to studying water's impact and influence, and the libraries of literature available about water.

I have used these endnotes for three basic purposes. First, where I have done anything more than basic arithmetic to come up with a particular way of capturing water's impact in statistical or numerical terms, I have tried to provide that mathematical reasoning so others can understand how you figure out how many molecules of water there are in a single blood cell, or how you calculate that a single serving of rice requires 14.4 gallons of water. And, of course, so others can check my math.

Second, I have occasionally used the notes to expand on a point, where that explanation or additional information would have slowed down the main text, but where the information is relevant enough that curious readers might appreciate knowing more.

And I have used the notes for the traditional purpose of providing source references and credit to the work of others, which I have consulted, relied on, and quoted.

All the online references were checked and current as of October 1, 2010. Newspaper and magazine stories for which there are only standard citations, and no online links, are not available online.

For ease of use, these notes will also be available online at www.thebigthirst.com.

Most of what I've learned about water has come from conversations with a wide range of people, some quoted in the text and some not. For a more complete list of people interviewed, by chapter, see the Acknowledgments.

1. THE REVENGE OF WATER

1. The first shuttle mission, the test flight of Columbia (STS-1), was launched April 12, 1981, and succeeded without the water sound-suppression system. But the shock waves recorded during launch caused NASA to install the system before the second flight.

    The peak noise at the pad comes 5 seconds after launch, according to NASA, with the shuttle at 300 feet above the pad and the full force of its five engines reflecting off the pad's surface. By the time the shuttle is rising through 1,000 feet, the noise on the pad is falling off rapidly.

    Scientists with payloads on the shuttle are required to harden them to withstand 145 decibels, inside the shuttle's payload bay. NASA says the cushion of water keeps the sound down to 142 decibels inside the shuttle's payload bay. (The noise of a chain saw is about 110 decibels.)

    A description of the sound suppression system at Kennedy Space Center's launch-pads 39-A and 39-B is below (scroll down to "sound suppression water system"). NASA calls the on-pad water delivery valves "rain birds." http://science.ksc.nasa.gov/shuttle/technology/sts-newsref/sts-lc39.html.

2. The U.S. Geological Survey report *Estimated Use of Water in the United States in 2005* says that total water use for all purposes in the U.S. is 410 billion gallons a day. Of that, electric power plants use 200 billion gallons (49 percent).

    Using the 2005 Census figure of 296 million Americans, electric power plants in the U.S. are using 676 gallons of water per person. As with water use, though, that doesn't mean the electricity that each of us uses requires 676 gallons of water per day—that includes electricity used for all industrial and commercial purposes.

    According to data from the U.S. Energy Information Administration, in a typical year, residential customers use 37 percent of electricity generated, commercial customers use 36 percent, and industrial customers use 27 percent. (Data from 2009 are a little anomalous because of the recession.)

    Electricity used in U.S. homes requires 37 percent of that 676 gallons per person per day, or 250 gallons per person per day.

    The full USGS water-use report for 2005 is here (PDF). http://pubs.usgs.gov/circ/1344/pdf/c1344.pdf.

    The EIA data on electricity use by broad sector are here. http://www.eia.doe.gov/cneaf/electricity/epm/table5_1.html.

3. *2008/2009 Sustainability Review*, Coca-Cola Company, 2009, p. 31 (PDF). http://www.thecoca-colacompany.com/citizenship/pdf/2008–2009_sustainability_review.pdf.

4. Women have less water, on average, because fat contains almost no water, and women in general have a higher percentage of body fat than men. Water weighs 8.33 pounds per gallon.

5. Peter Mayer, William DeOreo, Eva Opitz, et al., *Residential End Uses of Water*, 2000, Water Research Foundation.

    The executive summary is accessible here. http://www.waterresearchfoundation.org/research/topicsandprojects/execSum/241.aspx.

    More detail on the study is here. http://www.unep.or.jp/ietc/Publications/ReportSeries/IETCRep9/4.paper-F/4-F-nels1.asp.

    The EPA's presentation of the study is here. http://www.epa.gov/watersense/pubs/indoor.html.

6. Average residential water consumption in the United Kingdom is small compared with the U.S.—40 gallons (150 liters) per person per day, according to the Office for National Statistics.

    Estimated Household Water Consumption: Regional Trends 38, Office for National Statistics. http://www.statistics.gov.uk/STATBASE/ssdataset.asp?vlnk=7812.

    With a total population of 61.4 million people (2008), Brits use just 2.5 billion gallons of water at home.

    Average residential water consumption in Canada is closer to U.S. use—91 gallons (343 liters) per person per day.

    Factsheet: Water Use & Consumption in Canada, Program on Water Governance (PDF). http://www.watergovernance.ca/factsheets/pdf/FS_Water_Use.pdf.

    With a population of 34 million (2008), that's 3 billion gallons of water per day.

    UK and Canadian households, together, use 5.5 billion gallons of water a day for all purposes.

7. Water losses by U.S. utilities are calculated by the American Society of Civil Engineers (ASCE), among others, at 7 billion gallons a day, which comes to 15.8 percent of the total "utility supply" of 44.2 billion gallons a day, as calculated by the USGS.

    Drinking Water: Report Card for America's Infrastructure, ASCE, 2009. http://www.infrastructurereportcard.org/fact-sheet/drinking-water.

    Estimated Use of Water in the United States in 2005, USGS, 2009 (PDF). http://pubs.usgs.gov/circ/1344/pdf/c1344.pdf.

8. *VEWA Survey: Comparison of European Water and Wastewater Prices*, Metropolitan Consulting Group, May 2006, p. 4 (PDF). http://www.bdew.de/bdew.nsf/id/DE_id100110127_vewa-survey---comparison-of-european-water-and-wastewater-pr/$file/0.1_resource_200 6_7_14.pdf.

9. David M. Cutler and Grant Miller, "The Role of Public Health Improvements in Health Advances: The Twentieth-Century United States," *Demography*, vol. 42, no. 1, February 2005, pp. 1–22. http://muse.jhu.edu/login?uri=/journals/demography/v042/42.1cutler.html.

    The full text of the study, including charts and tables, is available online (PDF). http://www.economics.harvard.edu/faculty/cutler/files/cutler_miller_cities.pdf.

10. Cutler and Miller, PDF file at economics.harvard.edu, p. 18.

    In 1900, infectious diseases, often carried by water, were responsible for 44 percent of U.S. deaths. By 1940, they were responsible for only 18 percent of deaths. (Ibid., p. 6.)

11. Ibid., p. 4.

12. *Estimated Use of Water in the United States in 2005*, USGS, p. 19 (PDF). http://pubs.usgs.gov/circ/1344/pdf/c1344.pdf.

13. *Estimated Use of Water in the United States, 1955*, USGS, 1957. http://pubs.er.usgs.gov/publication/cir398.

14. No one routinely gathers data on the average monthly water bill. But the American Water Works Association (AWWA) has used usage and fee surveys to estimate that the monthly bill is $34 per household in the U.S. (just for water, not including sewer service).

    The average monthly cable TV bill in the U.S. in 2009 was $70; the average monthly cell phone bill in 2009 was $93, according to a study by the research firm Centris.

Communications Services Spending Increasing, Centris, January 26, 2010 (PDF). http://www.centris.com/Docs/PR/Nov%20Insights%20Report%20final.pdf.

15. The average price of 1,000 gallons of tap water for residential customers in the U.S. was $3.24, according to the 2008 survey conducted by the AWWA.

16. Al Goodman, "Spain Suffers Worst Drought," CNN, April 18, 2008. http://www.cnn .com/2008/WORLD/europe/04/18/spain.drought/index.html.

17. "Drought-Stricken Barcelona Ships In Water," Associated Press, May 16, 2008. http:// www.msnbc.msn.com/id/24629154/.

   Barcelona uses 220 million gallons of water a day—152,777 gallons a minute. So the *Sichem Defender*'s 5 million gallons of water lasted 32 minutes.

   The *Contester Defender* carried 9.5 million gallons of water, which lasted 62 minutes. That figure is from the French TV news account below:

   "Barcelona's Unprecedented Drought," France 24, May 26, 2008. http://www .france24.com/en/node/1939450/%252F2.

   Britain's *Guardian* newspaper has a good account of the arrival of the first water ship, but showed the quantity of water as different from other accounts:

   Graham Keely, "Barcelona Forced to Import Emergency Water," *Guardian*, May 14, 2008. http://www.guardian.co.uk/world/2008/may/14/spain.water.

   In terms of supplying water by supertanker, the largest supertankers carry about 3 million barrels of crude oil—at 42 gallons a barrel, that's 126 million gallons. Barcelona uses 220 million gallons of water a day. And of course, even if you could muster a continuously flowing fleet of two supertankers a day into Barcelona, you'd also have to find enough water to fill those supertankers.

   Supertanker information:

   Tanker Information, Pacific Energy Partners, May 2005 (PDF). http://www .pacificenergypier400.com/pdfs/TANKERS/TankerBusEmissions.pdf.

18. Thomas Catán, "Barcelona Relies on Water by Ship to Slake Its Thirst Amid Drought," *Times*, May 14, 2008. http://www.timesonline.co.uk/tol/news/world/europe/article 3927283.ece.

19. The story of Orme, Tennessee, running out of water is based on these news accounts:

   Drew Jubera, "Tennessee Town Rations Water," *Atlanta Journal-Constitution*, October 21, 2007.

   Dick Cook, "Rain Lacking; Troubles Welling," *Chattanooga Times Free Press*, September 29, 2007.

   "Tennessee Town Runs Out of Water in Southeast Drought," Associated Press, November 1, 2007. http://www.foxnews.com/story/0,2933,307437,00.html.

   Rusty Dornin, "Town Has Water Just Three Hours a Day," CNN, November 8, 2007. http://www.cnn.com/2007/US/11/08/dry.town/index.html.

   "Water Flows in Town Where Drought Dried Up Spring," Associated Press, *Knoxville News-Sentinel*, December 11, 2007.

20. Steve Helling, "The Town Without Water," *People*, December 3, 2007. http://www .people.com/people/archive/article/0,,20170838,00.html.

21. *Human Development Report, 2006: Beyond Scarcity: Power, Poverty and the Global Water Crisis*, UN Development Programme. http://hdr.undp.org/en/reports/global/ hdr2006/.

   1.1 billion people don't have access to drinking water: p. 2.

   700 million people live on $2 a day or less: p. 7.

(The full UN report on water and poverty is available from the page above as a PDF file, but it is 422 pages.)

22. *Human Development Report, 2006.*
      1.8 billion people whose water is within 1 km: p. 35.

23. Ibid.
      1.1 billion people use just 5 liters per day: p. 5.

24. Ibid.
      1.8 million children die annually from water-related disease: p. 6.

25. Florida population, Table S0901, "Children Characteristics," U.S. Census, 2006–2008. http://factfinder. census.gov/servlet/STTable?-geo_id = 04000US12 &-qr_nane = ACS _2008_1YR_G00_S0901&-ds_name=ACS_2008_1YR_G00_.
      As of 2008, Florida had approximately 1.3 million children between the ages of 6 and 11, according to the U.S. Census.

26. *Human Development Report, 2006.*
      1.2 billion more people by 2025, 2.4 billion more people by 2050: p. 138.

27. Ibid.
      Population up by factor of 4, water use up by factor of 7 since 1900: p. 137.

28. Most people find it surprising that only about 6 percent of the people who have ever lived are alive now. About 10 billion people were born between 1900 and 2010, which means in the years stretching back before 1900, 90 billion people were born and died.
      For a fascinating discussion and analysis of the long history of human birth and death, see Carl Haub, "How Many People Have Ever Lived on Earth?" *Population Today*, Population Reference Bureau, 2002. http://www.prb.org/Articles/2002/How ManyPeopleHaveEverLivedonEarth.aspx.

29. If we just consider drinking water for humans, 3 liters per day for 30 years equals 32,850 liters per person.
      32,850 liters per person × 100 billion people = 3.3×10^{15} liters.
      Three liters per day of drinking water as the basic human requirement comes from Peter Gleick, "Basic Water Requirements for Human Activities: Meeting Basic Needs," *Water International*, vol. 21 (1996), pp. 83–92 (PDF). http://www.pacinst.org/ reports/basic_water_needs/basic_water_needs.pdf.

30. Animals alone outnumber people *at least* 1,000:1.
      Scientists have identified nearly 2 million different species on Earth, of which 1.4 million are animals. (Plants, of course, contain and use water as well.)
      Liz Osborn, *Number of Species Identified on Earth*, Current Results, 2010. http:// www.currentresults.com/Environment-Facts/Plants-Animals/number-species.php.
      Perhaps the most vivid example of how modest a piece of the biomass humans on Earth represent is the fact that the weight of ants, alone, roughly equals the weight of human beings.
      Bert Hölldobler and Edward O. Wilson, *Journey to the Ants* (Cambridge: Belknap Press of Harvard University Press, 1998), p. 1.

31. Sid Perkins, "Trackway Site Shows Dinosaur on the Go," *Science News*, October 26, 2002.
      The fossilized, bathtub-shaped depression near La Junta, Colorado, measured

3 meters by 1.5 meters by about 0.25 meters, and so might have held 1.125 cubic meters of liquid, or 300 gallons.

32.   The total water consumption for all creatures that have ever lived cannot realistically be calculated. But just to get a sense of the scale, start with the water consumption of all human beings, from note 29, above:

3.3×10^{15} liters.

If animals outnumber people in terms of mass by 1,000:1, and animals have been on Earth at least 500 million years, which is 10,000 times longer than humans:

$(1 \times 10^3) \times (1 \times 10^4) = 1 \times 10^7$ (which is 10 million times the water consumption of humans).

That gives you a figure of:

$(3.3 \times 10^{15}$ liters$) \times (1 \times 10^7) = 3.3 \times 10^{22}$ liters of water.

That figure is, if anything, conservative.

For comparison, at any given moment, the USGS says, there are just 1.1×10^{19} liters of liquid fresh water on Earth.

The Water Cycle, USGS. http://ga.water.usgs.gov/edu/watercyclesummary.html. The chart of water distribution is at the very bottom of the page.

33.   Yes, dinosaurs had kidneys.

Dinosaurs: Anatomy & Evolution, Smithsonian National Museum of Natural History. http://paleobiology.si.edu/dinosaurs/info/everything/gen_anatomy.html.

34.   Water use to make steel:

Mark Ellis, Sara Dillich, and Nancy Margolis, *Industrial Water Use and Its Energy Implications*, U.S. Department of Energy (PDF). http://www1.eere.energy.gov/industry/steel/pdfs/water_use_rpt.pdf.

Water use to cool nuclear power plants:

Got Water? Nuclear Power Plant Cooling Water Needs, issue brief, Union of Concerned Scientists, December 4, 2007, p. 4. http://www.ucsusa.org/nuclear_power/nuclear_power_technology/got-water-nuclear-power.html.

35.   According to the Beverage Marketing Corporation, a New York market research firm that closely tracks the bottled-water and wider beverage market, total bottled-water consumption in the U.S. in 2009 was 8.5 billion gallons. There are 8,760 hours in a year, so that's almost exactly 1 million gallons of bottled water an hour in the U.S.

"Bottled Water Confronts Persistent Challenges," Beverage Marketing Corporation, July 2010. http://beveragemarketing.com/?section=pressreleases.

36.   The USGS reports for water use in the U.S. for 2005, for 1980, and for every five years back to 1950, can be accessed at http://water.usgs.gov/watuse/50years.html.

37.   GDP figures for the U.S. economy, in annual dollars and adjusted for inflation, are easily accessible at MeasuringWorth.com, a Web site created by two economists to make national economic data more readily available and easier to use. http://www.measuringworth.com/.

38.   *Agricultural Productivity in the United States*, USDA Economic Research Service. http://www.ers.usda.gov/Data/AgProductivity/. See the chart, in Excel format, under the heading "National Tables, 1948–2008," table 1.

39.   Martin Wanielista, *Stormwater Reuse: A Summary*, 2006, University of Central Florida Stormwater Management Academy (PDF). http://www.stormwater.ucf.edu/research/publications/Stormwater%20Reuse%20A%20Summary.doc.

40. The Holy Quran, Surah 21, Al-Anbiya (The Prophets), 21:30.

41. The Holy Quran, Surah 25, Al-Furqan (The Criterion, The Standard), 25:54.

42. Genesis 1:1–4.

43. The way water appears in the English language is a remarkable indicator of our some-
     what unconscious attitude about water. Consider, for instance, Shakespeare's use of
     water imagery.

     Water makes an appearance in every one of William Shakespeare's 37 plays. In
     some, like *Macbeth*, water consistently fails to cleanse—in this case, it is unable to wash
     the blood from either the hands or the consciences of Macbeth and Lady Macbeth. In
     act 2, scene 2, Lady Macbeth exhorts, "Go get some water, / And wash this filthy wit-
     ness from your hand." And a little later in the same scene, she says to Macbeth again, "A
     little water clears us of this deed."

     Macbeth is not so sanguine. In the same scene, he replies to his wife: "Will all
     great Neptune's ocean wash this blood / Clean from my hand? No, this my hand will
     rather / The multitudinous seas in incarnadine, / Making the green one red."

     Mostly, water in Shakespeare is a comic foil, or a symbol of inadequacy, deception,
     or impermanence.

     In *Much Ado About Nothing*, the character Leonato says:

     > I pray thee, cease thy counsel,
     > Which falls into mine ears as profitless
     > As water in a sieve (5.1).

     In *King Lear*, Shakespeare writes, "[W]hen brewers mar their malt with water."
     (3.2). In *Henry VI, Part 1*, "[G]lory is like a circle in the water, / Which never ceaseth to
     enlarge itself / Till by broad spreading it disperse to nought" (1.2).

     At least two expressions involving water that survive to this day come from
     Shakespeare. In *Henry VI, Part 2*, Shakespeare writes, "Smooth runs the water where
     the brook is deep." That's the Shakespearean version of "still waters run deep." And as
     the rest of that moment in the play shows, despite two more water puns, Shakespeare is
     using water to make a point not about hydrology but about human character:

     > Smooth runs the water where the brook is deep;
     > And in his simple show he harbours treason.
     > The fox barks not when he would steal the lamb.
     > No, no, my sovereign; Gloucester is a man
     > Unsounded yet and full of deep deceit (3.1).

     And in *Henry VIII*, a Shakespeare character observes, "Men's evil manners live in
     brass; their virtues / We write in water" (4.2).

     The line is inscribed in a sculpture that runs along the banks of the Thames
     River, near the London reproduction of Shakespeare's Globe Theater. Most famously,
     it echoes the melancholy line John Keats asked to have inscribed on his tombstone:
     "Here lies one whose name was writ in water."

     In terms of taking a direct slap at water, though, nothing Shakespeare wrote quite
     matches a line from *Othello*. At the climax of the play, as Othello is trying to justify
     having just smothered his wife, Desdemona, he says of her, "She's, like a liar, gone to
     burning hell"; "She'd turn to folly, and she was a whore." And then, the final insult:
     "She was false as water" (5.2).

     She was false as water. Meaning, in Shakespeare's era, she was changeable, volatile, unreliable. Desdemona could not be trusted. (Othello, of course, thought mistakenly that his wife had been unfaithful.)

Shakespeare wasn't typically thinking about water, of course—and Othello wasn't maligning water, he was denouncing his wife. Water was just the linguistic tool. In some ways, that makes the pattern all the more interesting, all the more revealing. Shakespeare's reflexive attitude about water was wariness.

2. THE SECRET LIFE OF WATER

1. The oldest rocks found to date were formed when Earth was just 300 million years old.
   Here is an account from the *New York Times*:
   Kenneth Chang, "Rocks May Be Oldest on Earth, Scientists Say," *New York Times*, September 25, 2008. http://www.nytimes.com/2008/09/26/science/26rock.html.
   The discovery is reported in the journal *Science*:
   Richard A. Kerr, "Geologists Find Vestige of Early Earth—May Be World's Oldest Rock," *Science*, September 26, 2008, vol. 321, no. 5897, p. 1755. http://www.sciencemag.org/cgi/content/short/321/5897/1755a.
   The technical paper in *Science* appears in the same issue.
   Jonathan O'Neill, et al., "Neodymium-142 Evidence for Hadean Mafic Crust," *Science*, September 26, 2008, vol. 321, no. 5897, pp. 1828–1831. http://www.sciencemag.org/cgi/content/short/321/5897/1828.

2. How many molecules of water are there on the surface of the Earth?
   There are 1.4 billion cubic km of water on Earth, according to the USGS reference *The Water Cycle*. http://ga.water.usgs.gov/edu/watercyclesummary.html.
   To get from cubic km of water to molecules of water, we're going to use basic chemistry—moles of water, and Avogadro's number, the number of molecules in a mole of a substance, 6.02×10^{23} molecules per mole.

   $(1.4 \times 10^9$ cubic km of water$) \times (2.64 \times 10^{11}$ gallons per cubic km$) =$
   3.7×10^{20} gallons of water on Earth.
   $(3.7 \times 10^{20}$ gallons of water$) \times (8.33$ pounds per gallon$) =$
   3.1×10^{21} pounds of water on Earth.
   $(3.1 \times 10^{21}$ pounds of water$) \times (4.54 \times 10^2$ grams per pound$) =$
   1.4×10^{24} grams of water on Earth.
   $(1.4 \times 10^{24}$ grams of water$) \div (18$ grams per mole of water$) =$
   8×10^{22} moles of water on Earth.
   $(8 \times 10^{22}$ moles of water$) \times (6.02 \times 10^{23}$ molecules per mole$) =$
   4.8×10^{46} molecules of water on Earth.

   If our interstellar cloud were forming 1 million molecules of H_2O per second (1×10^6), that comes to 4.6×10^{40} seconds to create all the water on Earth, which is 1.5×10^{33} years—far older than the universe itself. So the water molecules are popping into existence very quickly.

3. How does one even begin to conceive of a space 420 times the size of our solar system in human terms?
   ⌐   is to use the journey of a spaceship launched by humans.

Voyager 2, which was the first spacecraft to visit Neptune and is on track to leave the solar system, flies through space at 42,000 miles an hour, fast enough to circle Earth in 35 minutes. It took *Voyager 2* twelve years to cover about half the total width of the solar system; to travel across the width of Orion's "water factory," *Voyager 2* would have to fly for 10,080 years, which is to say, it would have to fly for more than all of recorded human history.

Voyager: The Interstellar Mission, Neptune, NASA's Jet Propulsion Laboratory. http://voyager.jpl.nasa.gov/science/neptune.html.

4. *Report of the Task Group on Reference Man.* International Commission on Radiological Protection (Oxford: Pergamon Press, October 1974), p. 364.

5. The curb weight of a 2009 Honda Odyssey minivan is 4,400 pounds. If you take 0.025 percent of that weight, you get 1.1 pounds. As it happens, a half-liter of water weighs exactly 1.1 pounds.

6. The Earth-to-apple comparison works like this.

    The average depth of the oceans is 4 km, and the diameter of Earth is 12,756 km. The ratio of ocean cover to Earth's diameter is 0.00031.

    A medium apple is about 3 inches in diameter; the average thickness of an apple skin is 60 microns. Three inches is 76,200 microns, so the ratio of skin thickness to diameter of an apple is 0.00079—the apple skin is 2.5 times as thick relative to the apple as the oceans are compared to the Earth.

    The thickness of apple skins is from:

    I. Homutova and J. Blazek, "Differences in Fruit Skin Thickness Between Selected Apple Cultivars," *Horticultural Science* (Prague), vol. 33, no. 3, 2006, pp. 108–113 (PDF). http://www.cazv.cz/userfiles/File/ZA%2033_108–113.pdf.

7. As it happens, the U.S. each day uses just about a cubic kilometer of water for all purposes—drinking, cooking, farming, power plants. The precise number is 1.55 km^3. The volumes of water on the surface of the planet are in the millions and billions of cubic kilometers—10.5 million km^3 in fresh ground water, 24 million km^3 of water frozen in polar ice, 1.3 billion km^3 in the oceans.

    The volumes, and other data about the movement of water around Earth, are available near the bottom of the USGS Water Cycle Web site. http://ga.water.usgs.gov/edu/watercyclesummary.html.

    Even a single cubic kilometer of water is not an intuitively understandable unit.

    1 km^3 = 1 trillion liters, enough to give every person on Earth 300 half-liter bottles of Evian.

    Just the amount of humidity in the atmosphere at any given moment, small compared with volumes like the water in the polar ice caps or the oceans, is 12,900 km^3. That's enough water to fill a cube bigger than a mountain: 14 miles on each side, and 14 miles high.

    Daily U.S. water consumption, 410 billion gallons of water for all purposes, comes from *Estimated Use of Water in the United States in 2005*, USGS (PDF). http://pubs.usgs.gov/circ/1344/pdf/c1344.pdf.

8. The American Museum of Natural History, New York, NY, had an exhibit about water, Water: H_2O = Life, from November 2007 to May 2008, that presented a range of water data, including the evaporation rate for an acre of trees. http://www.amnh.org/exhibitions/water/.

NASA reports that a one-acre corn field can evaporate 4,000 gallons of water a day.

The Water Cycle: A Multi-Phased Journey, Earth Observatory, NASA. http://earthobservatory.nasa.gov/Features/Water/water_2.php.

9. The nine days a molecule spends floating in the atmosphere before returning to Earth as precipitation—"residence time" in the atmosphere is the phrase scientists use—is part of the data presented here:

The Water Cycle—a Climate Change Perspective, Windows to the Universe, National Earth Science Teachers Association. http://www.windows.ucar.edu/tour/link=/earth/Water/water_cycle_climate_change.html.

10. Water: H_2O = Life, American Museum of Natural History. http://www.amnh.org/exhibitions/water/.

Having read the observation that a particular cloud doesn't usually last more than an hour, I've begun to watch clouds differently. In watching a single cloud for just a few minutes many times, I've seen what I never noticed before, that clouds are more dynamic than we realize, and that their shapes are almost never static.

11. "Rain: A Valuable Resource," *Water Science for Schools*, USGS. http://ga.water.usgs.gov/edu/earthrain.html.

12. Biological water volume is listed in the USGS Water Cycle Web site chart. http://ga.water.usgs.gov/edu/watercyclesummary.html.

The volume of the Great Lakes is here:

Great Lakes Fact Sheet, EPA. http://www.epa.gov/greatlakes/factsheet.html.

13. The math on the human proportion of biological water works like this.

If, for purposes of a rough estimate, we say that the average person on the planet weighs 80 pounds—including men, women, and children—and that the average person contains 57.5 percent water, then the average person contains:

80 pounds × .575 = 46 pounds of water = 5.5 gallons.

With 6.9 billion people, that's 38 billion gallons of water contained inside people.

In terms of the average weight of a person, 30 percent of the people in the world are under the age of 15.

2008 World Population Data Sheet, Population Reference Bureau, Washington, DC, 2008 (PDF). http://www.prb.org/pdf08/08WPDS_Eng.pdf.

14. Igor Shiklomanov, who was born in 1939, is director of the Russian State Hydrological Institute in St. Petersburg and a regular member of international scientific panels, including the Intergovernmental Panel on Climate Change. He did not reply to numerous e-mail inquiries requesting an interview.

A brief biography is here, from the UNESCO International Hydrological Programme. http://www.unesco.org/water/ihp/cvshiklomanov.shtml.

His original chart, "Water Reserves on the Earth," is in Peter H. Gleick, ed., *Water in Crisis* (USA: Oxford University Press, 1993), p. 13.

15. The world's deepest borehole was drilled in the Kola Peninsula, near Finland, as a research effort by the Soviet Union between 1970 and 1994, and reached 12,262 meters. The peninsula is still part of Russia.

Pamela J. W. Gore, *The Interior of the Earth*, Georgia Perimeter College. http://facstaff.gpc.edu/~pgore/geology/geo101/interior.htm.

16. The title of the world's deepest mine passed in 2008 to South Africa's TauTona gold mine, owned by AngloGold. The mine is 3.9 km deep; the deepest tunnels require air conditioning to make mining possible, bringing the air temperature down from 131°F to 82°F. The working rock face at TauTona is 140°F.

TauTona, Anglo Gold, South Africa, Mining-Technology.com. http://www.mining-technology.com/projects/tautona_goldmine/.

17. Don Murray, "Percy Spencer and His Itch to Know," *Reader's Digest*, August 1958, p. 114. http://www.softslide.com/volumes/v2/t3/history/readers_digest.htm.

"Percy Spencer, Inventor, Dead; Retired Raytheon Executive, 76," *New York Times*, September 8, 1970.

18. Murray, "Percy Spencer and His Itch to Know," p. 114.

Percy Spencer's grandson, George (Rod) Spencer, confirms the details of his grandfather's discovery of microwaves' cooking ability in an e-mail exchange. Although many accounts report that Spencer was carrying a chocolate candy bar, Rod Spencer says it was a peanut cluster bar.

19. Of the 130 patents Percy Spencer was awarded, the one for microwave cooking is No. 2,495,429, Method of Treating Foodstuffs. It is just 2½ pages long, and one of them is a simple, full-page diagram. It's not clear why Spencer used the word "treating" in the title—50 years later, people worry that microwave ovens can somehow "irradiate" them; in fact, they produce no radioactivity, inside or out. But Spencer corrects himself in the patent's opening sentence: "My present invention relates to the treatment of foodstuffs, and more particularly to the cooking thereof through the use of electromagnetic energy."

Percy L. Spencer, Method of Treating Foodstuffs, U.S. Patent No. 2,495,429, October 8, 1945. http://www.google.com/patents?id=x_tuAAAAEBAJ.

The U.S. Census provides figures on the number of appliances in U.S. homes. The latest data, released in November 2009, show 96.4 percent of all homes have a microwave, 90.6 percent have a landline telephone, 67.1 percent have a computer.

"Homes with Cell Phones Nearly Double in First Half of Decade," U.S. Census, November 19, 2009. http://www.census.gov/newsroom/releases/archives/income_wealth/cb09–174.html.

The data tables are here: "Extended Measures of Well-being: Living Conditions in the United States, 2005, Detailed Tables," U.S. Census. http://www.census.gov/population/www/socdemo/extended-05.html.

The Popcorn Board reports that, as of 2008, Americans bought 966 million pounds of unpopped corn a year—3 pounds for every man, woman, and child in the U.S., the equivalent of 15 regular-size bags of microwave popcorn per person each year. Of that, at least 70 percent is cooked in a microwave.

Industry Facts, Popcorn Board, Chicago. http://www.popcorn.org/Encyclopedia Popcornica/WelcometoPopcornica/IndustryFacts/tabid/108/Default.aspx.

Percentage of U.S. popcorn that is microwave popcorn comes from: *Popcorn Profile*, Ag Marketing Resource Center, Iowa State University, 2010. http://www.agmrc.org/commodities_products/grains_oilseeds/corn/popcorn_profile.cfm.

20. This elegant—and sticky—metaphor comparing water molecules to socks in a dryer comes from the American Museum of Natural History's 2007–2008 exhibit Water: H_2O = Life. http://www.amnh.org/exhibitions/water.

21. The calculations for how many water molecules would fit in the interior of single red blood cell are rough approximations, but work like this:

    A typical red blood cell is:

    8 micrometers (μm) long and 2 micrometers (μm) high.

    If you simply assume that a red blood cell is a cylinder, its volume is: 3.14×16 micrometers $\times$ 2 micrometers = 100 micrometers^3 (μm^3).

    100 μm^3 = 1×10^{-13} liters.

    Now, how many molecules of water are in 1 liter of water?

    1 liter = 1,000 grams of water.

    (1,000 gm of water) ÷ (18.015 grams / mole of water) =
    55.51 moles of water.

    55.51 moles of water $\times$ 6.022×10^{23} molecules / mole =
    3.34×10^{25} molecules of water in one liter.

    So in a cell with a volume of 1×10^{-13} liters, the number of molecules is:

    (3.34×10^{25} molecules of water/liter) ÷ (1×10^{-13} liters) =
    3.34×10^{12} molecules.

    So 3.34 trillion water molecules would fit inside a single red blood cell.

22. The comparisons that follow assume a suburban sidewalk that is 3 feet (0.914 meters) wide.

    The microchip pathway is 90 nm (nanometers) wide; the sidewalk is 914 million nm wide—10 million times wider than the chip pathway. So each item lying on the chip—on the metaphoric sidewalk—is also 10 million times bigger than life-size.

    A human hair is between 50 microns and 150 microns thick (0.00005 to 0.0001 meters)—500 to 1,000 meters in relative terms, between 1,640 and 3,280 feet high across the ordinary sidewalk.

    A single red blood cell is about 8 microns wide (8×10^{-6} meters)—80 meters wide on the sidewalk.

    A single particle of flu virus is about 130 nanometers long (1.3×10^{-7} meters)—1.3 meters wide on the sidewalk.

    A single water molecule is about 275 picometers long (2.75×10^{-10} meters)—about 3 mm on the sidewalk.

    Here are two good sites for understanding the relative size of very small objects:

    Cell Size and Scale, Learn.Genetics, Genetic Science Learning Center, University of Utah. http://learn.genetics.utah.edu/content/begin/cells/scale.

    Exploring the Nanoworld, Intro to Size and Scale, Materials Research Science and Engineering Center, University of Wisconsin Madison. http://mrsec.wisc.edu/Edetc/nanoscale/index.html.

23. Michael Graham Richard, *How Many Atoms Encode the Humane Genome?* April 6, 2008. http://michaelgr.com/2008/04/06/how-many-atoms-to-encode-the-human-genome/.

3. DOLPHINS IN THE DESERT

1. Las Vegas temperature and precipitation data—72 days a year at 100 degrees or more, 19 days of precipitation—come from the climate data available online from the National Climatic Data Center. Data through 2009—some going back 30 years, some going back more than 50 years—are in:

    Comparative Climatic Data for the United States Through 2009, National Climatic Data Center, National Oceanic and Atmospheric Administration (NOAA), Asheville, NC (PDF). http://www1.ncdc.noaa.gov/pub/data/ccd-data/CCD-2009.pdf.

    The data for days over 100 degrees come from a slightly older analysis, for Las Vegas specifically, archived here:

    Climatography of the United States, No. 20, 1971–2000, Station: Las Vegas, National Climatic Data Center, NOAA, Asheville, NC (PDF). http://cdo.ncdc.noaa.gov/climate normals/clim20/nv/264436.pdf.

2. What does it mean to say Las Vegas is the driest city in the U.S.?

    The National Climate Data Center publication *Comparative Climatic Data* has data on 274 major U.S. cities and weather reporting stations, going back at least three decades. According to that compilation, the lowest-precipitation cities in the U.S. are Barrow, Alaska; Yuma, Arizona; and Las Vegas. But Yuma and Barrow are small compared with Las Vegas.

    Yuma averages 3.01 inches of rain a year, with 16 days of precipitation. The city of Yuma has about 90,000 people, and the larger Yuma metro area has 190,000 people, according to U.S. Census data (2009). Yuma is 1/10th the size of Las Vegas.

    Barrow, Alaska, is the only other city with less precipitation than Las Vegas, with 4.16 inches a year, on 74 days a year. Barrow, according to the census, has a population of 4,091 (2009), less than the population in many of the individual hotels on the Las Vegas Strip on a typical night.

    Comparative Climatic Data for the United States Through 2009 (PDF). http://www1.ncdc.noaa.gov/pub/data/ccd-data/CCD-2009.pdf.

    Annual number of days of precipitation is in a table that begins on p. 37.

    Average annual precipitation is in a table that begins on p. 136.

    A separate, slightly older analysis of precipitation data from the U.S. Census uses population data from 2000 and precipitation data from 1961 to 1990: "Cities with 100,000 or More Population in 2000 Ranked by Annual Precipitation," table C-7, *County and City Data Book: 2000*, U.S. Census Bureau. http://www.census.gov/statab/ccdb/cit7140r.txt.

    The U.S. Census list of 280 cities with populations of 100,000 or greater is here. http://www.census.gov/popest/cities/SUB-EST2009.html.

3. Details about Lake Mead's size and capacity are here:

    Hoover Dam: Frequently Asked Questions, U.S. Bureau of Reclamation, Lower Colorado Region. http://www.usbr.gov/lc/hooverdam/faqs/lakefaqs.html.

    A list of the largest reservoirs in the U.S., by water capacity, from Stanford University's civil and environmental engineering department:

    Largest U.S. Reservoirs, National Performance of Dams Programs, Department of Civil and Environmental Engineering, Stanford University. http://npdp.stanford .edu/damlarge.html.

A somewhat different list—which still puts Lake Mead at the top—of the largest reservoirs by water capacity, from the U.S. Society on Dams, is here:

"Largest Manmade Reservoirs in the United States," *Dam, Hydropower and Reservoir Statistics*, United States Society on Dams. http://www.ussdams.org/uscold_s.html.

According to the U.S. Geological Survey, U.S. water utilities supply 44.2 billion gallons a day to homes and businesses, about 11 percent of the water the nation uses each day if you include electricity generation and irrigation. That 44.2 billion gallons comes to 136,000 acre-feet of water, so the 28.5 million acre-feet in a full Lake Mead would last 210 days.

USGS water-use statistics are here:

Estimated Use of Water in the United States in 2005, USGS, 2009 (PDF). http://pubs.usgs.gov/circ/1344/pdf/c1344.pdf.

4. Las Vegas's formal allocation from Lake Mead is about 300,000 acre-feet of water a year. Las Vegas takes about 450,000 acre-feet, because it returns 180,000 acre-feet of treated water back to the lake, for a total "use" of 270,000 acre-feet in recent years. But even at 450,000 acre-feet, the 28.5 million acre-feet in Lake Mead would last Las Vegas 63 years.

When it's full, Lake Mead is 157,900 acres, so if Las Vegas takes 270,000 acre-feet a year, that would lower a full lake 1.7 feet. The surface area of the lake shrinks as it falls, so Las Vegas's 270,000 acre-feet in net withdrawals typically lower the lake between 2 and 3 feet a year.

5. Details of Las Vegas's and Nevada's water entitlements can be found in the most current strategic plan of the Las Vegas area's water authority, the Southern Nevada Water Authority (SNWA): *Water Resource Plan 09*, Southern Nevada Water Authority, 2009, p. 15 (PDF). http://www.snwa.com/html/wr_resource_plan.html.

The "Law of the River," a series of laws, court cases, and agreements among the states and the federal government, fixes how much water the various users of the Colorado River are allowed to take.

The law around use of the water is baroquely complex. The basic allocations for the three states tapping the lower Colorado, using Lake Mead, are: California, 4.4 million acre-feet a year; Arizona, 2.85 million acre-feet a year; Nevada, 0.3 million acre-feet a year.

6. The population of metro Las Vegas (Las Vegas and Clark County) in 1980 was 462,000. Thirty years later, it was 2 million. This chart, from the Las Vegas Convention and Visitors Authority, provides population figures going back to 1970:

Population Trends, Las Vegas Convention and Visitors Authority (PDF). http://www.lvcva.com/getfile/241/Population%202009.pdf.

The SNWA says that per-capita, per-day water use in Las Vegas in 2009 was 240 gallons, the lowest it has been in the last 14 years. What 240 gallons per person per day means is that each new resident requires 87,600 gallons of water a year.

The number of hotel rooms in metropolitan Las Vegas was 45,815 in 1980. In 2009, it was 148,941.

Historical Las Vegas Visitor Statistics (1970–2009), Las Vegas Convention and Visitors Authority (PDF). http://www.lcva.com/getfile/80/Historical%201970%20to%202009.pdf.

7. Lake Mead has only been as low as it was in August 2010 twice before—for a few months in 1964–65, and for a few months in 1956. Lake Mead's historical water levels, going all the way back to February 1935, are here, month by month:

Lake Mead at Hoover Dam, Elevation (Feet), U.S. Bureau of Reclamation, Lower Colorado Region. http://www.usbr.gov/lc/region/g4000/hourly/mead-elv.html.

In terms of how much water has been lost in Lake Mead, according to the Bureau of Reclamation figures for July 2010, Lake Mead is holding 10.5 million acre-feet of water, or 41 percent of capacity (level, 1,088.8 feet). Using that figure, Lake Mead's usable capacity is 25.5 million acre-feet, meaning 15 million acre-feet have disappeared since January 2000.

The Strip is conventionally defined as being 4.2 miles long (from Sahara Avenue on the north to Russell Road on the south). If, generously, you define the Strip's boundaries to extend a mile on each side of Las Vegas Boulevard, then the area of the Strip is an oblong shape of roughly 8.4 square miles, which is 5,376 acres.

Fifteen million acre-feet of water would cover 5,376 acres of land to a depth of 2,790 feet, slightly more than half a mile.

8. Figures for the number of visitors, and the percentage of visitors from the U.S.:

    2009 Las Vegas Visitor Profile Study, Las Vegas Visitors and Convention Authority, GLS Research, p. 82 (PDF). http://www.lvcva.com/getfile/107/2009%20Las%20Vegas%20Visitor%20Profile.pdf.

9. The average historic level of Lake Mead is 1,173 feet, and Intake 1 stops being usable when the lake falls to 1,050 feet. Here is a good graphical representation of the history of Lake Mead water levels:

    Lake Mead Water Levels—Historical and Current. http://www.arachnoid.com/NaturalResources/.

10. Pat Mulroy has told the story of trying to get rid of the smaller fountains on several occasions, including directly to me. Most of this account comes from interviewing Mulroy, and I've used contemporary newspaper accounts to check her version. But this quote, comparing the fireboat at New York New York and the canals at the Venetian, is from an oral history with Mulroy in the book:

    Corinne Platt and Meredith Ogilby, *Voices of the American West* (Golden, CO: Fulcrum Publishing, 2009), p. 268.

11. It's important to understand that the figures for gallons of water used per person per day (GPCD—gallons per capita per day) do not reflect how much water each person in Las Vegas (or any U.S. city) uses. They are simply total water consumption in the metro area, divided by total population. They include water used at hotels and hospitals, at factories and restaurants. But GPCD is a good measure of changing use overall in a community. The figures for gallons of water used per person come from the Southern Nevada Water Authority.

12. Robert Reinhold, "Battle Lines Drawn in Sand as Las Vegas Covets Water," *New York Times*, April 23, 1991. http://www.nytimes.com/1991/04/23/us/battle-lines-drawn-in-sand-as-las-vegas-covets-water.html.

13. The "volume discount" line comes from a profile of Mulroy written when she had been on the job just five months:

    Jamie McKee, "Conserve Now or Pay More, Water Manager Warns," *Las Vegas Business Press*, April 19, 1990.

14. Both the county commissioner's quote, and Mulroy's, are from:

    Reinhold, "Battle Lines Drawn in Sand as Las Vegas Covets Water." http://www.nytimes.com/1991/04/23/us/battle-lines-drawn-in-sand-as-las-vegas-covets-water.html.

15. Barbosa says that during the slowest part of the 2008–2009 recession, work at Mission's Plant 50 was cut back from 22 hours to 16 hours a day, and water-use volumes fell by almost half. By summer 2010, business in Las Vegas's hotels and at Mission's laundry plants was picking back up.

16. Angel Park course superintendent Bill Rohret provided the figures on Angel Park's water use and rounds of play. The calculations of water use per round of golf work like this. Angel Park has two large, 18-hole courses, and a smaller, 12-hole, par-3 course.

    Rohret says the two big courses average 130,000 rounds of golf a year. At 18 holes per round, that's 2.34 million holes played.

    The par-3 course averages 30,000 rounds of golf. At 12 holes per round, that's 0.36 million additional holes.

    Total holes played per year: 2.7 million.

    Converted to 18-hole equivalents, that comes to 150,000 rounds.

    In 2009, the course used 376 million gallons of water.

    376 million gallons ÷ 150,000 rounds = 2,507 gallons / round.

    According to the American Water Works Association (www.awwa.org), the average water use in U.S. homes is 350 gallons per day, although that varies widely.

17. For purposes of its turf-removal program, popularly known as "cash for grass," the SNWA calculates that the average homeowner uses 73 gallons of water per square foot of grass. That comes to 3.18 million gallons for every acre of lawn, or 9.8 acre-feet of water for every acre of lawn.

    The SNWA currently restricts golf courses to using 6.3 acre-feet of water per acre of turf per year—36 percent less water per acre.

18. The latest census figures (2009) show the Orlando metropolitan area with 2.1 million people, compared with Las Vegas's 1.9 million. (Clark County's own population figure for 2010 is 2 million in the metro area.) http://www.census.gov/popest/metro/tables/2009/CBSA-EST2009-01.csv.

    The number of visitors to the Orlando–Orange County region, which includes Walt Disney World, comes from the Orlando Orange County Convention and Visitors Bureau, and is reported to be 48 million, compared with the 36 million visitors Las Vegas had in 2009.

    Orlando statistics here:

    Frequently Asked Questions, Orlando/Orange County Convention & Visitors Bureau. http://www.orlandoinfo.com/media/orlando/faqs.cfm#Visitors.

    Las Vegas statistics here:

    Historical Las Vegas Visitor Statistics (1970–2009), Las Vegas Convention and Visitors Authority (PDF). http://www.lvcva.com/getfile/80/Historical%201970%20to%202009.pdf.

    The normal precipitation in Orlando is 48 inches; normal precipitation in Las Vegas is 4.5 inches:

    Comparative Climatic Data for the United States Through 2009 (PDF). http://www1.ncdc.noaa.gov/pub/data/ccd-data/CCD-2009.pdf.

19. The SNWA water restrictions for golf courses are outlined here:

    "Golf Course Water Budgets," *Drought & Restrictions: Conservation Measures*, Southern Nevada Water Authority. http://www.snwa.com/html/drought_restrictions_golf.html.

    The 6.3 acre-feet of water for each acre of turf comes to 2.1 million gallons of water, per acre, per year. That's 5,753 gallons of water per acre per day.

The Imperial Valley and its use of Lake Mead water is discussed at greater length in chapter 9, "It's Water. Of Course It's Free," starting on page 267.

The average Imperial Valley farmer uses 6.0 acre-feet of water per acre of irrigated farmland, according to the Imperial Valley Irrigation District's reports:

2005 Annual Water Report, Imperial Irrigation District (PDF). http://www.iid .com/Media/2005IIDWaterAnnualReport.pdf.

20. Lawn watering is a huge consumer of water in the U.S. Even in Florida, half the water delivered to homes is used for lawn watering:

Martin Wanielista, *Stormwater Reuse: A Summary*, 2006, University of Central Florida Stormwater Management Academy (PDF). http://www.stormwater.ucf.edu/ research/publications/Stormwater%20Reuse%20A%20Summary.doc.

According to the SNWA, 44.5 percent of all water delivered in the Las Vegas area goes to single-family homes, and 70 percent of that water ends up outside, which comes to 31 percent of water pumped used outdoors at single-family homes.

The breakdown of water use by category and customer is in the *Water Resource Plan 09*, Southern Nevada Water Authority (PDF), p. 16. http://www.snwa.com/html/ wr_resource_plan.html.

21. The SNWA provided figures for total withdrawals from Lake Mead, total water returned, and consumptive use (withdrawals minus returns) for 25 years. All data are in acre-feet.

Year	Withdrawals	Returns	Consumptive Use	% Returned
1985	175,711	74,002	101,709	42
1986	194,168	81,951	112,217	42
1987	201,427	92,564	108,863	46
1988	232,407	102,987	129,420	44
1989	273,052	116,839	156,213	43
1990	294,795	116,685	178,110	40
1991	298,576	118,352	180,224	40
1992	305,669	128,118	177,551	42
1993	335,561	131,159	204,402	39
1994	361,293	135,465	225,828	37
1995	359,858	144,140	215,718	40
1996	390,508	142,006	248,502	36
1997	404,933	162,888	242,045	40
1998	401,173	156,663	244,510	39
1999	456,570	167,055	289,515	37
2000	485,870	166,014	319,856	34
2001	489,554	175,617	313,937	36
2002	500,679	175,452	325,227	35
2003	491,434	193,042	298,392	39
2004	475,761	192,755	283,006	41
2005	502,651	210,873	291,778	42
2006	522,157	229,293	292,864	44
2007	517,165	216,853	300,312	42
2008	479,974	210,320	269,654	44
2009	457,963	209,350	248,613	46

22. The SNWA explanation of how to drain your Las Vegas swimming pool into the city's sanitary sewers is at two sites.

Step-by-step instructions are here:

"How to Drain a Pool or Spa," *Conservation & Rebates: Pools & Spas*, Southern Nevada Water Authority. http://www.snwa.com/html/cons_pools_drain.html.

Tips and warnings are here:

"Pool Draining Tips," *Conservation & Rebates: Pools & Spas*, Southern Nevada Water Authority. http://www.snwa.com/html/cons_pools_draintips.html.

23. Lake Lanier's level, like that of other federal reservoirs, is measured in feet above sea level. The lake, formally named Lake Sidney Lanier, is managed by the Army Corps of Engineers. Lanier's usable water ends at 1,035 feet. Below that is the "dead pool" of water, which is inaccessible to water intakes without additional pumps. Lanier is considered full at 1,071 feet.

Lanier reached its all-time operational low on December 26, 2007, at 1,050.79 feet, just 16 feet above the dead-pool level. It hadn't been that low since February 6, 1958, when the reservoir was first being filled.

The history of Lanier's levels, along with other data on its operations, is here:

ACF Historic Project Data, U.S. Army Corps of Engineers: Mobile District. http://water.sam.usace.army.mil/gage/acfhist.htm.

24. Georgia's grab for a little slice of Tennessee River water by moving its border north got a fair amount of media attention.

Andrea Jones and Ben Smith, "By Wide Margin, Resolutions Pass to Seek Border Change," *Atlanta Journal-Constitution*, February 21, 2008. Includes the details of the vote, the singing of "This Land Is Your Land," and the quote from Tennessee state representative Gerald McCormick. (It is no longer available free online.)

Governor Sonny Perdue did not sign the border bill until May 14, 2008:

Jim Galloway, "Fetch Your Buckets! Perdue Signs Up for a Border War with Tennessee," *Atlanta Journal-Constitution*, May 15, 2008. http://www.ajc.com/metro/content/shared-blogs/ajc/politicalinsider/entries/2008/05/15/fetch_your_buckets_perdue_sign.html.

A *New York Times* account, before the final vote, is here:

Shaila Dewan, "Georgia Claims a Sliver of the Tennessee River," *New York Times*, February 22, 2008. http://www.nytimes.com/2008/02/22/us/22water.html.

The story of Chattanooga's mayor sending a truckload of water south is here:

James Baird, "Chattanooga Mayor Pokes Fun at Southeastern Water Crisis," *Tennessee Journalist*, February 27, 2008. http://tnjn.com/2008/feb/27/chattanooga-mayor-pokes-fun-at.

25. The population of Atlanta in 2009 was 5.475 million, according to the census. http://www.census.gov/popest/metro/tables/2009/CBSA-EST2009-01.csv.

The Atlanta real estate consulting firm Haddow & Company has an excellent table of Atlanta's population growth, since 1960, using census data:

Population Trends—Atlanta MSA, Haddow & Company (PDF). http://www.haddowandcompany.com/marketdata/Population%20Trends%20Atlanta%20MSA%20-%202008.pdf.

26. Details of Lake Lanier's size and operation can be found at the Army Corps of Engineers site for the lake:

"Map Room," *Lake Sidney Lanier*, U.S. Army Corps of Engineers: Mobile District. http://lanier.sam.usace.army.mil/MapRoom.htm.

Lake Mead's size and capacity are described here:

Hoover Dam: Frequently Asked Questions, U.S. Bureau of Reclamation, Lower Colorado Region. http://www.usbr.gov/lc/hooverdam/faqs/lakefaqs.html.

Lake Lanier's historical levels are here:

ACF Historic Project Data. http://water.sam.usace.army.mil/gage/acfhist.htm.

27. For a profile of the Atlanta area's water system, see:

Water Supply and Water Conservation Management Plan—May 2009, Metropolitan North Georgia Water Planning District. http://www.northgeorgiawater.com/html/88.htm.

28. Stacy Shelton, "Metro Atlanta's Need for Water: Three Months from a Mudhole," *Atlanta Journal-Constitution*, October 11, 2007.

29. Atlanta mayor William Hartsfield's letter is excerpted in a federal court decision from July 2009 in which Atlanta's use of Lake Lanier water was ruled illegal. The full text of U.S. District Court judge Paul Magnuson's decision, in what is called the Tri-State Water Rights Litigation, is below. Mayor Hartsfield's letter is on p. 13 of the PDF.

"Memorandum and Order," *In re: Tri-State Water Rights Litigation*, U.S. District Court, Middle District of Florida, p. 13 (PDF). http://www.dep.state.fl.us/mainpage/acf/files/statements/071709_magnuson_ruling.pdf.

30. Representative Phil Gingrey's quote is here:

"Georgia Delegation Introduces Legislation to Alleviate Water Crisis," October 16, 2007. http://isakson.senate.gov/press/2007/101607water.htm.

31. As part of its drought coverage, the *Atlanta Journal-Constitution* produced a map that shows the major water users throughout the 542-mile length of the Chattahoochee-Apalachicola river system:

Dale E. Dodson, "Heavy Demands on Our Water," *Atlanta Journal-Constitution*. http://www.ajc.com//metro/content/metro/stories/2007/10/26/watermap.html.

32. Bill Rankin, "Federal Judge Rules Against Ga. in Water Litigation," *Atlanta Journal-Constitution*, July 17, 2009. http://www.ajc.com/news/federal-judge-rules-against-94051.html.

33. Congressman Gerald Ford's question, during hearings over the construction of Lake Lanier, is excerpted (p. 16) in the federal court decision from July 2009 in which Atlanta's use of Lake Lanier water was ruled illegal.

"Memorandum and Order," *In re: Tri-State Water Rights Litigation*, U.S. District Court, Middle District of Florida, p. 16 (PDF). http://www.dep.state.fl.us/mainpage/acf/files/statements/071709_magnuson_ruling.pdf.

34. The full text of U.S. District Court judge Paul Magnuson's decision in the Tri-State Water Rights Litigation makes compelling reading for anyone interested in a compressed history of the southeastern water wars.

"Memorandum and Order," *In re: Tri-State Water Rights Litigation*, U.S. District Court, Middle District of Florida (PDF). http://www.dep.state.fl.us/mainpage/acf/files/statements/071709_magnuson_ruling.pdf.

35. Ibid., p. 93.

36. Ibid., pp. 94–95.

37. Ibid., pp. 93–94, 95.

38. Ibid., pp. 93, 96.

39. Dan Chapman and Leon Stafford, "Will Water Ruling Dry Up Growth?" *Atlanta Journal-Constitution*, July 24, 2009. http://www.ajc.com/news/will-water-ruling-dry -99500.html.

40. *Water Contingency Planning Task Force: Findings and Recommendations*, State of Georgia, December 21, 2009, p. 4 (PDF). http://gov.georgia.gov/vgn/images/portal/ cit_1210/59/57/154449884Water%20Contingency%20Planning%20Task%20Force %20Final%20Report.pdf.

Governor Perdue's quote is in the following. Jeremy Redmon, "Perdue: Lake Lanier Georgia's Best Option for Drinking Water," *Atlanta Journal-Constitution*, December 11, 2009. http://www.ajc.com/news/perdue-lake-lanier-georgias-238353.html.

41. There is ongoing coverage in the *Atlanta Journal-Constitution* of the negotiations over the allocation of water among Georgia, Alabama, and Florida. Stories in late 2009 and 2010 include:

Jeremy Redmon, "Three Governors Say a Water-Sharing Agreement Is in the Works," December 15, 2009. http://www.ajc.com/news/georgia-politics-elections/ three-governors-say-a-241962.html.

Bob Keefe, "Tri-State Water Talks Bog Down," May 31, 2010. http://www.ajc .com/news/georgia-politics-elections/tri-state-water-talks-538766.html.

Patrick Fox, "Lanier Ruling Anniversary Finds Perdue, Groups at Odds," July 16, 2010. http://www.ajc.com/news/atlanta/lanier-ruling-anniversary-finds-572203 .html.

42. The text of Georgia's Water Stewardship Act of 2010 is at the link below (PDF). The exemptions from the watering limitations for home vegetable gardens, golf courses, and athletic fields are on p. 6. http://www.legis.ga.gov/legis/2009_10/pdf/hb1094.pdf.

43. Jeremy Redmon, "Corps to Tighten Spigot at Lake Lanier in Wake of Judge's Ruling," *Atlanta Journal-Constitution*, November 18, 2009. http://www.ajc.com/news/corps-to -tighten-spigot-203870.html.

44. Henry Brean, "Third Straw: Water Authority Digs Deep for Third Intake Pipe at Lake Mead," *Las Vegas Review-Journal*, December 13, 2009. http://www.lvrj.com/ news/water-authority-digs-deep-for-third-intake-pipe-at-lake-mead-79158322.html.

45. Gallons per capita per day (GPCD) figures for the Las Vegas metro area for the last 21 years were supplied by the SNWA:

Year	GPCD
1989	348
1990	347
1991	344
1992	339
1993	337
1994	329
1995	327
1996	329
1997	322
1998	317
1999	315
2000	315

2001	318
2002	314
2003	294
2004	274
2005	269
2006	264
2007	255
2008	248
2009	240

46.   Water Utility Climate Alliance. http://www.wucaonline.org/html.

Members include: Denver Water, the Metropolitan Water District of Southern California, the New York City Department of Environmental Protection, the Portland Water Bureau, the San Diego Utilities Commission, Seattle Public Utilities, and the Southern Nevada Water Authority.

4. WATER UNDER WATER

1.   The *Wall Street Journal*'s "Washington Wire" noted at the time how fresh developments were elbowing equally dramatic events out of the limelight:

Evan Perez, "Washington Wire: Palin Overshadows Hurricane Recovery, Lehman Talks," *Wall Street Journal*, September 14, 2008. http://blogs.wsj.com/wash wire/2008/09/14/palin-overshadows-hurricane-recovery-lehman-talks.

The dates of the series of events in September 2008:

September 1: Republican vice presidential candidate Sarah Palin's daughter Bristol is revealed to be pregnant.

September 7: Mortgage companies Freddie Mac and Fannie Mae are placed under federal control.

September 11: Galveston residents are ordered to evacuate as Hurricane Ike approaches.

September 14: AIG seeks its first emergency bailout of $40 billion from the federal government; Lehman Brothers files for bankruptcy.

September 16: The U.S. government agrees to loan AIG $85 billion and take an 80 percent ownership stake in the insurer.

September 20: The Bush administration proposes its first $700 billion financial bailout.

September 24: Senator John McCain suspends his presidential campaign to return to Washington and consult on solutions to the financial crisis.

September 26: The FDIC seizes Washington Mutual bank, the largest bank failure in U.S. history.

September 29: The U.S. House rejects the Bush administration's first financial bailout proposal; Citigroup purchases Wachovia's banking operations.

2.   At the time, Hurricane Ike was described as the largest-ever Atlantic basin hurricane, in terms of width, but Dennis Feltgen of the National Hurricane Center says while Ike turned out to be one of the largest, both Hurricane Donna (1960) and Hurricane Betsy (1965) were bigger across.

Hurricane Ike's wind field is described here. "Hurricane Ike 2008," *Hurricane*

History, National Hurricane Center. http://www.nhc.noaa.gov/HAW2/english/history.shtml.

3. The text of the "certain death" warning from the local Houston/Galveston National Weather Service office is archived at this link. http://www.srh.noaa.gov/images/hgx/projects/ike08/HGX_Products/HLS/HLSHGX_091208_1000Z.txt.

    The unusually blunt nature of the "certain death" warning was noted in the media, even as Hurricane Ike approached, and was not without controversy.

    The *Dallas Morning News* called it "the storm warning heard 'round the world."

    Jeffrey Weiss, "Weather Forecasters Say Hurricane Warnings Can Make the Difference Between Life, Death," *Dallas Morning News*, September 19, 2008. http://www.dallasnews.com/sharedcontent/dws/dn/latestnews/stories/091908dnmetcertaindeath.1677b70.html.

    National Geographic News wrote two stories about it.

    Willie Drye, "Hurricane Ike's 9-Foot Floods to Bring 'Certain Death,'" *National Geographic News*, September 12, 2008. http://news.nationalgeographic.com/news/2008/09/080912-hurricane-ike.html.

    Drye, "Why Hurricane Ike's 'Certain Death' Warning Failed," *National Geographic News*, September 26, 2008. http://news.nationalgeographic.com/news/2008/09/080926-hurricane-ike-evacuation.html.

4. The San Luis Resort's construction on the site of Fort Crockett is described here:

    The San Luis Resort History, San Luis Resort. http://sanluisresort.com/about-us/history/index.cfm.

5. The water that drowned the motors in 30th Street Station turned out to have come up into the motor pits through the drains in the floor.

    Says Eric Wilson: "Should I have known about the drains? Yes. I have no issue accepting that blame."

6. How much water may have filled up Galveston's 59th Street Pump Station?

    The building is roughly 20 meters long and 7.7 meters wide (66 feet by 25 feet). The water rose at least 2.4 meters (8 feet) inside—and probably more like 2.75 meters.

    With a depth of 2.4 meters of water inside, that's 370 cubic meters of water—97,740 gallons.

7. *Mississippi River Water Quality and the Clean Water Act: Progress, Challenges, and Opportunities* (Washington, DC: The National Academies Press, 2008), p. 102. http://www.nap.edu/catalog.php?record_id=12051.

8. Ronald Schuyler, an expert on wastewater treatment with the Denver engineering firm Tetra Tech (www.tetratech.com), provided details on what kinds of bacteria help eat and digest routine waste in wastewater treatment plants, via e-mail. The most surprising thing is how little we know about how our own wastewater treatment plants operate.

    "Most of the bacteria are typical soil bacteria, bacteria right out in your backyard. In fact, a typical well-digested wastewater treatment sludge ready for land disposal will have an 'earthy' odor," he wrote. "I have read that we have only been able to identify about 30 percent of the bacterial species inhabiting activated sludge-type treatment processes, mainly because those are the only ones we can grow in the laboratory. Thus, we only know how to grow the other 70 percent within the wastewater treatment process, but cannot identify them yet. Most of our systems do quite well without knowing the 70 percent, or for that matter, the 30 percent. We do not have enough time or money to identify them. All we need to know is that they will function properly when the system is controlled properly."

9. A large tanker truck can hold 9,000 gallons of water, so 5,000 tankers would yield 45 million gallons of water a day, or about 9 gallons for each of 5 million Atlanta-area residents.

10. Major Daren Payne's comment about there being no plan for what to do if Atlanta ran out of water is from an Associated Press story by Greg Bluestein, "No Backup if Atlanta's Faucets Run Dry," October 19, 2007, which was widely published, although not in the *Atlanta Journal Constitution*: http://www.breitbart.com/article .php?id=D8SCHTI00&show_article=1.

11. The *Clarion-Ledger* in Jackson, Mississippi, provided thorough coverage of the city's water-main breaks. Just a few of the paper's stories are cited below. The paper archives older stories, but requires payment to read them online.
    Chris Joyner, "Water Repairs Continue," January 13, 2010.
    LaRaye Brown, "Businesses Coping Without Water," January 13, 2010.
    Gary Pettus, "Jackson Water Still Not OK to Drink," January 16, 2010.

12. Jackson mayor Harvey Johnson's quote "We have a disaster. It's just not one you can see" is from Joyner, "Water Repairs Continue."

13. The National Hurricane Center's official report on Ike calculates that the storm surge across this bayside part of Galveston Island was between 10 feet and 15 feet.
    Robbie Berg, *Tropical Cyclone Report: Hurricane Ike*, National Hurricane Center, pp. 6–7 (PDF). http://www.nhc.noaa.gov/pdf/TCR-AL092008_Ike_3May10.pdf.

14. Greg Bluestein, "No Backup if Atlanta's Faucets Run Dry," October 19, 2007. http:// www.breitbart.com/article.php?id=D8SCHTI00&show_article=1.

15. This quote is from an interview that a group of Galveston city officials gave to the Austin weekly newspaper a year after Hurricane Ike hit Galveston.
    Kate X Messer, "Q&A: Austin's Coastal Neighbors," *Austin Chronicle*, September 11, 2009. http://www.austinchronicle.com/gyrobase/Issue/story?oid=oid%3A842282.

16. Figures for states with more than 60 percent of homes without complete indoor plumbing in 1950 come from the U.S. Census Bureau. The census defines complete plumbing facilities as "hot and cold piped water, a bathtub or shower, and a flush toilet."

State	% Without Complete Indoor Plumbing
Alabama	68
Arkansas	71
Georgia	63
Kentucky	64
Mississippi	74
North Carolina	65
North Dakota	66
South Carolina	65
South Dakota	61
Tennessee	63

"Plumbing Facilities," *Historical Census of Housing Tables*, U.S. Census Bureau. http://www.census.gov/hhes/www/housing/census/historic/plumbing.html.

17. Statistics for 1960 and 1970 indoor plumbing are from the census:
     "Plumbing Facilities." http://www.census.gov/hhes/www/housing/census/
     historic/plumbing.html.
          In 1960, 88 percent of U.S. households had televisions, according to the U.S. Census Bureau.
          U.S. Bureau of the Census, *Statistical Abstract of the United States: 1961*, Washington, DC, 1961, p. 516 (PDF). http://www2.census.gov/prod2/statcomp/documents/
     1961–09.pdf.
          In 1970, 95 percent of U.S. households had televisions, according to the U.S. Census.
          U.S. Bureau of the Census, *Statistical Abstract of the United States: 1971*, Washington, DC, 1971, p. 487 (PDF). http://www2.census.gov/prod2/statcomp/documents/
     1971–05.pdf.
          The census makes the texts of the *Statistical Abstracts* available online. http://
     www.census.gov/prod/www/abs/statab1951–1994.htm.

18. No one routinely gathers data on the average monthly water bill. But the American
     Water Works Association (AWWA) has used usage and fee surveys to estimate that
     the monthly bill is $34 per household in the U.S. (just for water, not including sewer
     service).
          The figure of $260 per family each year for water infrastructure upkeep comes
     from the total infrastructure spending on water systems, compiled by the American
     Society of Civil Engineers (ASCE). The ASCE calculates that from 2009 to 2014, U.S.
     governments will spend $29.2 billion a year maintaining water and wastewater systems; the U.S. has 112 million families.
          "Estimated 5-Year Investment Needs in Billions of Dollars," *Report Card for America's Infrastructure*, American Society of Civil Engineers. http://www.infrastructure
     reportcard.org/report-cards.

19. Officials at the water company Aqua America provided the data on construction costs
     to lay replacement water pipe.
          Desalination plant costs in Australia come from:
          Norimitsu Onishi, "Arid Australia Sips Seawater, but at a Cost," *New York Times*,
     July 10, 2010. http://www.nytimes.com/2010/07/11/world/asia/11water.html.

5. THE MONEY IN THE PIPES

1. There are roughly 73 million sheep in Australia—three sheep for every person. Each
    sheep produced an average of 10.3 pounds of greasy wool in 2009–2010.
          Australia produced 21.5 percent of the greasy wool grown in the world in 2008,
    followed by China (19 percent), New Zealand (10 percent), and Argentina (4 percent).
    Greasy wool is wool weighed before it has been cleaned.
          Worldwide and Australian wool production figures from:
          Australian Wool Production 2009/10. Australian Wool Innovation Ltd. http://
    www.wool.com/Media-Centre_Australian-Wool-Production.htm.

2. Michael Grealy, "Michell Wool Dynasty Shrouded in Mystery," *Sun-Herald* (Sydney,
    Australia), October 29, 1989, quoting David Coombes, then executive director of the
    Wool Council of Australia.

3. At home, each pound of laundry requires about two gallons of water to clean and rinse. Conventional home washing machines (top-load) hold 12–16 pounds of laundry and use 30 gallons of water.

    "Washing Machines: Types of Washing Machines," ConsumerReports.org. http://www.consumerreports.org/cro/appliances/laundry-and-cleaning/washing -machines/washing-machine-buying-advice/washing-machine-types/washing -machine-types.htm.

    "How Much Does It Cost to Run a Washing Machine?" Michaelbluejay.com. http://michaelbluejay.com/electricity/laundry.html.

4. How does a half-gallon-a-minute change in showerhead flow add up to millions of gallons in water savings a year?

    The showerhead in each room in Aria, down from 2.5 gallons per minute to 2 gpm, saves 0.5 gpm. If the average shower is 8 minutes, each shower uses 4 fewer gallons of water than it otherwise would. (The Water Research Foundation study of U.S. water use, *Residential End Uses of Water*, cited in chapter 1, found that people with low-flow showerheads took showers that were 8.5 minutes, on average.)

    The Aria has 4,004 rooms. If, at the low end, there are 6,000 guest showers a day at the Aria (75 percent occupancy, with two people per room showering), that saves 24,000 gallons of water a day. Over the year, that's a savings of 8.8 million gallons of water. With a few more showers—that is, with a couple months of the higher occupancy typical of Las Vegas (average hotel occupancy is 90 percent), savings could easily be 10 million or 12 million gallons of water a year.

5. The CityCenter project has earned six gold ratings for environmentally conscious design from the U.S. Green Building Council.

    The development's environmental efforts and design innovations are outlined in this press release, and detailed in a separate Web site:

    "City of Gold: Vegas' CityCenter Earns Six LEED Gold Certifications," City-Center Press Room, November 20, 2009. http://www2.citycenter.com/press_room/ press_room_items.aspx?ID=778.

    "CityCenter—Environment: The Nature of Luxury," CityCenter. http://www2 .citycenter.com/environment/.

6. As the calculations above show, every 0.5 gpm reduction in water used by the showerhead at the Aria hotel saves a minimum of 24,000 gallons a day—1,000 gallons an hour. Going from 2.5 gpm to 1.5 gpm saves 48,000 gallons a day, 2,000 gallons an hour.

    Water consumption in Las Vegas, overall, is 400 million gallons a day, of which 59 percent is residential, or 236 mgd. As of 2008, according to the U.S. Census, there are 676,617 housing units in Clark County, which comes to 349 gallons of water per household per day. So 48,000 gallons a day would supply 138 homes.

7. Press accounts routinely describe Monsanto as the largest seller of seeds in the world:

    Donald L. Barlett and James B. Steele, "Monsanto's Harvest of Fear," *Vanity Fair*, May 2008. http://www.vanityfair.com/politics/features/2008/05/monsanto200805.

    Jack Kaskey, "Monsanto 'Warrior' Grant Fights Antitrust Accusations, Critics," Bloomberg, March 4, 2010. http://www.bloomberg.com/apps/news?pid=newsarchive &sid=axVdNmPtSgts&pos=7.

    Monsanto reports its seed sales each quarter. In 2009, seed sales were $4.5 billion out of total revenue of $11.7 billion.

    Fourth-Quarter 2009 Financial Results, Monsanto, October 7, 2009, pp. 4–5. http:// www.monsanto.com/pdf/investors/2009/10_07_09.pdf.

8. "Water Conservation," *2009 Intel Corporate Responsibility Report*, Intel, pp. 41–45. http://www.intel.com/about/corporateresponsibility/report/build/index.htm.

   Intel's detailed water-use figures, and water use per chip produced ("normalized water use"), along with basic financial data for comparison purposes, are available at the same link.

   Intel's water productivity is even worse in per-chip terms. From 2005 to 2009, the water required to make a single chip increased 58 percent.

   Still, the very existence of such detailed figures, reported voluntarily, is significant. Intel says in its 2009 CRR report that the increase in water use was due in part to the recession ("low manufacturing levels"), and in part to "the increasing complexity of our manufacturing processes" (p. 42).

9. The Muhtar Kent comment comes from his appearance on *Charlie Rose*, PBS, June 9, 2009.

   Coke says it provides 1.6 billion servings of soft drinks a day—584 billion a year, or 86 for every person on the planet. A "serving," according to Coke's figures, is 8 ounces, so at 86 per person, Coke is serving 688 ounces of beverages to each person on the planet—57 twelve-ounce cans.

   Coke says that it uses 313 billion liters of water a year, which is 83 billion gallons a year, or 227 million gallons a day. At the standard U.S. citywide consumption rate of 150 gallons per person, per day, that 227 million gallons would support a city of 1.5 million Americans.

   This data is from:

   2008/2009 Sustainability Review, Coca-Cola Company (PDF). http://www.the cocacolacompany.com/citizenship/pdf/2008–2009_sustainability_review.pdf.

   Basic water-use data: p. iv.

   Water-use discussion: pp. 31–33.

   Servings per day: p. 4.

10. Here are the calculations of Coke's water productivity, compared with that of IBM and GE.

    All figures are from 2008, the most recent year for which the water-use numbers are available.

    > Coke: 83 billion gallons of water per year.
    > $31.9 billion in revenue.
    > GE: 12.3 billion gallons of water per year.
    > $183 billion in revenue.
    > IBM: 13.4 billion gallons of water per year (just microchip manufacturing).
    > $104 billion in revenue.
    > Each $1 of revenue for Coke requires 2.6 gallons (333 ounces).
    > Combined revenue for GE and IBM is $287 billion. Combined water use for GE and IBM is 25.7 billion gallons.
    > Each $1 of revenue for GE and IBM requires 0.09 gallons (11.5 ounces).
    > GE water-use data from:
    > *2008 Citizenship Report: Resetting Responsibilities*, GE, p. 41 (PDF). http://files.gecompany.com/gecom/citizenship/pdfs/ge_2008_citizen ship_report.pdf.

IBM water-use data from:

IBM and the Environment: 2008 Annual Report, IBM, p. 20 (PDF). http://www.ibm.com/ibm/environment/annual/IBMEnvReport_2008 .pdf.

IBM reports on p. 20 that in 2008, "microelectronics manufacturing operations achieved a 2.4 percent savings [in water use]. This translates to an annual conservation savings of 1,214 thousand cubic meters of water." That is, 2.4 percent savings equaled 1,214,000 cubic meters of water.

By calculation, in 2008, IBM used 50.6 million cubic meters of water, or 13.4 billion gallons.

11. *2008/2009 Sustainability Review*, Coca-Cola Company, p. 31.

12. Coke's SEC filings back to 1994 are online here. http://ir.thecoca-colacompany.com/ phoenix/zhtml?c=94566&p=irol-sec.

The "Raw Materials" section of Coke's 2002 annual report begins on p. 10, *Form 10-K Annual Report for the Fiscal Year Ended Dec. 31, 2002*, Coca-Cola Company, filed March 26, 2003.

The "Raw Materials" section of Coke's 2009 annual report begins on p. 9, the "Risk Factors" section begins on p. 12, *Form 10-K Annual Report for the Fiscal Year Ended December 31, 2009*, Coca-Cola Company, filed February 26, 2010.

13. Intel's SEC filings are online here. http://www.intc.com/financials.cfm.

14. *A Product Lifecycle Approach to Sustainability*, Levi Strauss & Co., San Francisco, March 2009, pp. 11, 15, 18 (PDF). http://levistrauss.com/sites/default/files/librarydoc ument/2010/4/Product_Lifecycle_Assessment.pdf.

15. The figures on ice and water savings come from Scott Steenrod, director of food and beverage operations, Celebrity Cruises.

If each ship saves 7,500 pounds of ice a day, that's 52,500 pounds of ice a week. At 8.3 pounds per gallon of water, that comes to 6,300 gallons of water per ship per week not required to make ice.

Celebrity's total fleet carries about 23,000 passengers a week; the ships together save about 55,000 gallons of water a week.

16. IBM Burlington uses 3.2 million gallons of water a day, and gathers 400 million data points about that water a day. So it gathers an average of 133 bits of data about every gallon of water.

17. IBM Burlington's chip production in 2008 was 33 percent higher than it had been in 2000. But in 2009, chip production fell sharply, so while water use in 2009 continued to fall, the "water productivity" of the plant in 2009 wasn't as good as in 2008.

18. IBM describes, somewhat superficially, its effort to revolutionize desalination with new filters and solar power:

Steve Hamm, "Solar Power + Water Desalination = Rivers in the Desert," Building a Smarter Planet, IBM, April 5, 2010. http://asmarterplanet.com/blog/2010/04/ solar-power-water-desalination-rivers-in-the-desert.html.

19. Whole Foods spokeswoman Kate Lowery says that in the grocery category, yogurt was the No. 1–selling item by volume in 2007, 2008, and 2009. Water was the No. 2 item in volume in 2007, 2008, and 2009. By mid-2010, water as a product had slipped to No. 3 at Whole Foods, behind salty snacks.

20. The total U.S. wholesale bottled-water market in 2009 was $10.6 billion, according to the industry's research leader, the Beverage Marketing Corporation. The markup on bottled water is typically 100 percent between wholesale and retail, so total sales to consumers in 2009 were roughly $21 billion. (Sales in 2009 were down 2.7 percent from 2008 in gallons of water sold, and down 5.2 percent in revenue.)

   "Bottled Water Confronts Persistent Challenges," Beverage Marketing Corporation, July 2010.

   Beverage Marketing makes a wealth of data available about U.S. beverage consumption at http://www.beveragemarketing.com/?section=pressreleases.

   Apple breaks down sales by product and by category in detail in its annual 10-K filing with the SEC.

   In 2009 total iPhone sales were $6.8 billion; total iPod sales were $8.1 billion; total sales from iTunes were $4 billion. Together, iPhone, iPod, and iTunes sales were $18.9 billion.

   Apple sold 21 million iPhones and 54 million iPods.

   Form 10-K Annual Report for the Fiscal Year Ended September 26, 2009, Apple Inc., filed October 27, 2009, p. 41 (PDF). http://phx.corporate-ir.net/External.File?item=U GFyZW50SUQ9MTglOTB8Q2hpbGRJRD0tMXxUeXBlPTM=&t=1.

21. Total 2009 U.S. bottled-water consumption of 8.4 billion gallons comes from the Beverage Marketing Corporation. http://www.beveragemarketing.com/?section= pressreleases.

   The American Society of Civil Engineers estimates that U.S. municipal water systems leak about 7 billion gallons a day, about 16 percent of what they pump.

   Drinking Water: Report Card for America's Infrastructure, ASCE, 2009. http:// www.infrastructurereportcard.org/fact-sheet/drinking-water.

22. Tap water is regulated in the U.S. by the Environmental Protection Agency (EPA), under the Safe Drinking Water Act. Bottled water that crosses state lines is regulated as a food product by the Food and Drug Administration (FDA). And while the FDA has adopted the EPA's drinking water standards for bottled water, the actual regulation of bottled water amounts mostly to a system of voluntary compliance and trust, because enforcement rules and factory inspections are both minimal.

   For instance, the EPA requires that any significant violation of tap water standards be reported, both to the EPA and to the public, within 24 hours.

   Bottled-water companies are not required to report violating water standards to either the public or the FDA—ever.

   U.S. water utilities serving more than 100,000 people are required to test their water for bacterial contaminants every few hours.

   Bottled-water companies, which have facilities that produce millions of bottles of water a day, are only required to test for bacterial contamination once a week, and only required to test for some contaminants once a year.

   The U.S. Government Accountability Office (GAO) found in a 2009 study that, at best, bottled-water facilities are only inspected every 2 to 3 years—but that the EPA doesn't have a comprehensive list of bottling companies, so "we could not determine the percentage of bottled water facilities inspected" (p. 10). What's more, because bottled-water companies are only required to maintain records of their testing for two years (compared with 5 to 10 years for water utilities), "FDA would most likely not be aware that a contamination problem existed if a facility was not inspected within a 2-year time frame" (p. 9).

Drinking bottled water amounts to a leap of faith in the company whose water you're purchasing and consuming.

The GAO's 2009 study is:

Bottled Water: FDA Safety and Consumer Protections Are Often Less Stringent Than Comparable EPA Protections for Tap Water (GAO-09–610), Government Accountability Office, Washington, D.C.: June 2009 (PDF). http://www.gao.gov/new.items/d09610.pdf.

23. During the Boston water-main break and outage from May 1 to May 4, 2010, even doctors were ordered to use bottled water, to scrub in before surgery.

Tracy Jan, "Residents, Businesses Race to Adapt; Water Vanishes from Stores," *Boston Globe*, May 2, 2010. http://www.boston.com/news/local/massachusetts/articles/2010/05/02/residents_businesses_race_to_adapt_water_vanishes_from_stores.

24. The American Society of Civil Engineers (ASCE) issues an "infrastructure report card" each year, to focus attention on how the U.S. is maintaining, or failing to maintain, vital systems like roads, bridges, airports, air traffic control, schools, and water systems. The ASCE report says that actual spending on water and wastewater treatment systems (excluding dams) over the five years from 2009 to 2014 will be $146.4 billion—including the 2009 federal stimulus funds. That comes to $29.2 billion a year.

"Estimated 5-Year Investment Needs in Billions of Dollars," *Report Card for America's Infrastructure*, American Society of Civil Engineers. http://www.infrastructurereportcard.org/report-cards.

25. "S.F. Mayor Bans Bottled Water at City Offices," Associated Press, June 25, 2007. http://www.msnbc.msn.com/id/19415446.

Sharon Plan Chan, "City of Seattle Won't Buy Bottled Water," *Seattle Times*, March 13, 2008. http://seattletimes.nwsource.com/html/localnews/2004280866_webwater13m.html.

Jennifer 8. Lee, "City Council Shuns Bottles in Favor of Water from Tap," *New York Times*, June 17, 2008. http://www.nytimes.com/2008/06/17/nyregion/17water.html.

Several universities, including Washington University in St. Louis, the University of Portland, and DePauw University in Greencastle, Indiana, have banned sales of bottled water on campus.

26. *2008 Report on Postconsumer PET Container Recycling Activity: Final Report*, National Association for PET Container Resources, p. 4 (PDF). http://www.napcor.com/pdf/2008_Report.pdf.

27. The Beverage Marketing quote about tap water comes from "Liquid Refreshment Beverage Market," Beverage Marketing Corporation, March 24, 2010. http://www.beveragemarketing.com/?section=pressreleases.

28. Gary Hemphill, senior vice president at Beverage Marketing, says FIJI Water was the No. 1 imported brand in 2008, but fell during the recession; San Pellegrino was the No. 1 import in 2009.

29. "President-Elect Obama Drinks FIJI Water on Election Night," FIJI Blog, FIJI Water, November 21, 2008. http://www.fijiwater.com/blog/2008/11/president-elect-obama-drinks-fiji-water-on-election-night/.

Susan Donaldson James, "Gym Rat in Chief? Obama's Fitness Regimen," ABC News, December 4, 2008. http://abcnews.go.com/Health/President44/Story?id=6387559&page=1.

30. The GE ecomagination commercial "Clouds," produced by BBDO New York, is available online on YouTube. http://www.youtube.com/watch?v=SWJ7iVbKRj8.

31. GE does not break out revenue for GE Water separately. The 2009 revenue figure of $2.5 billion comes from:

    Scott Malone, "GE Sees Tide Coming In for Water Business," Reuters, August 11, 2009. http://www.reuters.com/article/idUSN2235851820090811.

32. In order, references for GE Water's work at the Virginia coal mine, the Algiers desalination plant, the Sydney golf course, and China's Lake Taihu:

    "Turning Mine Water Into a Useful Resource," GE Water Press Center, March 4, 2010. http://www.ge-energy.com/about/press/en/2010_press/030410.htm.

    "Hamma Seawater Desalination Plant," GE Water Press Center, February 24, 2008. http://www.gewater.com/who_we_are/press_center/vpr/hamma.jsp.

    "Pennant Hills Golf Course Goes Green," Sydney Water, May 30, 2008. http://www.sydneywater.com.au/whoweare/MediaCentre/MediaView.cfm?ID=470.

    Peter Ford, "Pollution Puts Chinese Lake Off Limits," *Christian Science Monitor*, June 4, 2007. http://www.csmonitor.com/2007/0604/p07s02-woap.html.

    "GE Wastewater Treatment Solution to Help Restore Health of China's Third Largest Lake," GE Water Press Center, June 24, 2008. http://www.gewater.com/who_we_are/press_center/pr/06242008.jsp.

33. The GE Web site lists Schaefer's discovery of cloud seeding in its timeline "The Science of Improvement," under the title "The Rainmaker." http://www.ge.com/innovation/timeline/eras/science_of_improvement.html.

    Schaefer's papers are archived at SUNY Albany, and the online guide to those papers includes a biographical sketch of the scientist. http://library.albany.edu/speccoll/findaids/ua902.010.htm.

34. Steven Prokesch, "How GE Teaches Teams to Lead Change," *Harvard Business Review*, January 2009, p. 7 (PDF). http://www.ge.com/pdf/innovation/leadership/hbr_crotonville.pdf.

35. GE CEO Jeffrey Immelt spoke briefly about GE's water business at the GE investors conference on December 15, 2009. An analyst asked Immelt, "Tell us what you learned from the things which were not successes. I don't want to pick on water. But it might be a good example."

    Immelt replied, "Look, water was—we paid too much for growth that was hard to materialize. And we had no foundational point in the company to plug it into. So we ran it as a freestanding business, having paid 14 or 15 times EBITDA. So what do you learn? Don't pay so much. And put things inside core businesses, right?"

    GE Water is now a part of the larger GE division, GE Energy.

    The exchange is in this presentation transcript, p. 22 (PDF). http://www.ge.com/pdf/investors/events/12152009/ge_annualoutlook_transcript_12152009.pdf.

36. Andrew C. Revkin, "Dredging of Pollutants Begins in Hudson," *New York Times*, May 15, 2009. http://www.nytimes.com/2009/05/16/science/earth/16dredge.html.

    In April 2010, GE released an accounting of its costs for the initial dredging effort on the Hudson of $561 million.

    Michael Hill, "GE Says Hudson Dredging Cost $561M," Associated Press, April 30, 2010. http://abcnews.go.com/Business/wireStory?id=10523307.

    On its own Web site, GE said that since 1990, the company had spent a total of $830 million on the Hudson River cleanup.

"GE Reports Cost of First Phase of Dredging," *The Hudson River Dredging Project*, GE Corporate Environmental Program. http://www.hudsondredging.com/phase_one_costs.

37. Delta's Paul Patton sent me a sample of the Aria's custom-designed four-hole shower-head, and we installed it in one of our bathrooms. It was a pleasure to use—the water did stay warmer, and the four holes created a cone of spray much wider, denser, and more even than from showerheads with five or ten times the number of holes. Without knowing in advance, you'd never guess that the Aria showerhead was using 25 percent or 40 percent less water than typical.

6. THE YUCK FACTOR

1. Carenda Jenkin, "Two Die of Thirst in Bush Tragedy," *Centralian Advocate*, January 15, 2008.
     The trio whose Mitsubishi Pajero ran out of water were never identified in the Australian media.

2. Paige Taylor and Victoria Laurie, "How a Simple Flat Tyre Killed Artist and Bush-man," *Australian*, January 16, 2007. http://www.theaustralian.com.au/news/nation/how-a-simple-flat-tyre-killed-artist-and-bushman/story-e6frg6nf-1111112837559.

3. Lindsay Murdoch, "How a Desert Claimed Two Ill-Prepared Travellers," *Age*, April 13, 2005. http://www.theage.com.au/news/National/How-a-desert-claimed-two-illprepared-travellers/2005/04/12/1113251629492.html.
     The men were identified as Bradley John Richards, 40, his nephew Mac Bevan Cody, 21, and their dog, VB.

4. Bellinda Kontominas, "Triple-0 Review Urged by Coroner as Iredale Inquest Ends," *Sydney Morning Herald*, May 8, 2009. http://www.smh.com/au/national/triple0-review-urged-by-coroner-as-iredale-inquest-ends-20090507-awla.html.

5. Katie Finn, "St. Luke's Packed as City Prays for Rain," *Chronicle* (Toowoomba), April 22, 2005.

6. "Toowoomba Takes Out Top Tidy Town Award," ABC News (Australia), April 20, 2008. http://www.abc.net.au/news/stories/2008/04/20/2221872.htm.

7. A brief history of Toowoomba is at the official Toowoomba Regional Council Web site. The precise origins of Toowoomba as the city's name are murky—as it happens, while the Aboriginal word for swamp is *tawampa*, two other possibilities are equally intriguing.
     The Aboriginal phrase *woomba woomba* means "reeds in the swamp." And the area was known for a melon that grew abundantly, which the natives called *toowoom* or *choowoom*.
     "About Council: History: Toowoomba," Toowoomba Regional Council. http://www.toowoombarc.qld.gov.au/index.php?option=com_content&view=article&id=111%3Atoowoomba&catid=6%3Ahistory&Itemid=18.

8. "Population Estimates by Local Government Area, 2001 to 2009," Catalog No. 3218.0, Australian Bureau of Statistics. http://www.abs.gov.au/AUSSTATS/abs@.nsf/DetailsPage/3218.02008-09?OpenDocument.

9. Leisa Scott, "Beaten, Bloody Well Unbowed," *Courier-Mail* (Brisbane), September 2, 2006. http://www.couriermail.com.au/news/beaten-bloody-well-unbowed/story -e6frer7x-1111112173213.

    This is a rollicking profile of Mayor Di Thorley. The word "shit" is in the first quote from Thorley.

10. Australians have routinely used the phrase "big dry" to refer to periods of drought, although the period from 2001 to 2009 has clearly supplanted previous Big Dry periods.

    In 1986, the *Courier-Mail* in Brisbane, writing about a mid-1980s drought and its impact on cattle ranchers, wrote, "Most of south-east Queensland also had been hit hard by the big dry." ("Double Blow as Prices, 'Dry' Hit QLD Cattlemen," *Courier-Mail*, April 9, 1986.)

    Similarly, during a drought in the early 1990s, headlines and stories relied on the phrase: Åsa Wahlquist, "Big Dry Claims NSW Wheat Exports," *Sydney Morning Herald*, October 2, 1991.

    The current dry period was being referred to as the "Big Dry" as early as 2002: Anna Merola, "The Big Dry: Lack of Rainfall Drying Up Hope," *Sunday Mail* (Adelaide), September 8, 2002.

11. Susan Searle, "The Plan to Save Our City," *Chronicle* (Toowoomba), July 2, 2005.

12. Searle, "Watering Cans Sell Out as Restrictions Tighten," *Chronicle* (Toowoomba), August 9, 2005.

13. Australia uses the metric system, and all Australian water authorities measure large water volumes in megaliters—1 megaliter is 1 million liters (264,172 gallons, or 0.8 acre-foot).

    Toowoomba, a town of 120,000 people, uses about 27 megaliters of drinking water a day (about 7 million gallons—60 gallons per person). Toowoomba requires roughly 10,000 megaliters of water a year; under tight water restrictions, the city has been using 9,000 megaliters a year, about 24 megaliters a day.

    Toowoomba's three reservoirs hold 127,000 megaliters of water when full—a 10-year supply. But the capacity is actually relatively small; the reservoirs would only serve 1.2 million people over that 10 years. The total volume of Toowoomba's reservoirs, when full, would only serve the needs of Las Vegas for three months.

    In the U.S., the basic measure of large volumes of water is the acre-foot—the amount of water covering an acre of space (43,560 square feet), to a depth of one foot. 1 acre-foot is 325,841 gallons.

    1 acre-foot is equal to 1.2 megaliters.

    Toowoomba uses 27 megaliters of water a day—Las Vegas uses 1,520, New York uses 4,164 megaliters of water a day.

14. In order of appearance, here are references for information on the Occoquan Reservoir, in Fairfax County; the Orange County, California, recycling facility; and Singapore's NEWater effort:

    Rob Davis, "Where Water Reuse Isn't a Dirty Word," Voice of San Diego, January 7, 2009. http://www.voiceofsandiego.org/environment/article_068e2ad1-1b57-5313 -936c-ebd65aca818a.html.

    Randal C. Archibold, "From Sewage, Added Water for Drinking," *New York Times*, November 27, 2007. http://www.nytimes.com/2007/11/27/us/27conserve.html.

    NEWater: Plans for NEWater, PUB: Singapore's National Water Agency. http:// www.pub.gov.sg/newater/plansfornewater/Pages/default.aspx.

15. The annual list of the richest Australians is compiled by the Australian business magazine *BRW*. As of the May 2010 list, Clive Berghofer was listed at No. 118, with an estimated net worth of A$340 million (US$307 million).

    He was No. 110 in 2009.

    BRW posts the lists online, but a subscription is required for access: "Rich," *BRW*. http://brw.com.au/lists/rich.

    Below is the story from the Toowoomba *Chronicle* about the 2010 list:

    Jim Campbell, "Berghofer on BRW Rich List," *Chronicle*, May 28, 2010. http://www.thechronicle.com.au/story/2010/05/28/340m-wealth-berghofer-brw-list/.

16. The venues in Toowoomba bearing Clive Berghofer's name:

    - Clive Berghofer Arena, Toowoomba
    - Clive Berghofer Events Center, Toowoomba
    - The Berghofer Pavilion, Toowoomba
    - Clive Berghofer Stadium, Toowoomba
    - Clive Berghofer Recreation Centre, at the University of Southern Queensland

    And, at St. Vincent's Hospital, Toowoomba, the Clive Berghofer Intensive Care Unit.

    In Brisbane, the capital of Toowoomba's state of Queensland, Berghofer donated money to the Queensland Institute of Medical Research, which now has the Clive Berghofer Cancer Research Centre.

17. "Clive Casts Doubts on Water Plans," *Chronicle*, July 27, 2005.

    I interviewed Clive Berghofer in person. The quotes in this paragraph come not from the initial story about his opposition from the *Chronicle*, above, but from a magazine story on Toowoomba's water issues that ran during the recycling battle:

    Roy Eccleston, "Bottoms Up—Aqua Blue," *Australian*, July 29, 2006.

    The story is no longer available on the *Australian*'s Web site, but its full text is here:

    "Bottoms Up—Aqua Blue," Travestonswamp.info, July 29, 2006. http://www.travestonswamp.info/forum/viewtopic.php?t=737&sid=071a0e066b4d086ce3b9e55291f13150.

18. Wendy Frew, "The Yuk Factor," *Sydney Morning Herald*, September 5, 2005. http://www.smh.com.au/news/national/the-yuk-factor/2005/09/04/1125772411914.html.

19. Searle, "No Sign of Scientists to Answer Questions," *Chronicle* (Toowoomba), August 23, 2005.

20. Peter McCutcheon, "Residents Oppose Toowoomba Recycled Water Proposal," *7:30 Report*, ABC, March 22, 2006. http://www.abc.net.au/7.30/content/2006/s1598458.htm.

    "Chronicle: Oct. 8, 2005: Petition—MP Asks Council to Delay Water Plans," Water Futures, October 8, 2005. http://waterfutures.blogspot.com/2005/10/chronicle-oct-8-2005-petition-mp-asks.html.

    The text of the Toowoomba petition to parliament and to MP Ian Macfarlane is here:

    "Recycled Sewage Water for Drinking," Closed E-Petition, Queensland Parliament Petitions. http://www.parliament.qld.gov.au/view/EPetitions_QLD/ClosedEPetition.aspx?PetNum=528&1Index=-1.

21. The data on what's in purified recycled water in Australia come from an "expert advisory panel" of the Queensland Water Commission, appointed to provide technical guidance on purified recycled water.

The panel reported in a letter to the Queensland Water Commission, with a full technical report attached:

Paul Greenfield, Chairman, Expert Advisory Panel, Vice Chancellor, University of Queensland, letter to Elizabeth Nosworthy, Chairwoman, Queensland Water Commission, February 20, 2009 (PDF). http://www.qwc.qld.gov.au/myfiles/uploads/purified%20recycled%20water/Interim%20water%20quality%20report/Interim%20water%20quality%20report.pdf.

The concentration of acetaminophen is described on p. 5. (Acetaminophen is known in Australia as paracetamol.)

The amount of bisphenol A that a typical person would consume per day is from:

"Monograph on the Potential Human Reproductive and Developmental Effects of Bisphenol A," NIH Publication No. 08–5994, National Toxicology Program, Center for the Evaluation of Risks to Human Reproduction, U.S. Department of Health and Human Services, September 2008, p. 5 (PDF). http://cerhr.niehs.nih.gov/evals/bisphenol/bisphenol.pdf.

22. In the U.S., there is no cabinet secretary exclusively for water at the national level. Of the 50 state governments, just two have a cabinet-level water official—Arizona and Idaho both have a director of water resources. Minnesota has an advisory group called the "water cabinet," composed of state officials from several departments.

During 2008, Australian Minister of Water Penny Wong was mentioned in 1,007 newspaper stories and radio broadcasts, versus 1,102 stories for then Defence Minister Joel Fitzgibbon (using a Nexis search of stories in which the name of each appeared at least three times). Fitzgibbon resigned in June 2009 as a result of questions about travel costs.

In the year after his departure, from July 2009 to July 2010, Penny Wong appears in 704 stories; the new defense minister, John Faulkner, appeared in 1,281 stories.

23. The A$30 billion of spending in Australia just on water projects related to the Big Dry is likely conservative. As of 2010, the *New York Times* reported that Australian cities had A$15 billion (US$13 billion) in desalination plants either built or under construction.

The A$30 billion comes from Australia's water utility association, the Water Services Association of Australia (WSAA), cited in this story:

Åsa Wahlquist, "Water, Special Report: Policy Hinges on Data Pool," *Weekend Australian*, March 7, 2009.

Norimitsu Onishi, "Arid Australia Sips Seawater, but at a Cost," *New York Times*, July 10, 2010. http://www.nytimes.com/2010/07/11/world/asia/11water.html.

24. The Australian Bureau of Statistics calculates that 85 percent of Australians live within 50 km (31 miles) of the Australian coast.

"Census of Population and Housing: Population Growth and Distribution, Australia, 2001," Catalog No. 2035.0, Australian Bureau of Statistics. http://www.abs.gov.au/ausstats/abs@.nsf/mf/2035.0.

Professor Leon van Shaik, of the Royal Melbourne Institute of Technology, says in the article below that half the nation's homes are within 8 miles of the coast.

"With the Blenburn House in Rural Australia, Sean Godsell Perfects an Eco-

friendly Prototype," *Architectural Record*, April 2008. http://archrecord.construction
.com/residential/recordHouses/2008/08Glenburn.asp.

25. Climate, rainfall, runoff, and river flow data are from a remarkable reference document, *The Australian Water Map*. Some of the charts on the four-foot-wide map are
available online:
    The Australian Water Map, Earth Systems, Melbourne, Australia: 2003. http://
www.earthsystems.com.au/mapwater/index.htm.

26. The land area of Western Australia is 2.5 million km².
    The total land area of the Western European nations is 2.05 million km²—UK
(0.24 million km²), France (0.64 million km²), Spain (0.51 million km²), Germany
(0.36 million km²), and Italy (0.30 million km²).

27. Data on changes in Perth's rainfall and runoff are from officials of the Water Corpora-
tion, the water utility for Western Australia. The data are summarized in the Water
Corporation's most recent strategic planning document:
    Integrated Water Supply Scheme: Source Development Plan, 2005–2050, Water
Corporation, April 2005, p. 2 (PDF). http://www.watercorporation.com.au/_files/
PublicationsRegister/22/SourcePlan_2005_Summary.pdf.

28. Tim Flannery's original observation about Perth comes from:
    Paul Sheehan, "The Flannery Eaters," *Sydney Morning Herald*, June 5, 2004.
www.smh.com.au/articles/2004/06/04/1086203632052.html.
    The prediction was reported in Perth's *West Australian* newspaper, based on its
interview with Flannery, June 25, 2004.
    Carmelo Amalfi, "Perth Will Die, Says Top Scientist," *West Australian*, June 25,
2004.

29. Adelaide residents have dramatically reduced their total water consumption, even
as the city has grown. In 2002, the city used 194 gigaliters of water. In 2009, it used
137 gigaliters—a reduction of 57 gigaliters. The Perth desalination plant produces
50 gigaliters of drinking water a year; Adelaide's planned desalination plant was originally designed to produce 50 gigaliters of water a year; its size has been doubled during
planning. See the release below.
    "Water Consumption Remains Below Target," SA Water, January 2, 2010
(PDF). http://sawater.com.au/NR/rdonlyres/2FC06879–3EE4–4A3A-90FB-6EB87
6233504/0/waterconsumption.pdf.
    "Adelaide to Double Size of Its Desalination Plant," SA Water, May 13, 2009
(PDF). http://sawater.com.au/NR/rdonlyres/E47D55A8-91F9-4029-A9A6-79A3EC
21BE62/0/MedRelDesalMay09.pdf.

30. Melbourne's swimming-pool filling restrictions are detailed here:
    "Stage 3 Water Restrictions: Frequently Asked Questions, Pools and Spas," Tar-
get 155. http://www.target155.vic.gov.au/stage-3-water-restrictions/frequently-asked
-questions.
    Trucking in water to fill swimming pools is described below:
    "Having Trouble Filling the Swimming Pool? There's a Solution," *Age*, Octo-
ber 8, 2007. http://www.theage.com.au/articles/2007/10/07/1191695744867.html.

31. Historical figures for Melbourne's water use:
    "Water Use," Melbourne Water. http://www.melbournewater.com.au/content/
water_conservation/water_use/water_use.asp.

Current water use figures for Melbourne:

"Water Consumption," Target 155. http://www.target155.vic.gov.au/water-supply
-and-use/water-consumption.

32.   "Seawater Desalination Plant," Melbourne Water. http://www.melbournewater.com
.au/content/current_projects/water_supply/seawater_desalination_plant/seawater_
desalination_plant.asp.

Victorian premier John Brumby's "pray for rain" comment is from Rick Wallace,
"Desal Project to Defy Slump," *Australian*, July 31, 2009.

33.   Basic information about the size and significance of the Murray-Darling Basin is from
the Murray-Darling Basin Authority (MDBA):

"About the Basin," Murray-Darling Basin Authority. http://www.mdba.gov.au/
water/about_basin.

River flow data from 2007, 2008, and 2009 is from the MDBA drought updates
(the "river year" runs from June 1 to May 31).

The Murray's long-term average flow for one year: 8,840 gigaliters.

The Murray's total flow for the three years 2007–2009: 5,040 gigaliters.

"River Murray System: Drought Update," Murray-Darling Basin Authority, Issue No. 19, June 2009, p. 2 (PDF). http://www.mdba.gov.au/system/files/drought
-update-June-2009.pdf.

34.   In order, references for the loss of fruit trees, the shrinking of the sheep herd, and the
dramatic falloff in the rice harvest are:

Debra Jopson, "Murray Towns 'Are Living Hand to Mouth,'" *Sydney Morning
Herald*, March 9, 2009. http://www.smh.com.au/environment/water-issues/murray
-towns-are-living-hand-to-mouth-20090308-8sgi.html.

"Sheep and Lamb Numbers Fall to Their Lowest Levels Since 1905: Agricultural
Commodities, Australia, 2008–09," Catalog No. 7121.0, Australian Bureau of Statistics, April 9, 2010. http://www.abs.gov.au/ausstats/abs@.nsf/mediareleasesbytitle/D79
3AD9EE6BCF107CA257456001F1839?OpenDocument.

"Agricultural Commodities, 2007–08," Catalog No. 7121.0, Australian Bureau
of Statistics, May 22, 2009, p. 12 (PDF). http://www.ausstats.abs.gov.au/Ausstats/sub
scriber.nsf/0/0B7FE368D21623E0CA2575BD001CE264/$File/71210_2007-08.pdf.

35.   Lauren Novak, "Water: End of the Sprinkler," *Advertiser* (Adelaide), April 8, 2010.

36.   "Water Pollution Americans' Top Green Concern," Gallup Poll, March 25, 2009. http://
www.gallup.com/poll/117079/Water-Pollution-Americans-Top-Green-Concern
.aspx.

Gallup has polled about concern over drinking water going back to 1990, and the
trend is, in fact, toward less worry. In 1990, 65 percent of Americans worried about
drinking water "a great deal." The number peaked in 2000 at 72 percent. The 2009
level of concern, 59 percent worried "a great deal," is the highest since 2001.

Only 5 percent don't worry about drinking water at all.

37.   The issue of micropollutants is just starting to get both the scientific and public policy
attention that it requires. Alongside the question of micropollutants, there is a whole
new wave of industrial water pollutants that are just coming under scrutiny and
regulation—including those from factory farms, and from natural gas drilling where
huge volumes of high-pressure water, mixed with a secret brew of chemicals, are used
to crack open rock deep under ground and release gas supplies.

The U.S. EPA announced plans in 2010 to dramatically overhaul and update drinking water standards to account for the micropollutants, without providing any of the crucial specifics. But there is plenty of research and reporting for those curious about micropollutants.

In 2009, the *New York Times* produced an ambitious series called "Toxic Waters" that focused on the increasing pollution of U.S. water supplies and drinking water. The stories included several extraordinary cases of dangerous drinking water pollution, especially from coal mines and farms.

The series is available at http://www.nytimes.com/water.

The impact of the natural gas industry's hydraulic fracturing technique on water supplies has been written about in great detail by the investigative reporters at *ProPublica*. http://www.propublica.org/series/buried-secrets-gas-drillings-environmental-threat.

For those interested in the more detailed scientific research involving micropollutants and emerging water-supply contaminants, the lab that Shane Snyder now codirects, the Arizona Laboratory for Emerging Contaminants, has links to its own current research and to the work of others at http://www.alec.arizona.edu.

And finally, one of the early pioneers trying to understand the impact of endocrine-disrupting chemicals on humans and animals is Dr. Frederick vom Saal at the University of Missouri–Columbia. He and his team, the Endocrine Disruptors Group, maintain a Web site with current research and an archive of studies on endocrine-disrupting substances. http://endocrinedisruptors.missouri.edu/vomsaal/vomsaal.html.

38. According to Toowoomba's Alan Kleinschmidt, the new pipeline is expected to provide 10 gigaliters of water per year, or 27.4 megaliters per day.

27.4 megaliters = 7.2 million gallons.

7.2 million gallons × 8.33 pounds per gallon = 60 million pounds.

The pipeline is sending 60 million pounds of water up the hill a day—that's 400,000 people at an average weight of 150 pounds each.

7. WHO STOPPED THE RAIN?

1. The Imperial Valley gets just 3 inches of rain a year, on average.

Our County . . . Our Community, Imperial County. http://www.co.imperial.ca.us/.

The Imperial Valley is the 11th most productive agricultural producing county in the U.S., according to the USDA:

2007 Census of Agriculture: County Profile: Imperial County, California, National Agricultural Statistical Service, USDA (PDF). http://www.agcensus.usda.gov/Publications/2007/Online_Highlights/County_Profiles/California/cp06025.pdf.

The number of golf courses in Phoenix, Arizona:

"Golf," VisitPhoenix.com. http://www.visitphoenix.com/media/media-kit/golf/index.aspx.

2. In the growing season that ended in June 2010, Murray River farmers produced rice at an average of 11 metric tons per hectare, which is 4.9 tons per acre. The world average is 3.9 metric tons per hectare, 1.7 tons per acre.

Åsa Wahlquist, "Perfect Weather Yields a Rice Bowl Record," *Australian*, July 12, 2010. http://www.theaustralian.com.au/news/nation/perfect-weather-yields-a-rice-bowl-record/story-e6frg6nf-1225890495021.

3. How do you calculate that each dinner-plate serving of rice Laurie Arthur grows requires 14.4 gallons of irrigation water?

One pound of dry rice cooks up to provide roughly 10 servings.

Arthur produces 10 metric tons of rice per hectare, which is 4.5 tons of rice per acre.

1 acre = 9,000 pounds of rice.

Arthur uses 12 megaliters of water per hectare of rice =

4.9 megaliters of water per acre = 1.3 million gallons per acre.

9,000 pounds of rice = 1.3 million gallons of water.

1 pound of rice = 144 gallons.

1 serving = 14.4 gallons of water.

4. What's the math behind Laurie Arthur growing enough food to feed 100,000 people for a year?

In a good year (which was typical until 2006), his land produced 9,000 metric tons of high-calorie grain (mostly rice, with some barley and wheat), along with 2,000 sheep.

9,000 metric tons of grain = 9 million kg.

9 million kg = 90 kg of rice for each of 100,000 people.

90 kg = 200 pounds of rice per person.

That's roughly 4 pounds of rice per person, per week.

1 pound of cooked rice contains roughly 3,200 calories, so the rice provides 12,800 calories a week, or 1,800 calories per day per person.

"I'm not saying you're going to get fat on it," says Arthur, "but it's enough to survive."

Of course, you wouldn't want to live on just rice, but Arthur's point is that the water he uses produces enough calories to feed an entire city.

5. *Detailed Historic Timeline of the Australian Rice Industry*, SunRice (PDF). http://www .sunrice.com.au/uploads//documents/education/Detailed_History_of_the_Austra lian_Rice_Industry.pdf.

6. The Murray-Darling Basin is 410,000 square miles. The five states come to 408,000 square miles: Kansas, 82,000 square miles; Missouri, 69,700; Iowa, 56,300; Nebraska, 77,400, Oklahoma, 69,900; and Arkansas, 53,100.

7. "Water and the Murray-Darling Basin: A Statistical Profile, 2000–01 to 2005–06," Catalog No. 4610.0.55.007, Australian Bureau of Statistics, August 15, 2008. http:// www.abs.gov.au/ausstats/abs@.nsf/mf/4610.0.55.077.

8. Laurie Arthur says 2006–2007 was the first zero-allocation year for Murray Irrigation Limited.

9. The Australian Bureau of Statistics says, as of 2006, 175,100 Australians identified themselves as "farmers" out of a population of 21.9 million—0.8 percent, or 1 out of 125.

"Agriculture in Focus: Farming Families, Australia, 2006," Catalog No. 7104. 0.55.001, Australian Bureau of Statistics, August 12, 2008. http://www.abs.gov.au/ ausstats/abs@.nsf/Latestproducts/7104.0.55.001Main%20Features32006?open document&tabname=Summary&prodno=7104.0.55.001&issue=2006&num=& view=.

10. "Watering Down an Agricultural Dilemma," *Age*, February 21, 2003.

11. In order, citations for water use by Adelaide, by Melbourne, and by the Murray Basin farmers are:

    "How Much Do We Use?" Water for Good. http://www.waterforgood.sa.gov.au/using-and-saving-water/how-much-do-we-use/.

    "Water Use," Melbourne Water. http://www.melbournewater.com.au/content/water_conservation/water_use/water_use.asp.

    "Water and the Murray-Darling Basin—A Statistical Profile, 2000–01 to 2005–06," Catalog No. 4610.0.55.007, Australian Bureau of Statistics, August 15, 2008. http://www.abs.gov.au/ausstats/abs@.nsf/mf/4610.0.55.007.

    For the number of farmers, see "Farmers in the Murray-Darling Basin," from the "Contents" page, Catalog No. 4610.0.55.007.

12. "River Murray System: Drought Update," Murray-Darling Basin Authority, Issue No. 19, June 2009, pp. 1–2 (PDF). http://www.mdba.gov.au/system/files/drought-update-June-2009.pdf.

13. Fred Pearce, "Fertile Crescent Will Disappear This Century," *New Scientist*, July 27, 2009. http://www.newscientist.com/article/dn17517-fertile-crescent-will-disappear-this-century.html.

14. Mundaring Weir overflowed on September 12, 1996.

    The Water Corporation's description of the dam is below, with the notation that the spill rate is 1,020 meters^3 per second, which is 269,455 gallons per second, or 16 million gallons of water per minute.

    "Mundaring Weir," Water Corporation. http://www.watercorporation.com.au/D/dams_mundaring.cfm.

    Pictures of the dam overflowing are here:

    "1996 Mundaring Weir Overflow," Mundaring Weir Hotel. http://www.mundaringweirhotel.com.au/over.html.

15. "Barnett's a Canal Too Far," On Line Opinion, February 25, 2005. http://www.onlineopinion.com.au/view.asp?article=3079.

    Matt Liddy and Nadia Farha, "The Poll Vault: How the West Was Won," ABC, February 28, 2005. http://www.abc.net.au/elections/wa/2005/weblog.

    Barnett's proposed canal was 3,700 km, which is 2,300 miles. The distance from Niagara Falls to Las Vegas is 2,264 miles.

16. The Tampa Bay, Florida, desalination plant's woes have been extensively covered by the *St. Petersburg Times*. Here's a small sampling of stories:

    Craig Pittman, "Tampa Bay Water Desal Plant Isn't Running Often Enough for Swiftmud," *St. Petersburg Times*, August 25, 2010. http://www.tampabay.com/news/environment/water/tampa-bay-water-desal-plant-isnt-running-often-enough-for-swiftmud/1117191.

    Pittman, "More Problems for Tampa Bay Water Desalination Plant," *St. Petersburg Times*, March 17, 2009. http://www.tampabay.com/news/environment/water/article984409.ece.

    Pittman, "Cost to Fix Desalination Plant Jumps by Millions," *St. Petersburg Times*, August 7, 2004. http://www.sptimes.com/2004/08/07/Tampabay/Cost_to_fix_desalinat.shtml.

17. Here is an online water pressure calculator by depth:

    "Pressure Calculator," A to Z Diving. http://www.atozdiving.co.nz/waterpressure.htm.

Safe operating depths of U.S. submarines are not made public, but are discussed in this publication from the Federation of American Scientists:

Run Silent, Run Deep, Federation of American Scientists, December 8, 1998. http://www.fas.org/man/dod-101/sys/ship/deep.htm.

18. Jim Gill's pee example already assumes one conservation measure that is hardly universal: 6 liters of toilet water is just 1.6 gallons, which is a low-flow toilet. Although 3.5-gallon-per-flush toilets haven't been sold in the U.S. since 1994, many U.S. homes and businesses have fixtures that are sixteen years old, which use 3.5 gallons of water per flush. That's 13 liters of water to get rid of half-a-liter of pee—a dilution ratio of 26:1. And, in fact, very few people, even adults, pee as much as 0.5 liters a time.

19. One of the striking things about water management in Australia is how many women there are in senior positions, in contrast to the male-dominated arena of water in both the U.S. and the rest of the world. As of October 2010:

At the federal level, Penny Wong is the Australian minister for water. Three of seven members of the National Water Commission are women, and one of those, Chloe Munro, is also chairman of the consortium building Victoria's large desalination plant.

In the state of South Australia, Karlene Maywald was minister for water security for six years, until March 2010; Robyn McLeod is currently the commissioner for water security.

Sue Murphy is CEO of the Water Corporation, the water utility for Perth and Western Australia; Anne Howe is CEO of SA Water, the water utility for Adelaide and South Australia; Kerry Schott is managing director of Sydney Water; Eleanor Underwood is chairman of the board of Melbourne Water; Judith King is chairman of Power and Water, the water utility for the Northern Territory, including Darwin.

Jennifer Westacott is head of the Australia water practice for the large consulting company KPMG.

20. *Desalination in Australia*, CSIRO (Commonwealth Scientific and Industrial Research Organization), February 2009 (PDF). http://www.csiro.au/files/files/ppcz.pdf.

Norimitsu Onishi, "Arid Australia Sips Seawater, but at a Cost," *New York Times*, July 10, 2010. http://www.nytimes.com/2010/07/11/world/asia/11water.html.

8. WHERE WATER IS WORSHIPPED, BUT GETS NO RESPECT

1. The survey of water service in the 35 largest Indian cities comes from a World Bank project-planning document:

"India—Delhi Water Supply and Sewerage Project," Report No. 36065, Project Information Document, World Bank, March 15, 2006. http://www-wds.worldbank .org/external/default/WDSContentServer/WDSP/IB/2006/05/03/000012009_20060 503095630/Rendered/INDEX/36065.txt.

Detail on the water utilities in 20 Indian cities is in:

2007 Benchmarking and Data Book of Water Utilities in India, Asian Development Bank and the Ministry of Urban Development of the Government of India (PDF). http://www.adb.org/documents/reports/Benchmarking-DataBook/2007-Indian -Water-Utilities-Data-Book.pdf.

2. Divya Datt and Shilpa Nischal, eds., *Looking Back to Change Track*, TERI: The Energy and Resources Institute, New Delhi: November 2009, p. 5.

Unfortunately, the TERI report is not available online.

3. The numbers for water-related illnesses and deaths come from a sobering 2009 UNICEF/World Health Organization report devoted to diarrhea, and the toll diarrhea still takes in the developing world. *Diarrhoea: Why Children Are Still Dying and What Can Be Done* says that in India, there are 386,000 deaths of children under 5 each year (p. 7).

According to the WHO data, 88 percent of those deaths—931 per day, 39 each hour just in India—are caused by dirty water, dirty hands, and poor sanitation. The report also points out that diarrhea deaths in India and the developing world are utterly unnecessary—diarrhea is no more difficult to treat in India than in England.

Diarrhoea: Why Children Are Still Dying and What Can Be Done, UNICEF/World Health Organization, New York, 2009 (PDF). http://whqlibdoc.who.int/publications/2009/9789241598415_eng.pdf.

4. *Looking Back to Change Track*, p. 5.

5. Ibid., p. 60.

The figures for incidence of diarrhea and cost are from 2008.

6. The history of urban water systems in India is a little hard to get details on. This comes from an interview with V. S. Chary, an expert on urban water systems in India and director, Centre for Energy, Environment, Urban Governance, and Infrastructure Development at the Administrative Staff College of India, Hyderabad, India. Another book with some history is:

John Briscoe and R.P.S. Malik, *India's Water Economy* (New Delhi: Oxford University Press and the World Bank, 2006).

7. The Grand hotel's lobby, with the gardens beyond it, is pictured online.

The Grand, New Delhi. http://www.thegrandnewdelhi.com/.

8. *The "Bird of Gold": The Rise of India's Consumer Market*, McKinsey Global Institute, May 2007 (PDF). http://www.mckinsey.com/mgi/reports/pdfs/india_consumer_market/MGI_india_consumer_full_report.pdf.

Figures for cutting the number of poor people in half: pp. 10–12.

Figures for the movement of Indians into the middle class: p. 13.

McKinsey says that in 2005, 50 million Indians qualified as middle class. By 2025, McKinsey estimates, 583 million Indians will be middle class, an increase of 533 million new middle-class Indians in 20 years, or 73,014 per day.

McKinsey's estimates assume an average growth rate of 7.3 percent for India's GDP over the 20-year period, as explained on p. 165 of the report.

9. "The World's Billionaires," *Forbes*, March 10, 2010. http://www.forbes.com/lists/2010/10/billionaires-2010_The-Worlds-Billionaires_Rank.html.

The Indians in the top 50 are:

#4 Mukesh Ambani (Mumbai, Reliance Industries)
#5 Lakshmi Mittal (London, ArcelorMittal)
#28 Azim Premji (Bangalore, Wipro)
#36 Anil Ambani (Mumbai, Reliance)
#40 Shashi Ruia, Ravi Ruia (Mumbai, Essar Group)
#44 Savitri Jindal (Delhi, Jindal Group)

The rank of nationalities among the top 50 billionaires:

United States: 20
India: 6

Hong Kong/China: 4
Russia: 4
Germany: 3
United Kingdom: 1
Saudi Arabia: 1

10. "Fuel Used for Cooking," Census of India, 2001. http://www.censusindia.gov.in/
Census_Data_2001/Census_data_finder/HH_Series/Fuel_used_for_cooking.htm.
"Source of Drinking Water (Including Availability of Electricity and Latrine),"
Census of India, 2001. http://www.censusindia.gov.in/Census_Data_2001/Census_
data_finder/HH_Series/Source_of_drinking_water.htm.
The main page for the Census of India is http://www.censusindia.gov.in.

11. The adult literacy rate refers to Indians 15 and over, and comes from the CIA *World
Factbook*, derived from the Census of India data:
"India: People," *The World Factbook 2009,* Central Intelligence Agency. https://
www.cia.gov/library/publications/the-world-factbook/geos/in.html.
The CIA *Factbook* also reports that 70 percent of Indians are 15 or older, and that
the total of those "adults" who are women is 391.5 million.

12. The announcement of the original discovery of water molecules on the Moon:
"NASA Instruments Reveal Water Molecules on Lunar Surface," Release
No. 09–222, NASA, September 24, 2009. http://www.nasa.gov/home/hqnews/2009/
sep/HQ_09–222_Moon_Water_Molecules.html.
Seth Borenstein, "It's Not Lunacy, Probes Find Water in Moon Dirt," Associ-
ated Press, ABC News, September 23, 2009. http://abcnews.go.com/Technology/wire
Story?id=8655085.
The announcement of the discovery of large amounts of ice in Moon craters:
"NASA Radar Finds Ice Deposits at Moon's North Pole; Additional Evidence
of Water Activity on Moon," Release No. 10–055, NASA, March 1, 2010. http://www
.nasa.gov/home/hqnews/2010/mar/HQ_10–055_moon_ice.html.
Tariq Malik, "Tons of Ice Found on Moon's North Pole," *Florida Today*, March 1,
2010.
How much ice is there in an iceberg? NASA found "at least" 660 million tons of
ice in 40 craters—16.5 million tons per crater.
NASA says large icebergs have a mass of about 1 million tons.
"Iceberg," *World Book at NASA*, NASA. http://www.nasa.gov/worldbook/
iceberg_worldbook.html.
The Geological Survey of Denmark and Greenland says iceberg masses can
range much higher, up to 10 million tons or more.
Environmental Oil Spill Sensitivity Atlas for the West Greenland Coastal Zone, Dan-
ish Energy Agency, Ministry of Environment and Energy, 2000, pp. 8-46, 8-47 (PDF).
http://www.geus.dk/departments/quaternary-marine-geol/oliespild_v_gr/PDF
files/Chapter8/Chapter8_7.pdf.

13. This sampling of Indian headlines on the water-on-the-Moon discovery:
"One Big Step for India, Giant Leap for Mankind," *Times of India*, September 25,
2009. http://timesofindia.indiatimes.com/india/One-big-step-for-India-a-giant-leap
-for-mankind/articleshow/5053202.cms.
Full coverage from the *Times of India:*

"Special Coverage: Chandrayaan Finds Water on Moon," *Times of India*. http://timesofindia.indiatimes.com/chandrayaan-finds-water-on-moon/specialcoverage/5361062.cms.

"Water on Moon Is India's Discovery, Says ISRO Chief," *Hindustan Times*, September 26, 2009. http://www.hindustantimes.com/News-Feed/india/Water-on-moon-is-India-s-discovery-says-ISRO-chief/Article1-458154.aspx.

"It's the Finding of the Millennium: Nair," *Times of India*, March 3, 2010. http://timesofindia.indiatimes.com/india/Its-the-finding-of-the-millennium-Nair/articleshow/5635381.cms.

14. *2007 Benchmarking and Data Book of Water Utilities in India*, Asian Development Bank and the Ministry of Urban Development of the Government of India, p. 3 (PDF). http://www.adb.org/documents/reports/Benchmarking-DataBook/2007-Indian-Water-Utilities-Data-Book.pdf.

The ADB *Data Book* sometimes overstates the utilities' performance. A survey by the Indian magazine *Business Today* in June 2009 found that 49 percent of Bangalore residents don't receive water even every day—although the *Data Book* says Bangalore offers water service 4.5 hours a day.

"Cover Story: Best Cities to Work, Play and Live: Cities in Numbers," *Business Today*, June 9, 2009. http://businesstoday.intoday.in/index.php?option=com_content&task=view&id=11649&sectionid=22&secid=0&Itemid=1&issueid=1166.

15. "Wipro and Thames Water: Build a Successful Long-Term Relationship," Wipro, October 1, 2009 (PDF). http://www.wipro.com/resource-center/library/pdf/wipro-article1.pdf.

16. The full text of the Indian Supreme Court's order to the central government to resolve the nation's water problem "on a war footing" is here:

M. K. Balakrishnan & Others v. Union of India & Others, Supreme Court of India, Civil Original Jurisdiction, "Writ Petition No. 230 of 2001: Order," April 28, 2009 (PDF). http://kgfindia.com/judgement-28april09.pdf.

The order got scant press coverage:

"Supreme Court Attempts to Resolve India's Water Woes," *Hindustan Times*, April 28, 2009.

17. The list of prominent Americans who are either Indian or of Indian origin is remarkable for its range of talent:

Film director M. Night Shyamalan, CNN journalist Fareed Zakaria, and conductor Zubin Mehta were all born in India. Indra Nooyi, the CEO of PepsiCo, C. K. Prahalad, the recently deceased Harvard management guru, and Vikram Pandit, the CEO of Citigroup—all also born in India.

Singer Norah Jones, Harvard surgeon and author Atul Gawande, and Amar Bose, founder of the company that makes Bose stereo equipment, are all children of Indian parents—as are six of the last ten U.S. National Spelling Bee champions.

18. The Indian head shake doesn't mean only "Yes, I agree with you." The gesture has a wide range of subtlety and interpretation. Kavita Pillay, a filmmaker based in Cambridge, MA, who went to India on a Fulbright scholarship, described it in a piece on World Hum this way:

"In its myriad iterations, the Indian head nod can mean 'Yes,' 'Nice to meet you,' and 'I agree to the price you have just mentioned.' It can also mean 'Maybe,' 'Hell no,' and 'You are the enemy of intelligence.' Interpreting the meaning requires time, practice, a little self-effacement, and a lot of humor."

Pillay's brief essay is highly entertaining:

Kavita Pillay, "How to Tilt Your Head Like an Indian," World Hum, January 30, 2006. http://www.worldhum.com/features/how-to/tilt_your_head_like_an_india _20060128/.

19. *State of Pollution in the Yamuna,* Centre for Science and Environment, May 2009 (PDF). http://www.indiaenvironmentportal.org.in/files/State%20of%20the%20 Yamuna_0.pdf.

20. Andrew Buncombe, "Unholy Water: Delhi's Rotting River," *Independent*, May 1, 2008. http://www.independent.co.uk/news/world/asia/unholy-water-delhis-rotting -river-818774.html.

Neha Lalchandani, "Pollution Cloud Looms Over Yamuna," *Times of India,* September 20, 2009. http://timesofindia.indiatimes.com/news/city/delhi/Pollution-cloud -looms-over-Yamuna/articleshow/5034865.cms.

21. Figures for the irrigation efficiency of India's farmers range from 25 percent to 40 percent. The 25 to 35 percent range comes from *Looking Back to Change Track,* TERI, p. 56.

22. Citations, in order, for continuous water service in Hanoi, Ho Chi Minh City, Phnom Penh, and Kampala:

Hanoi: "Base Data of Hanoi Water Business Company," *Benchmarking 2004,* Southeast Asian Water Utilities Network. http://www.seawun.org/benchmarking/ basedata/basedata.asp?sRespondent=Ha%20Noi.

Ho Chi Minh City: *Socialist Republic of Viet Nam: Preparing the Ho Chi Minh City Water Supply Project,* Asian Development Bank Technical Assistance report, June 2008, p. 2 (PDF). http://www.adb.org/Documents/TARs/VIE/41070-VIE-TAR.pdf.

Phnom Penh: "Proceedings of the Regional Consultation Workshop on Water in Asian Cities—The Role of Civil Society," *Water for All,* Asian Development Bank, October 14–16, 2002, p. 77 (PDF). http://www.adb.org/documents/books/water_for _all_series/Water_Asian_Cities/Annex.pdf.

Kampala: *Change Management for Achieving Continuous Water Supply for All in Urban Areas,* Administrative Staff College of India. http://www.asci.org.in/continuous -water-supply.asp.

23. The official from the Delhi Jal Board who wrote that when people proposed 24/7 water in India, "we laugh it off as an absurdity," was Ashish Kundra, then additional CEO of the Delhi Jal Board.

Ashish Kundra, "Water 24X7: Not Just a Pipe Dream," *Indian Express*, April 1, 2006. http://www.indianexpress.com/news/water-24x7-not-just-a-pipe-dream/1512/0.

24. The fact that most Indian water utilities do not recover their operating costs from water fees is detailed in *2007 Benchmarking and Data Book of Water Utilities in India*, p. 30. That the Delhi Jal Board's revenue only covers 60 percent of operating costs comes from a World Bank assessment of Delhi's water system from 2006:

India—Delhi Water Supply and Sewerage Project," Report No. 36065, Project Information Document, World Bank.

25. "Estimated 5-Year Investment Needs in Billions of Dollars," *Report Card for America's Infrastructure,* American Society of Civil Engineers (ASCE). http://www.infra structurereportcard.org/report-cards.

The ASCE also estimates, in the same chart, that of the $255 billion in water infrastructure needs in the U.S. through 2014, only $147 billion will be available.

26. The story of the *E. coli* deaths resulting from contaminated tap water in Hyderabad in May 2009 received extensive press coverage:

"Hyderabad Water Contamination Raises Toll to 9," Indo-Asian News Service, May 8, 2009. http://www.thaindian.com/newsportal/health1/hyderabad-water -contamination-raises-toll-to-nine_100190025.html.

"Four Die, 100 Taken Ill after Drinking Contaminated Water in Hyderabad," Indo-Asian News Service, May 5, 2009. http://sify.com/news/four-die-100-taken-ill -after-drinking-contaminated-water-in-hyderabad-news-hyderabad-jffvy8jddej.html.

"Four Die, 100 Sick after Drinking Contaminated Water in Hyderabad," May 5, 2009, Indo-Asian News Service. http://www.thaindian.com/newsportal/health1/ four-die-100-sick-after-drinking-contaminated-water-in-hyderabad-lead_100188724 .html.

27. Augustin Maria, "Urban Water Crisis in Delhi. Stakeholders Responses and Potential Scenarios of Evolution," *Idées Pour le Débat*, No. 6, 2008, Paris: Institut du Développement Durable et des Relations Internationales, p. 10 (PDF). http://www.iddri.org/ Publications/Collections/Idees-pour-le-debat/Id_0806_Maria_Urban-Crisis-Water -Delhi.pdf.

Maria's research concluded that just 43 percent of Delhi's population lived in settlements where they were entitled to a permanent water connection (p. 10). He also reports that, as of the most recent data he could obtain, from 2004, Delhi had 11,533 public standposts—water spigots (p. 6).

Maria is now a water economist for the World Bank.

28. For the Census of India, the definition of "away" for purposes of getting water is on this page of charts about "amenities at home":

"Availability of Eminities [*sic*] and Assets," table 16, Census of India, 2001. http:// censusindia.gov.in/Census_And_You/availability_of_eminities_and_assets.aspx.

Within the three distance categories—"within premises," "near premises," and "away"—the census asks whether the water comes from a tap, a hand pump, a tube well, a well, or some other source.

The total number of households in India, and the average household size, are here:

"Data Highlights," Census of India, 2001 (PDF). http://censusindia.gov.in/Data_ Products/Data_Highlights_Data_Highlights_link/data_highlights_hh1_2_3.pdf.

29. According to 2008 U.S. Census estimates, 177 million people live in the states east of the Mississippi River; 126 million live west of the Mississippi River.

30. There is some dispute about the minimum amount of water each person needs for basic needs. The WHO and UNICEF set 20 liters per day per person as the minimum level necessary for drinking and personal hygiene.

The UN's 2006 report *Beyond Scarcity: Power, Poverty and the Global Water Crisis* uses the 50 liters (13 gallons) per person per day standard as a "water poverty line," as discussed on p. 34 of the report.

People routinely receiving less than 50 liters per person each day live in "water poverty," limited in their ability to bathe, cook, and keep their clothes clean. (The report is a PDF file 422 pages long.)

Human Development Report, 2006: Beyond Scarcity: Power, Poverty and the Global

Water Crisis, UN Development Programme. http://hdr.undp.org/en/reports/global/hdr2006/.

A more passionate discussion, "Diminishing Standards: How Much Water Do People Need?" comes from Les Roberts, an epidemiologist who has worked for the Centers for Disease Control and the International Rescue Committee, and who is now an associate clinical professor of population and family health at Columbia University.

Writing in 1998, Roberts pointed out that at that point the "standards" for basic water requirements were essentially random guesses, not based on any science. One limited study of refugees did find that those with more water had dramatically less diarrhea. Writing about victims of war or natural disaster, Roberts asks rhetorically why bare minimum standards for providing water are often not met. "Unfortunately, the answer will most often be because someone, somewhere, with a flush toilet and hot shower, does not think that the extra investment to provide sufficient water is really worth it."

Les Roberts, "Diminishing Standards: How Much Water Do People Need?" *Forum: War and Water*, Geneva, 1998: International Committee of the Red Cross. http://www.icrc.org/Web/eng/siteeng0.nsf/htmlall/57JPL6.

31. It is a small but irresistible irony: A grown woman can reasonably carry 8 gallons of water (67 pounds). As it happens, that is almost exactly the amount of water that a grown woman's body contains. Women are 55 percent water by weight, so the body of a 120-pound woman contains 66 pounds of water—8 gallons.

32. One of the most popular grocery store items in the U.S. these days is the 24-pack of half-liter bottles of water, shrink-wrapped into an ungainly block. The package of water weighs 26 pounds.

     If you want a feel for carrying your own water, prop it on your head as you leave the grocery, and balance it there as you walk through the parking lot to your car.

     The women and girls in Jargali were carrying between 41 pounds and 67 pounds of water on their heads, over the 2 km trip back home from the well.

33. There is very little agreement on the boundaries of the world's largest metropolitan areas, and so very little consensus on their precise populations and the rankings among them.

     One well-regarded source puts the 2010 population of urban Delhi at 23.2 million, ranking it fifth. According to City Population, which includes a list of adjacent areas encompassed in its population calculation, the top 10 metropolitan areas are

1.	Tokyo, Japan	34.0 million
2.	Guangzhou, China	24.2 million
3.	Seoul, South Korea	24.2 million
4.	Mexico City, Mexico	23.4 million
5.	Delhi, India	23.2 million
6.	Mumbai, India	22.8 million
7.	New York, U.S.	22.2 million
8.	São Paulo, Brazil	20.9 million
9.	Manila, Philippines	19.6 million
10.	Shanghai, China	18.4 million

Thomas Brinkhoff, "The Principal Agglomerations of the World," City Population. http://www.citypopulation.de/world/Agglomerations.html.

34. Garima Sharma, N. K. Mehra, R. Kumar, "Biodegradation of Wastewater of Najaf-garh Drain, Delhi Using Autochthonous Microbial Consortia," *Journal of Environmental Biology,* October 2002, vol. 23 no. 4, pp. 365–71. http://www.ncbi.nlm.nih.gov/pubmed/12674375.

Sharma et al. report the flow from the first enormous drain into the Yamuna as 1,668 million liters a day—441 million gallons a day, or 1 million gallons every 3.25 minutes.

35. "State of Pollution in the Yamuna," Centre for Science and Environment, May 2009, p. 8 (PDF). http://indiaenvironmentportal.org.in/files/State%20of%20the%20Yamuno_0.pdf.

This is a clear explanation of the U.S. EPA *E. coli* standards from the Willamette Riverkeeper:

"*E. coli* Monitoring," Willamette Riverkeeper. http://www.willamette-river keeper.org/programs/ecoli/e_colimain.htm.

36. The number of buses and auto rickshaws in Delhi comes from a 2007 study of the CNG fuel conversion, for EMBARQ, an NGO focused on environmentally sustainable urban transportation:

Monica Bansal, "Clean It Up, Don't Throw It Away: Greening Delhi's Para-transit," EMBARQ, World Resources Institute, Washington, DC, 2007 (PDF). http://www.embarq.org/sites/default/files/Monica_Bansal_Delhi_Paratransit.pdf.

The number of taxis in New York City:

Annual Report 2009, New York City Taxi and Limousine Commission, p. 9 (PDF). http://www.nyc.gov/html/tlc/downloads/pdf/tlc_annual_report_2009.pdf.

37. Vijay Singh, "Top Celebs, VIPs Get Notices from Navi Mumbai Cess Dept," *Times of India,* January 25, 2010. http://timesofindia.indiatimes.com/city/mumbai/Top-celebs -VIPs-get-notices-from-Navi-Mumbai-cess-dept/articleshow/5496470.cms.

9. IT'S WATER. OF COURSE IT'S FREE

1. Many hotels provide their own house brand of bottled water, as if a half-liter bottle of water with the Ritz-Carlton lion-crest logo on it somehow confers a richer hotel experience, or a richer water experience.

The most exotic version of this I've ever encountered was at the Hilton Hotel in Clear Lake, Texas, down the street from NASA's Johnson Space Center. Years before water was actually discovered on the Moon, the Hilton was offering "Luna" water, with a crescent moon logo, a beautiful blue glass bottle, and the story of the Clear Lake Hilton on the back. (The hotel's restaurant is named Luna.) "Luna" water costs $8 for a one-liter bottle and, as of summer 2010, was still for sale at the Hilton Clear Lake.

2. Starwood Four Points hotels explains free bottled water in hotel rooms here. http://www.starwoodhotels.com/fourpoints/index.html#/quad1/water/.

3. The average U.S. monthly water bill—just water, not including sewer service—has been hard to calculate, but the American Water Works Association has used usage and fee surveys to estimate that it is about $34 per household.

The average price of 1,000 gallons of tap water for residential customers in the U.S. is $3.24, according to the 2008 AWWA survey—10 gallons of water costs about 3 cents.

If you flush a 3-gallon-per-flush toilet 100 times, you've used 300 gallons of water—97 cents' worth in the typical U.S. home.

For comparison, if you wanted to flush the toilet once with bottled water, you'd need a full 24-pack of Deer Park half-liters (3 gallons). One flush would cost you at least $4, as opposed to the single penny a typical flush costs.

Water in other developed countries is more expensive than in the U.S., although still inexpensive compared with either bottled water or, say, cell phone service.

In Sydney, Australia, in the midst of that nation's water crisis, municipal water cost homeowners A$18.86 (US$16.97) for 1,000 gallons, five times what it cost in the U.S. A single U.S. penny would still buy four half-liter bottles of water.

In Germany, the average price of water was 6.99 euros (US$8.92) for 1,000 gallons. A single U.S. penny would buy eight half-liter bottles of water.

Thames Water has 8.5 million water customers, out of a total UK population in 2010 of 61 million people—14 percent of the population. The company posts its average water bills here:

"Our Business: Facts and Figures," Thames Water. http://www.thameswater.co.uk/cps/rde/xchg/corp/hs.xsl/4625.htm.

4. Municipal water rates are accessible for most places online. You just need to be careful about units of water being measured. Some utilities provide rates per 1,000 gallons; some use cubic feet (1 cubic foot = 7.48 gallons); some use an industry unit, CCF, that is, cost per 100 cubic feet (748 gallons).

All rates are current as of July 1, 2010.

The Las Vegas Valley Water Authority provided the average bill, at $23.62 per month.

The LVVWA has an online bill calculator that estimates your bill, based on water consumption. http://www.lvvwd.com/apps/rate_calculator/index.cfml.

Water rates for Atlanta come from the city of Atlanta's water department Web site, which also includes an online bill calculator. http://www.atlantawatershed.org/billcalculator.

The water rates for "suburban Philadelphia" are for my own home in Wyncote, PA, where we are customers of the company Aqua American, and the numbers come from our bills.

5. The Imperial Valley, in Imperial County, is the 11th most productive agricultural producing county in the U.S., and the 10th most productive in California.

2007 Census of Agriculture: County Profile: Imperial County, California. National Agricultural Statistical Service, USDA (PDF). http://www.agcensus.usda.gov/Publications/2007/Online_Highlights/County_Profiles/California/cp06025.pdf.

6. How do we know the water for a 3-pound bag of carrots from the Imperial Valley cost the farmer 1 penny?

The math works like this—references are at the end of the note.

Typical productivity of carrot farmers is 15 tons (30,000 pounds) per acre.

The Imperial Irrigation District reports that the district delivered 2.5 million acre-feet of water to its farmers in 2008, and that the district's farmers irrigate about 430,000 acres of farmland: 6 acre-feet of irrigation water for each acre of farmland.

The IID charges $19 per acre-foot, so the water for one acre of carrots costs $114.

Which means that the water to grow 30,000 pounds of carrots costs $114, which is one penny for three pounds.

Carrot farm productivity:

Joe Nuñez et al., "Carrot Production in California," *Vegetable Production Series*, UC Vegetable Research & Information Center, p. 1 (PDF). http://ucanr.org/freepubs/docs/7226.pdf.

Water per acre in the Imperial Valley:

2005 Annual Water Report, Imperial Irrigation District, p. 32 (PDF). http://www.iid.com/Media/2005IIDWaterAnnualReport.pdf.

2008 Annual Report, Imperial Irrigation District, p. 21 (PDF). http://www.iid.com/Media/iid_annual_08_web.pdf.

Water rates in 2010 for Imperial Valley farmers were $19 an acre-foot. Those rates are set to rise to $20 an acre-foot in 2011.

7. As noted above, the IID reports that the district delivered 2.5 million acre-feet of water to its farmers in 2008, and that the district's farmers irrigate about 430,000 acres of farmland: 6 feet of irrigation water for each acre of farmland.

The Southern Nevada Water Authority reports that total water demand for Clark County, Nevada, is 553,000 acre-feet of water.

Water Resource Plan 09, Southern Nevada Water Authority, 2009, p. 41 (PDF). http://www.snwa.com/html/wr_resource_plan.html.

Clark County is 5.1 million acres (8,012 square miles). http://www.accessclarkcounty.com/depts/public_communications/pages/About_clark_county.aspx.

So the county uses 0.1 acre-foot of water per acre of land per year.

The farmland uses 60 times as much water per acre as the resort land does.

8. Total value of agricultural products produced in Imperial County:

2009 Annual Agricultural Crop and Livestock Report for Imperial County Agriculture, Imperial Valley Agricultural Commissioner, June 2010, p. 1 (PDF). http://imperialcounty.net/ag/Crop%20&%20Livestock%20Reports/Crop%20&%20Livestock%20Report%202009.pdf.

In 2009, carrots produced for routine consumption (as opposed to for processing into other foods or for feed) were worth $54.6 million, and Imperial County farmers grew 242 million pounds, three-quarters of a pound of carrots for every person in the country, just from Imperial County (p. 4).

Total gaming revenue in Las Vegas is from the University of Nevada / Las Vegas Center for Business and Economic Research.

"Historical Economic Data for Metropolitan Las Vegas," Center for Business and Economic Research, University of Nevada, Las Vegas, 2010. http://cber.unlv.edu/snoutlk.html.

9. Adam Smith, *The Wealth of Nations*, ed. Edwin Cannan (New York: Modern Library, 1965), p. 28.

10. All the price comparisons for 1912, 1963, and 2010 use the U.S. Bureau of Labor Statistics online comparative "buying power" calculator. http://data.bls.gov/cgi-bin/cpicalc.pl.

The average spending on alcohol in 1960 comes from:

100 Years of U.S. Consumer Spending, 1960–61, U.S. Bureau of Labor Statistics, May 2006, p. 32 (PDF). http://www.bls.gov/opub/uscs/1960-61.pdf.

The complete *100 Years of U.S. Consumer Spending* is available here (PDF). http://www.bls.gov/opub/uscs/report991.pdf.

The cost of telephone service in 1963 is from:

"Historical Charges for Individual Residence Telephone Service," Appendix B-5, in *Local Telephone Rates: Issues and Alternatives*, staff working paper, Congressional

Budget Office, January 1984, table B-3 (PDF). http://www.cbo.gov/ftpdocs/109xx/doc10952/84doc01b.pdf.

The $5.65 for a single telephone line in 1963 equates to a cost of $40 a month today.

11. The Indianapolis resident who said she gets nothing for her water rates is quoted in:
Brendan O'Shaughnessy, "Fountain of Debt May Soak Water Users," *Indianapolis Star*, April 12, 2009.

The Indianapolis resident who said she pays "hellacious water rates" is quoted in:
Brendan O'Shaughnessy, "Bond Woes Could Soak Indy Water Users," *Indianapolis Star*, February 20, 2009.

12. The El Dorado Irrigation District water rate information is from:
El Dorado Irrigation District Proposed Rate Increase—Key Issues, El Dorado Irrigation District, January 2010 (PDF). http://www.eid.org/doc_lib/02_dist_info/2010_rate_increase_fact_sheet.pdf.

John Fraser, "Viewpoints: El Dorado Water Board Trying to Make Up for Past Neglect," *Sacramento Bee*, February 4, 2010. http://www.sacbee.com/2010/02/04/2511690/el-dorado-water-board-trying.html.

13. Washington, DC, rate increase information:
"DC WASA Board Approves 2011 Budget, Funding Critical Infrastructure and Environmental Protection," news release, District of Columbia Water and Sewer Authority, February 4, 2010. http://www.dcwater.com/site_archive/news/press_release430.cfm.

Charles Duhigg, "Saving U.S. Water and Sewer Systems Would Be Costly," *New York Times*, March 14, 2010. http://www.nytimes.com/2010/03/15/us/15water.html.

14. Plato, "Euthydemus," trans. Benjamin Jowett, Project Gutenberg. http://www.gutenberg.org/files/1598/1598-h/1598-h.htm.

15. "Household TV Trends Holding Steady: Nielsen's Economic Study 2008," Nielsen-Wire, Nielsen Company, February 24, 2009. http://blog.nielsen.com/nielsenwire/media_entertainment/household-tv-trends-holding-steady-nielsen's-economic-study-2008/.

16. For those skeptical that it is accepted wisdom that water use is inelastic in terms of price, below is a 2009 study from the California Energy Commission and the California Environmental Protection Agency.

The paper reviews the literature on the impact of price on water consumption, as well as doing its own analysis of California consumption and price data. It has fairly technical math, but here's one summary sentence, p. 4: "Studies of the impact of water price on residential water use suggest that water use is price inelastic . . . that a given percent change in water price elicits a relatively small change in water use."

And on p. 5: "Most studies suggest that water demand is inelastic."

Larry Dale et al., *Price Impact on the Demand for Water and Energy in California Residences*, California Climate Change Center, August 2009. http://www.energy.ca.gov/2009publications/CEC-500-2009-032/CEC-500-2009-032-F.PDF.

17. This article in *Water Efficiency*, a trade journal, says that water utilities use 3 percent of the electricity generated in the U.S.:
David Engle, "Controlling the Power," *Water Efficiency*, April 1, 2008, p. 1. http://www.waterefficiency.net/elements-2009/water-agency-costs.aspx.

This report from the California Energy Commission says 20 percent of the electricity used in the state goes to move and treat water:

Water Supply-Related Electricity Demand in California, Water and Energy Consulting and the Demand Response Research Center, California Energy Commission, November 2007, p. 3. (PDF). http://www.energy.ca.gov/2007publications/CEC-500-2007-114/CEC-500-2007-114.PDF.

The 2007 report below, from the Water Research Foundation, analyzing electricity use by U.S. water utilities, says water utilities are "the largest single user of electricity in the United States" (p. 7).

The report goes on to say that globally, water pumping and treatment consume 3 percent of total electricity generated worldwide; if agricultural uses are included, moving water consumes 7 percent of all electricity generated (p. 8).

Risks and Benefits of Energy Management for Drinking Water Utilities, Water Research Foundation, 2008. http://www.waterresearchfoundation.org/research/Topics andProjects/execSum/PDFReports/91200.pdf.

The total number of power plants in the U.S. is 5,400, of which 3 percent is 162.

Frequently Asked Questions—Electricity, U.S. Energy Information Administration. http://www.eia.doe.gov/ask/electricity_faqs.asp.

18. The water rates for Santa Fe, New Mexico, are here:

"Rate Schedule 1A," March 1, 2009, Sangre de Cristo Water Division, City of Santa Fe. http://www.santafenm.gov/DocumentView.aspx?DID=5269.

19. Thames Water is in the midst of a major campaign to install water meters, with the goal of having 50 percent of London-area residences on water meters by 2015.

"Your Water, Your Future," news release, Thames Water, May 7, 2008. http://www.thameswater.co.uk/cps/rde/xchg/corp/hs.xsl/6372.htm.

The statistic that just 22 percent of properties in Thames Water's service area have meters comes from here:

"Lower Charges on the Way for Thousands of Customers," news release, Thames Water, February 5, 2008. http://www.thameswater.co.uk/cps/rde/xchg/corp/hs.xsl/4367.htm.

20. Mike Young has a Web site loaded with links to articles and presentations about water, economics, and his "robust water system."

Mike Young and Jim McColl, "Water Droplets." http://www.myoung.net.au/water/index.php.

An accessible explanation of his framework for allocating water is:

Young and McColl, "A Future-Proofed Basin," University of Adelaide, 2008 (PDF). http://www.myoung.net.au/water/publications/A_future-proofed_Basin.pdf.

21. This 2010 *New York Times* story says that the flow of the Shatt al Arab is no longer enough to keep out salt water from the Persian Gulf, and that salt water has intruded almost 100 miles upriver from the ocean, devastating everything from fruit groves and fresh-water fisheries to drinking-water supplies.

Steven Lee Myers, "Vital River Is Withering, and Iraq Has No Answer," *New York Times*, June 12, 2010. http://www.nytimes.com/2010/06/13/science/earth/13shatt .htm.

22. The *Cleveland Plain Dealer*'s coverage of the missing water fountains at the Q Arena is archived here:

Glenn Baird, "Cleveland Cavaliers Will Reinstall Water Fountains at the Q to Comply With State Building Code," *Cleveland Plain Dealer*, February 26, 2010. http://blog.cleveland.com/metro/2010/02/cleveland_cavaliers_will_reins.html.

The Cavaliers' statement about the water fountain removal is here:

"Quicken Loans Arena Water Update," Cleveland Cavaliers, NBA.com. http://www.nba.com/cavaliers/news/qarena_100210.html.

The Cavaliers' 2009–2010 schedule, showing 29 home games through the end of February 2010, is here:

"Cavaliers Schedule & Results," Cleveland Cavaliers, NBA.com. http://www.nba.com/cavaliers/schedule.

23. The basic calculation for water required to grow food is that 1 calorie of food energy requires 1 liter of water to produce. So a daily 2,000-calorie diet requires 2,000 liters (528 gallons) of water to produce. See this 2007 report on water and food from the International Water Management Institute (p. 5):

David Molden, ed., *Water for Food, Water for Life*, International Water Management Institute, 2007 (PDF). http://www.iwmi.cgiar.org/Assessment/files_new/synthesis/Summary_SynthesisBook.pdf.

10. THE FATE OF WATER

1. It is hard to imagine geysers of water and ice crystals shooting from the surface of a moon with such force that they go out hundreds of miles, straight out into space.

Scientists haven't figured out what is driving the geysers on Enceladus. The moon is just 311 miles in diameter—meaning that the circumference around its equator is 977 miles. The circumference of our own Moon is 6,786 miles; the circumference of Saturn, which Enceladus orbits, is 235,297 miles.

Enceladus, named for one of the giants in Greek mythology, is one of 62 moons of Saturn. For a brief explanation of the jets of water vapor and ice coming off Enceladus, see:

"Enceladus," *About Saturn and Its Moons*, Cassini Equinox Mission, Jet Propulsion Laboratory, NASA. http://saturn.jpl.nasa.gov/science/moons/enceladus.

"Jets on Saturn's Moon Enceladus Not Geysers from Underground Ocean, One Group of Researchers Say," *ScienceDaily*, June 24, 2009. http://www.sciencedaily.com/releases/2009/06/090624152813.htm.

An image of an ice plume from Enceladus:

"Successful Flight Through Enceladus Plume," *Image of the Day Gallery*, NASA Images. http://www.nasa.gov/multimedia/imagegallery/image_feature_1510a.html.

2. The Burj Khalifa in Dubai is 2,716 feet high.

The Sears Tower, now known officially as the Willis Tower, is 1,450 feet tall, not including its radio masts.

The Empire State Building is 1,250 feet tall, not including its lightning rod.

So together, the Willis Tower and the Empire State Building come to 2,700 feet.

For a graphic comparison of the heights of the tallest buildings in the world, here is the Burj Khalifa's site:

"World's Tallest Towers," Burj Khalifa. http://www.burjkhalifa.ae/language/en-us/the-tower/worlds-tallest-towers.aspx.

And this is the list of the world's tallest buildings:

"100 Tallest Completed Buildings in the World," *CTBUH Tall Buildings Database*, Council on Tall Buildings and Urban Habitat. http://buildingdb.ctbuh.org/.

Here is the official fact sheet on the fountain:

Fact Sheet—The Dubai Fountain at Downtown Burj Dubai, Souk Al Bahar (PDF). http://www.soukalbahar.ae/fountain.pdf.

In terms of cost, Mark Fuller said the fountain cost "about" $220 million. The creators of the Burj Khalifa said in a press release that the cost was $218 million.

The Dubai Fountain, Dubai Information Guide. http://www.dubaifaqs.com/dubai-fountain.php.

Just for comparison, the cost to build the Empire State Building (in 1930) was $24.7 million. In 2010, the cost to build the fountain of the world's tallest building was nearly ten times that amount.

Even adjusting for inflation, the Dubai Fountain is expensive. The $24.7 million cost of constructing the Empire State Building in 1930 equals $322 million in 2010.

3. It's hard to appreciate the scale and wonder of a great fountain from a video, but for a taste, there are videos of many of WET Design's fountains posted on YouTube, available with a simple search, including videos of the fountains at the Bellagio, at Burj Khalifa, Lincoln Center, and in Branson, Missouri.

4. Mark Fuller holds more than 50 patents on water-related technology, including a patent on the air-powered shooters:

Mark Fuller and Alan Robinson, Air Powered Water Display Nozzle Unit, U.S. Patent No. 5,553,779, September 10, 1996. http://www.google.com/patents?q=%22mark+fuller%22&btnG=Search+Patents.

Jim Doyle is typical of the range of talents WET Design has attracted. An experienced special effects designer who is WET's director of new technology, Doyle won an Academy Award for technical achievement in 1991 for creating a particularly effective fog machine, employing liquid nitrogen, that has been widely used in Hollywood productions, including Michael Jackson's *Thriller* video.

5. Senator Arlen Specter spoke and took questions at Muhlenberg College on April 3, 2009. A portion of the appearance was uploaded to YouTube, and is available here:

Arlen Specter on Natural Gas Drilling in PA Drinking Water, April 7, 2009. http://www.youtube.com/watch?v=EHj90ZqImGY.

6. In 2009, there were only nine U.S. senators with more seniority than Arlen Specter, six Democrats: Robert Byrd, Daniel Inouye, Patrick Leahy, Max Baucus, Carl Levin, and Christopher Dodd; and four Republicans: Richard Lugar, Orrin Hatch, Thad Cochran, and Chuck Grassley.

7. As detailed in chapter 5, note 22, bottled water is much less closely regulated than tap water.

Briefly, tap water is regulated in the U.S. by the Environmental Protection Agency (EPA), under the Safe Drinking Water Act. Bottled water that crosses state lines is regulated as a food product by the Food and Drug Administration (FDA). And while the FDA has adopted the EPA's drinking water standards for bottled water, the actual regulation of bottled water amounts mostly to a system of voluntary compliance and trust, because enforcement rules and factory inspections are both minimal.

For additional detail, see chapter 5, note 22, and:

Bottled Water: FDA Safety and Consumer Protections Are Often Less Stringent Than Comparable EPA Protections for Tap Water (GAO-09–610), Government Accountability Office, Washington, D.C.: June 2009 (PDF). http://www.gao.gov/new.items/d09610.pdf.

8. "Water Your Body" is from a software company called FoWare. http://foware.com/water.html.

9. *Human Development Report, 2006,* UN Development Programme, p. 138. http://hdr.undp.org/en/reports/global/hdr2006/.

10. The cost to put 1 pound of cargo into space on the space shuttle is the subject of debate, and depends on how you allocate costs, but in the detailed 2002 study, linked below, from the Futron Corporation, the cost was $5,000 per pound to low Earth orbit, where the International Space Station flies ($6,300 in 2010). By that standard, it cost $41,650 per gallon to get water to the astronauts.

   Space Transportation Costs: Trends in Price Per Pound to Orbit, 1990–2000, Futron Corporation, September 6, 2002, p. 3 (PDF). http://www.futron.com/upload/wysiwyg/Resources/Whitepapers/Space_Transportation_Costs_Trends_0902.pdf.

   Detail about the space station's water recycling system is here:

   "International Space Station: Environmental Control and Life Support System (ECLSS)," NASA Facts, Marshall Space Flight Center, May 2008 (PDF). http://www.nasa.gov/centers/marshall/pdf/104840main_eclss.pdf.

   Seth Borenstein, "Drink Up: Space Station Recycling Urine to Water," Associated Press, *USAToday*, May 21, 2009. http://www.usatoday.com/tech/science/space/2009-05-21-space-urine_N.htm.

11. Water losses in Italy and the UK:

    VEWA Survey: Comparison of European Water and Wastewater Prices, Metropolitan Consulting Group, May 2006, p. 4 (PDF). http://www.bdew.de/bdew.nsf/id/DE_id100110127_vewa-survey---comparison-of-european-water-and-wastewater-pr/$file/0.1_resource_20 06_7_14.pdf.

    Water losses by U.S. utilities from the American Society of Civil Engineers (ASCE):

    Drinking Water: Report Card for America's Infrastructure, ASCE, 2009. http://www.infrastructurereportcard.org/fact-sheet/drinking-water.

    Water use by state, from:

    Estimated Use of Water in the United States in 2005, USGS, 2009, p. 6 (PDF). http://pubs.usgs.gov/circ/1344/pdf/c1344.pdf.

12. *Human Development Report, 2006,* pp. 137–138.

13. It's hard to get good measures for changes in overall agricultural productivity (as opposed to changes in the productivity of individual crops like corn or cotton or rice). How do you bundle improvements in production of rice with improvements in production of cotton? For purposes of this comparison in the U.S., the agricultural production figures come from the U.S. Bureau of Economic Analysis (BEA), specifically, the BEA's National Income and Product Accounts, table 7.3.3—Real Farm Sector Output. The chart, which you set up yourself using the BEA's Interactive Table construction page, shows the total value of U.S. farm output across the years you request, in constant dollars.

    This is the "keyword index" page for National Income account pages, for "F." http://www.bea.gov/national/nipaweb/IndexF.htm#F.

    Under "farm," table 7.3.3 is real output.

    The main BEA charting page is here. http://www.bea.gov/national/nipaweb/Index.asp.

With 2005 as the base year: 2005 = $100 billion in output. Farm output for 1980 is, comparatively, $62.2 billion.

So in constant dollars, the value of farm output rose from $62.2 billion to $100 billion, an increase of 61 percent.

The total water used for irrigation in the U.S. comes from the USGS's five-year water-use analysis, for 2005 and for 1980, the year irrigation withdrawals peaked.

The main USGS water-use page is here. http://water.usgs.gov/watuse/.

Older water-use reports, back to 1950, are accessible here (PDF only). http://water.usgs.gov/watuse/50years.html.

Irrigation withdrawals in 1980 were 150 billion gallons per day.

Irrigation withdrawals in 2005 were 128 billion gallons per day.

So farmers used 15 percent less irrigation water in 2005 than in 1980.

Since the dollar-value figures are constant, we can compare the productivity of the water used. One billion gallons of irrigation water a day in 1980 produced $410 million worth of agricultural products; 1 billion gallons of water a day in 2005 produced $780 million worth of agricultural products—the same amount of water produced 90 percent more crop, by value.

According to page 8 of the 2007 report *Water for Food, Water for Life*, from the International Water Management Institute, world food prices have fallen almost in half during that same time, so if you measured farm productivity by volume of crops produced, the water productivity of U.S. farms has risen even more dramatically than 90 percent.

14. In the U.S., consumption of carbonated soft drinks is 46 gallons per person, per year, according to 2008 figures from Beverage Marketing. That's 368 sixteen-ounce bottles of soda a year—7 a week for every man, woman, and child in the country. If you didn't drink your 7 bottles of soda last week, well, someone else drank 14.

"Smaller Categories Still Saw Growth as the U.S. Liquid Refreshment Beverage Market Shrank by 2.0% in 2008, Beverage Marketing Corporation Reports," news release, Beverage Marketing Corporation. http://beveragemarketing.com/?section= pressreleases.

15. Outdoor water use was put at 58.7 percent by the AWWA study from 2000:

Residential End Uses of Water, 2000, Water Research Foundation, executive summary. http://www.waterresearchfoundation.org/research/topicsandprojects/exec Sum/241.aspx.

Legal issues around use of gray water are discussed here:

R. Waskom and J. Kallenberger, *Graywater Reuse and Rainwater Harvesting*, Colorado State University, Extension, Natural Resource Publications, No. 6.702. http://www.ext.colostate.edu/pubs/natres/06702.html.

Guidelines for Water Reuse, U.S. EPA, September 2004. http://www.ehproject .org/PDF/ehkm/water-reuse2004.pdf.

16. Orange County, Florida, water rules were described to me by Bill Hurley, manager of Orange County's water reclamation division. They are detailed here:

Water Reclamation (Utilities), Orange County Government, Florida. http://www .orangecountyfl.net/YourLocalGovernment/CountyDepartments/Utilities/Water ReclamationDivisionUtilities/tabid/656/Default.as px.

Orange County has a brochure for homeowners about use of treated wastewater here:

Reuse: Orange County's Reclaimed Water Program, A Customer's Guide, Orange

County Utilities (PDF). http://www.orangecountyfl.net/Portals/0/Resources/Inter
net/DEPARTMENTS/Utilities/docs/ReuseBrochure.pdf.

Orange County's water-use figures come from the county Web site, here (adjusted for daily use):

Utilities Department Statistics, Orange County Government, Florida. http://
www.orangecountyfl.net/YourLocalGovernment/CountyDepartments/Utilities/
Statistics.aspx.

17.   National housing data are available from the Department of Housing and Urban Development:

"U.S. Housing Market Conditions," U.S. Department of Housing and Urban Development. http://www.huduser.org/portal/periodicals/ushmc.html.

Historical home construction and sales data for the nation, through 2009, is in this compilation of statistics. Total new home sales, by year, are on p. 24:

"Exhibit 6: New Single-Family Home Sales: 1970–Present," *Historical Data*, U.S. HUD, p. 69 (PDF). http://www.huduser.org/portal/periodicals/ushmc/spring10/hist
_data.pdf.

In the same 10 years, from 2000 to 2009, 2.75 million apartment units were finished in buildings with five units or more (p. 75). Apartment buildings are easier to double-pipe than single-family homes, if only because of the density.

18.   Smoking was banned by the FAA on U.S. flights of two hours or less on April 23, 1988.

"Smokefree Transportation Chronology," Americans for Nonsmokers' Rights. http://no-smoke.org/document.php?id=334.

19.   *2010 Corporate Social Responsibility Report*, Campbell Soup Company, 2010, p. 20. http://www.campbellsoupcompany.com/csr/.

20.   Felix Franks, ed., *Water: A Comprehensive Treatise* (New York: Plenum, 1972).

Leland C. Allen, "Physical Chemistry of Water," *Science*, vol. 184, April 1974, p. 152.

21.   Bushkill Falls is owned by the Peters family, the descendants of Charles E. Peters, who bought the land and opened the falls to the public in 1904. The family refers to it as "the Niagara of Pennsylvania," and the waterfalls are spectacular, although the flow is intimate compared to the majesty of Niagara.

The facility has predictably kitschy gift shops and exhibits, but the park itself, with miles of hiking trails, is immaculately maintained. It is open from late March to late November, depending on the weather and the condition of the footpaths and trails.

Admission is $10 for adults and $6 for children. http://www.visitbushkillfalls
.com.

WATER MEASUREMENTS OF ALL KINDS

1 gallon = 3.8 liters
1 gallon = 8.33 pounds

1 cubic meter = 264 U.S. gallons
1 cubic meter = 1,000 liters
1 cubic meter = 1 metric ton (2,200 pounds)

1 cubic foot = 7.48 gallons
1 cubic foot = 62 pounds

1 megaliter = 1,000,000 liters (1 million)
1 gigaliter = 1,000,000,000 liters (1 billion)

1 megaliter = 264,000 gallons
1 gigaliter = 264,000,000 gallons

1 acre-foot = 325,851 gallons
1 acre-foot = 1.2 megaliters
1 acre-foot = 7.48 gallons per square foot

Average daily water use per person in the U.S. = 99 gallons
Average daily water use per household in the U.S. = 262 gallons
Average daily per capita use in U.S. = 149 gallons
(Above usage figures as of 2005; for average per capita use, total water supplied by utilities divided by total population; excludes use by electric utilities and agriculture.)

ACKNOWLEDGMENTS

THIS BOOK STARTED AS A SUGGESTION, really a single drop of water, in the form of an e-mail. The day I finished writing my first book, *The Wal-Mart Effect*, my friend Eric Mlyn wrote in a note, "I want to see an article on bottled water."

A few months later, my family was staying in a hotel where the rooms were stocked with FIJI Water. We'd never seen FIJI Water before. My wife opened a bottle, took a long swallow, and declared FIJI Water superior to her favorite, Evian.

My editors at *Fast Company* magazine—Keith Hammonds, Mark Vamos, and Bob Safian—realized that with water coming, in shipping containers, all the way from Fiji to hotels and convenience stores in the United States, the business of bottled water was worth a story. They let me follow the reporting all the way to the north coast of Fiji's main island, Viti Levu. The *Fast Company* story, "Message in a Bottle," was published in August 2007; the response made it clear that readers wanted to know more about their water.

Raphael Sagalyn, my agent, showed his usual understated brilliance. Over dinner, he persuaded me that the book proposal I had labored over was inside out—the real book idea appeared in a single paragraph on page 8 of the seventeen pages I had written. Rafe has the ability to inspire the most ambitious work, while also keeping me firmly planted in reality.

Emily Loose, who edited *The Big Thirst* at Free Press, as she previously edited *The Wal-Mart Effect* at Penguin, is the kind of editor that authors hope for. Sometimes it seemed as if she was thinking about water as much as I was, and our dozens of conversations shaped my reporting and crystalized my thinking. Her text editing is always graceful and respectful, often inspired.

Learning about water has taken me literally around the world, and it has been eye opening and challenging, fun, and frequently humbling. I am consistently amazed at how willing people are to share their hard-won knowledge and their time. It was not possible to land in a country like Australia or India, where I knew no one, and be effective without a lot of help and hospitality.

In Australia, the following people provided advice and connections, insight and time; some also offered a bed to sleep in: Åsa Wahlquist, Mike Duffy, Liz and Robbie Burns, Tim Calkins, Phil Kneebone, Jim Gill, Sue Murphy, Jorg Imberger, Laurie and Deb Arthur, Ross Young, Wayne Meyer, Mike Young, Bruce Naumann, Kevin Flanagan, Alan Kleinschmidt, Stephanie Simms, and Ken Harnett.

Before I got to India, David Strelneck of Ashoka opened doors for me with Ashoka's Fellows on the ground, introductions that were indispensable. David Foster provided valuable advice.

In India, the following people took me to places, introduced me to people, and allowed me to experience things I never would have found on my own: Mehmood Khan, Jyoti Sharma, V. S. Chary, Vimlendu Jha, Venkatesh D, Adrien Couton, Amit Jain, Kushal, Sanjay Desai, and Mohan B. Dagaonkar. Bridget Wagner and Steve Matzie welcomed me into their apartment in Defence Colony, giving me a base and a place to call home. It was an island amid the swirl of Delhi. And they've been endlessly patient with questions about daily life in India.

No less vital were those back here in the United States: Professor Geri Richmond at the University of Oregon; Eric Wilson in Galveston, Texas; J. C. Davis at the Las Vegas Valley Water District, and Yvette Monet at MGM Resorts, in Las Vegas; Janette Bombardier at IBM in Burlington, Vermont; Mark Fuller at WET Design; Jeff Fulgham at GE Water; Jane Lazgin at Nestlé Waters North America; the staff at Aqua America in Bryn Mawr, Pennsylvania; and the staff at Campbell Soup in Camden, New Jersey.

At Free Press, I'm grateful for the confidence, patience, and help of Martha Levin, Dominick Anfuso, Edith Lewis, Nicole Kalian, and Alexandra Pisano. At the Sagalyn Literary Agency, I rely on Bridget Wagner, Shannon O'Neill, and Jennifer Graham. And I couldn't get by without the help of Cheryl Maynard, Renae Vaughn, Heather Craige, Ken Wiley, Lily Shapiro, and, of course, Myrtle Kearse.

You can't write a book like this without good friends, friends who have both patience and good humor: G. D. Gearino, Chuck Salter, Kevin Spear, Ruth Sheehan, Keith Hammonds, and John Dornan; as well as my siblings, Matthew, Betsy, and Andrew.

My parents, Sue and Larry Fishman, are always among my first and most enthusiastic readers; their support is vital.

Nicolas Fishman and Maya Wilson have offered insight, inspiration, enthusiasm, fun, music recommendations, and during one memorable car ride, a string of a dozen excellent title ideas. They make the sun come up each morning.

As always, two people deserve special mention.

No one is closer to the creation of this book than Geoff Calkins. Every day for months, he has heard chapters read aloud, and then offered the kind of editing most writers never get, let alone having it available on-demand, eighteen hours a day. He can spot the dull parts and enliven them; he can hear the tangled sections and disentangle them. He is never too tired, too busy, or too distracted by his own responsibilities to offer advice, support, and inexhaustible cheerfulness. Geoff is that rarest of people, a best friend for thirty years.

And finally, my wife, Trish Wilson, has really had two jobs the last two years—her own, and editing this book. Her editing is so true that once she's made a point—about a sentence, a paragraph, or the arc of a whole story—it's not possible to go with the lesser version. Her enthusiasm and her love are as resilient as water itself—providing both steady support and that essential splash of magic.

INDEX

Page numbers beginning with 316 refer to notes.

acetaminophen, residue in water, 163, 174,
    348
Adelaide, Australia, 112, 115, 166, 167, 169,
    171, 187, 192, 194, 196, 197, 202, 212,
    214, 215, 278, 349
Adelaide, University of, 215, 265, 278
Administrative Staff College of India, 235,
    263, 355
Aggrawal, Praveen, 231
agriculture:
    productivity in, 21, 320, 368–69
    water conflict and, 169, 187–88, 192
    water productivity in, 21, 232, 303–4, 320,
        368–69
    water use in, 20, 21, 119, 169, 171, 196, 232,
        267–69, 331, 342, 366, 369
    see also rice farming
Agriculture Department, U.S. (USDA), 12,
    320, 351, 362
AIDS/HIV, 163
air, paying for, 266
Alabama, 74, 75, 76, 77–78, 79, 303
algae blooms, 140
Algiers, 140
Allegheny River, 100

Amana, 40
Andrew, Hurricane, 92
Angel Park golf course, Las Vegas, 66–69,
    72
    water use by, 67, 330
Anjana (Indian girl), 241, 244
Antigua, 15
Apalachicola Bay, 75, 76, 197
Apalachicola River, 75, 76
Apple Inc., 130–31, 133, 299, 342
Aquafina, 47, 133, 134
aquifers, 20, 85, 115, 116, 154, 158, 204, 205–7
Aria hotel, 52, 53, 122
    water fixtures in, 118, 142–44, 309–10,
        339, 345
Arizona, 54, 190, 328, 348
Arizona Laboratory for Emerging
    Contaminants, 176
Army Corps of Engineers, 75, 76, 77, 100
Arthur, Deb, 182, 212–13
Arthur, Laurie, 182–89, 191–93, 195, 196,
    197, 202, 212–17, 352
    philosophy of, 183–84, 186, 188–89,
        213–17
    water use by, 185–86, 192, 216

Arthur, Neil, 217
Atlanta, Ga., 9, 11, 20, 73–80, 151, 197
  lawsuits against, 76–78
  Magnuson's ruling against, 77–80
  population growth of, 74, 77, 303
  2007–2008 drought in, 73–76, 100
  water mismanagement in, 74–75, 83, 102,
    234, 303
  water rates in, 267, 268
  water use in, 100
attitudes about water, *see* humans, in
    relationship with water; water culture
Australia, 83, 112–17, 140, 145–81, 182–217,
  265, 274, 278, 287, 338, 354
  average rainfall in, 167–68
  climate change and, 14–15, 166, 167, 168,
    172, 198, 205, 216–17, 280, 349
  dehydration deaths in, 145–47, 345
  desalination plants in, 47, 109, 116, 168,
    169, 199, 204–9, 278, 338, 348, 349
  drought in, *see* Big Dry
  economy of, 151, 171, 350
  rainfall bar graph in, 191–92, 199–203
  rice yields in, 184–85, 350, 351
  urban vs. rural water use, 169, 187–88,
    192–93, 194
  water conservation in, 113–17, 170, 209,
    309
  water consumption in, 170, 194, 209, 346,
    349–50, 353
  water culture in, 166, 168–70, 171–72, 189,
    192–93, 209, 210–12, 354
  water politics in, 152–53, 161–62, 166,
    203–7, 273
  water scarcity in, 114, 148, 149–52, 154,
    166–74, 187, 192, 198, 212, 273
  water spending in, 166–67, 179–80, 210,
    212, 348, 362
  see also specific cities
Australian, 146, 193
Australian bush, 183, 212–13

bacteriology, 6
Bangalore, India, 16, 226, 228, 235, 256, 302,
  357
Barbosa, Ralph, 64, 66, 330
Barcelona, Spain, 9–11, 12, 20, 302, 318
Barnes, Roy, 102
Barnett, Colin, 205, 353
Basdevi (Indian villager), 241, 244
Beattie, Peter, 159
Bellagio, 52, 63, 87, 118
  Fountains of, 26, 52, 56, 57–8, 69, 87, 295,
    367

Bening, Annette, 49
Berghofer, Clive, 159–60, 162, 164, 165, 178,
  347
Berkshire Hathaway, 123
Berliner, Eric, 48, 112, 126, 128
Beverage Marketing Corporation, 137, 320,
  342, 343, 369
Bhagat, Pradeep, 255, 256
Big Dry, 115, 116, 117, 146, 148, 149, 151–52,
  153, 168, 169–72, 183, 185, 188, 190–91,
  194, 197, 201–2, 217, 287, 346
biological water, 36, 324
bisphenol A (BPA), 163, 348
Blair, Steve, 47
block pricing of water, 277–78
blood, 46, 321
  in recycled water, 161, 163–64
  water content of, 3, 43–44, 326
Bombardier, Janette, 46, 48, 126, 127, 128,
  129, 132
Boston, Mass., 6, 135, 343
bottled water, 8, 19, 24, 90, 120, 132–38, 174,
  265–66, 270, 288–89, 298–99
  bans on, 137
  and economics of water, 289
  as indulgence, 136–37
  as necessity, 237
  regulation of, 299, 342–43, 367
  as sign of water economy failure, 138
  U.S. consumption of, 137, 138, 320, 342
Boyer Construction, 105, 107, 108, 110
Brady, Jack, 190
Braff, Zach, 49
Brancatelli, Anthony, 288
Branson, Mo., 296
Brantley, Bert, 80
Bridgeport, Ala., 11, 12
Brisbane, Australia, 151, 166, 167, 179, 212
Brita water pitchers, 177
Britain, *see* United Kingdom
Brooks, Cameron, 132
Bunnings, 153
Burj Khalifa, 295, 366–67
Burlington, Vt., 44–48, 125–29
Bushkill Falls, 310–11, 370

Caesars Palace, 53
Caica, Paul, 171–72
California, 54, 80, 84, 184, 190, 197, 267–69,
  277, 328
Cambodia, 233
camels, dying of thirst, 146
Campbell Soup, 117, 267, 276, 306–7, 380
Canada, 5, 85, 317

canals, 85
Canberra, Australia, 214
carbonated soda, 137, 369
carbon footprint, 18
Caribbean Sea, 15
Carper, Tad, 288
carrots, water required to grow, 268, 269,
   362, 363
car washing, 70, 236, 305
"cash for grass," 68–69, 70
Catalan Federation of Commerce, 10
Celebrity Cruises, 124–26, 127, 341
Census of India, 227, 240, 356, 359
Champlain Water District, 128
Chandrayaan-1 spacecraft, 227, 228, 357
Chapman, Jeff, 132
Chary, V. S., 235, 263, 355
Chattahoochee River, 75, 76–77, 197
Chattanooga, Tenn., 11, 73, 332
Chávez, Hugo, 15
child mortality, 7, 13, 223, 300, 355
China, 15, 16, 140, 226, 338
chlorination systems, 6–7
chondrites, 38
Christie, Steve, 208
Citibank, 91, 236
Citizens Against Drinking Sewage
   (CADS), 154, 160–61
CityCenter complex, Las Vegas, 51–52, 118
Clark County, Nev., 59, 268
clean water, 173–78, 304–05
   cost of, 177
   defining of, 174–78
   impact on mortality of, 6–8, 23, 300
   see also micropollutants
Clean Water Act (1972), 99, 336
Cleveland Cavaliers, 287–89, 366
Cleveland *Plain Dealer*, 287–88, 365
climate change, 14–15, 84, 102, 119, 166, 172,
   198, 205, 216–17, 285–86, 301, 312
Clooney, George, 49
clouds, 35, 138–39, 324
   seeding of, 140–41, 344
coal mines, water use by, 140, 155, 180
Coca-Cola, 2, 21, 47, 78, 117, 119, 120–22,
   134, 142, 231, 252
   10-K filings of, 121–22
   water productivity of, 120, 340
   water use by, 120, 121, 140, 340
Cockburn Sound, Western Australia, 205,
   206
"Colin's Canal," 205, 353
Colorado River, 20, 54, 55, 197, 267, 281, 303,
   328

Colorado River Water Users Association
   (CRWUA), 80
Comfort Inn, 88–89, 111
compact fluorescent lightbulbs, 19–20, 131
computer chips, 2, 44–48, 125, 126–27, 129,
   326, 340, 341
Congress, U.S., Lake Lanier and, 75, 76,
   77, 78
Contester Defender, 10, 318
continental drift, 2, 37, 39
continental freeboard, 39
cooking fuel, in India, 227, 356
corporations:
   water as analogous to, 284–85
   water consciousness in, 112–25
   water data gathered by, 119–22, 306–7
   water management by, 117–32
   see also specific corporations
cosmic water, 27, 29–32
cow dung, 227, 230, 245
Creedence Clearwater Revival, 138
Crockett, Fort, 92, 336
cruise ships, 123–25, 127, 341
cubic kilometers of water, 35–36, 323
Cutler, David, 7, 317

Dagaonkar, Mohan, 261, 262
"Dancing Cheek to Cheek," 296
Darling River, Australia, 170, 190
Dartmouth reservoir, Australia, 191
Dasani, 19, 24, 47, 120, 133, 134
Dawson, Nyakul, 146
deep water, 32–35, 37–39
Delhi, India, 16, 228, 229, 230, 232, 235, 264,
   256, 298
   air pollution in, 258–59
   Defence Colony neighborhood of,
      236–39, 251
   population with access to water, 225, 239,
      359
   public taps in, 239, 359
   rainwater harvesting in, 253, 255
   Rangpuri Pahadi neighborhood in,
      253–54, 256, 290–91
   Sainik Farms neighborhood in, 254–56
   Shivanand neighborhood in, 253
   Vasant Kunj neighborhood of, 218–21,
      225, 309
   Vasant Vihar neighborhood of, 221–22
   wastewater from, 257–58, 358, 361
   water delivery by truck or pipe in, 218–23,
      236–39, 252, 253, 256
Delhi Jal Board (DJB), 219–21, 222, 233,
   236, 238, 239, 252, 253, 256, 358

Delta Faucet Company, 118, 143–44, 345
Desai, Sanjay, 261
desalination plants, 10, 47, 84, 131–32, 140, 251, 341
 in Australia, 109, 116, 168, 169, 199, 204–9, 212, 278, 349, 354
 as climate independent, 166, 212
 costs of, 109, 207, 212, 338, 348
 problems with, 204–5, 353
 shipboard, 123–24
 in U.S., 47, 205, 207, 353
 wind-powered, 206
Detroit Metro Airport, fountain in, 293–94, 297
Devi, Bhawan, 254
diamonds, 270
diarrhea, 246, 355, 360
 cost of, 223
dinosaurs, pee from, 17, 319, 320
dolphins, in Las Vegas, 52–53
double-plumbing, 62, 306, 370
"drains," India, 230, 257, 258
drinking fountains, 133, 287–89
drought, 15, 102, 119, 197
 in Atlanta, 73, 75, 76, 77, 100, 102, 333
 in Australia, *see* Big Dry
 in Barcelona, 9–10, 12, 318
 in Caribbean, 15
 in China, 15
 in Las Vegas, 56, 70
 in North Carolina, 14
 in Orme, Tennessee, 11–12, 318
 prayer and, 148, 149, 154
 in Syria, 15
 in Venezuela, 15
drought-resistant crops, 119
Dubai Fountain, 295, 366–67
Dunaway, Faye, 49
Dust Bowl, 15

Earth, 27
 origins of water on, 28–29, 32
 total surface water volume of, 32, 36–37, 322, 323
 water cycle of, 35–36
 water in mantle of, 32–35, 37–39
Eckelbarger, Bob and Beverly, 270–72
E. coli, 238, 258, 359, 361
economic development, 16, 226–27
economics, of water, 23, 138, 143–44, 265–92
 corporate awareness of, 117–23, 126, 129, 143–44
 poor people and, 220–21, 290–91
 price as solution to water problems, 267

price sensitivity and, 272–74, 275–77, 364
 pricing in, 8, 23, 136, 143–44, 193, 210, 265–92, 361–62
 problems with "free," 144, 267, 274–75, 291
 proposed market system and, 279–87
 skewed nature of, 269–70
 true cost of supplying, 83
 value of every gallon, 144
education, of girls in India, 20, 225, 227, 245, 246, 248–50
eendhi, 240–41, 242
El Dorado (Calif.) Irrigation District, 273
electricity, 15, 90, 92, 232, 276–77
 water use and, 2, 5, 84, 129, 276–77, 316, 364–65
Emlenton Water Company (Pa.), 100–101, 102, 270–72
Enceladus, 294, 366
Endangered Species Act, 76
endocrine disruptors, 176, 351
Energy and Resources Institute (TERI), India, 223, 224, 229, 231, 354, 358
Environmental Protection Agency, U.S. (EPA), 4, 142, 258, 316, 342, 343, 350–51, 367
Euphrates River, 197, 281
"Euthydemus" (Plato), 275, 364
evaporation, 35, 36, 72, 323–24
Evian, 13, 17, 29, 50, 90, 133, 136, 138, 185, 333

Fairfax County, Va., 158, 162, 346
farming, *see* agriculture; food; rice farming
fat, water content of, 316
fat three-ridge mussel, 75
FEMA, 110, 111
feminization, from hormone residue, 160–61
Fertile Crescent, 197, 353
Fiji, 134, 138
FIJI Water, 29, 133, 138, 353
filters, 6–7, 47, 176, 177, 207, 208, 224
fire hydrants, 11, 111, 263
"first-glass" problem, 275, 281, 282
fit-for-purpose water, 116–17, 290, 305, 308
Flanagan, Kevin, 152, 154–55, 156, 157–58, 162, 163, 166, 180–81
Flannery, Tim, 169, 209, 349
Florida, 13, 21, 47, 76–78, 79, 139, 167, 303, 305, 319
Florida Panhandle, 74, 75
Foley, Jeff and Kathy, 101

food, water use in production and processing of, 21, 289, 303, 304, 306, 366, 369; *see also* agriculture; rice farming
Food and Drug Administration (FDA), 342, 367
Forbes, 226, 355
FORCE (Forum for Organized Resource Conservation and Enhancement), 252–53, 255
Ford, Gerald, 76
fountains:
decorative, 52, 56–58, 59, 87, 293–98, 329, 367
drinking, 133, 287–89
Four Points hotels, 265–66, 361
"fourth state" of water, 32–35
fracking, 24, 351
Fraile, Miguel Angel, 10
Fraley, Robert, 119
France, utility water loss in, 5
Franks, Felix, 307–8
Fulgham, Jeff, 140, 141
Fuller, Mark, 293, 294–98, 367

Gallop, Geoff, 206–7, 208–9
Gallup Poll, 174, 350
Galveston, Tex., 88–100, 101–2, 103–8, 109–11
Airport Pump Station in, 96, 103, 107, 110
boil-water order for, 89, 111
destruction of water system in, 89, 94–95, 97–99
59th Street Pump Station in, 97–98, 99, 104, 105–6, 107, 108, 110, 336
long-term water system upgrades in, 110–11
30th Street Pump Station in, 94–95, 99, 104, 105, 106–7, 108, 110, 336
wastewater treatment plant in, 89, 99–100, 108, 111
Galveston Bay, 97, 98, 99, 104
Galway Bay, 131
Ganges River (Ganga River), 218, 230, 231, 232, 233, 258
GE, 21, 117, 119, 120, 141–42, 300, 340, 344
"Clouds" commercial of, 138–39, 141, 344
GE Water division, 139–42
PCB pollution by, 142, 354
Geological Survey, U.S. (USGS), 7, 21, 36, 316, 317, 320, 322, 323, 324, 328, 369
Georgia, 73–80, 100, 102, 303, 332, 333, 334
Gill, Jim, 198–207, 209, 210, 211, 212, 214, 273, 354

Gingrey, Phil, 76, 333
Gleick, Peter, 36–37, 319, 324
"global water crisis," 300–301
gods and goddesses, and water, 22–23, 44, 148, 230–32, 295
golf courses, 59, 66–69, 72, 79, 83, 86, 140, 184, 305, 330, 334, 344
water use by, 66–67
Google, 229
Goolwa, Australia, 194, 195
Government Accountability Office, U.S. (GAO), 342–43
Grand hotel, Delhi, 225–26, 240, 355
grass, removal of, 68–69, 70, 86, 330
gray water, 305, 369; *see also* recycled water
Great Lakes, 85, 324
groundwater, 60–61, 82, 86, 120, 201, 215, 246, 255
gulf sturgeon, 75
Gurgaon, India, 229, 231, 239, 246, 247, 260

Hains, Steve, 115, 116, 117
Hanoi, Vietnam, 233, 358
Hartsfield, William B., 75, 76, 80, 333
"Have You Ever Seen the Rain?," 138
Henderson, Nev., 55, 69
hepatitis C virus, 163
Hinduism, 22, 230
Hindustan Times, 228, 366
Ho Chi Minh City, Vietnam, 233, 358
Hollis, Me., 132, 136
Hoover Dam, 14, 54, 85, 262, 268, 327
hormone residue, in water, 160–61, 176
Hudson River, 131, 142, 344–45
humans:
daily water requirements of, 13, 16–17, 243, 319, 359–60
in relationship with water, 2, 3, 6, 8–9, 22–24, 25–27, 28, 50, 117, 133, 136, 143–44, 157, 160, 164, 165, 172, 174, 189–90, 192–93, 209–10, 230–32, 266–67, 286, 289–92, 297–298, 306–9, 310–13, 321–22; *see also* water consciousness; water culture
total water consumed by, 16–17, 319, 320
water as basic right of, 290
water content of, 3, 36, 316, 324
Hume Dam, Australia, 190–91
hurricanes, 92; *see also specific hurricanes*
Hyderabad, India, 226, 228, 238, 256, 263, 359
continuous water service test in, 250–51
hydrated minerals, 33–34
hydrogen bonding, 41–43

hydrogen sulfide, 42
hydroxil (OH), 33–34, 38

IBM, 21, 44–48, 112, 117, 119, 120, 125–132,
    134, 155, 229, 340–41
  desalination technology of, 131–32
  "smart water" and, 130–32
  water management services of, 129–32
IBM Burlington, Vermont, 44–48, 125–29
  water bill at, 126, 127
  water conservation at, 128–29
  water cost savings at, 129
  water data gathered by, 125–26, 129, 341
ice, 124–25, 341
  cosmic, 31, 294, 366
  density of, 42–43, 44
  on the Moon, 227–28, 356–57
Ike, Hurricane, 88–89, 91–94, 97, 98, 103,
    106, 110–11, 335, 336, 337
Imberger, Jorg, 206
Immelt, Jeffrey, 141, 142, 344
Imperial Irrigation District, 331, 362–63
Imperial Valley, 69, 184, 197, 267–69, 276,
    277, 331, 351, 362–63
  economic impact of, 268–69
India, 58, 120, 218–264
  agriculture in, 232, 358
  child mortality in, 223, 355
  climate change and, 14
  continuous water supply (24x7 water) for,
    232–35, 250–51, 260–64
  cross-contamination of water supply and
    sewage in, 237–38, 256, 261–62
  economic growth in, 16, 223, 226–27, 229,
    355
  economic growth in hurt by water,
    223–25
  education in, 227, 245, 246–50, 356
  lack of safe water access in, 223–26, 228–
    229, 232–33, 236–39, 240–44, 250–52
  pollution in, 230–31, 233, 256–59, 358, 361
  poverty in, 226–27, 246
  public taps in, 224, 239, 359
  resistance to continuous water service in,
    233–34
  sacred rivers of, 230–32, 233, 256–59
  space program of, 227–28, 258, 356–57
  Supreme Court in, *see* Indian Supreme
    Court
  urban growth in, 232, 233
  village water culture in, 239–44, 245–50,
    251–52
  water culture in, 221–23, 228–29, 230–33,
    259, 263–64

water infrastructure of, 221–24, 233, 235,
    259, 354
water innovation in, 247–48, 250–56,
    260–64
water mismanagement in, 224, 228–29,
    234, 250, 259
water poverty in, 20, 218–21, 224–25, 233,
    246–50, 251–52, 263–64, 290–91
water quality in, 236–39
water utilities in, 233–34, 260–64, 354,
    357, 358
women and girls in, 20, 225, 227, 240–44,
    245, 248–50, 360
see also Delhi
Indianapolis, Ind., 272–73, 364
Indian Ocean, 116, 168, 194, 195, 197, 205,
    208
Indian Space Research Organization
    (ISRO), 227–28
Indian Supreme Court, 228–29, 258, 357
indirect potable reuse (IPR), 156
indoor plumbing, 109, 337–38
Infrared Space Observatory, 29–30
Intel, 21, 117
  10-K filings of, 122, 341
  water productivity of, 119–20, 340
International Space Station, 8, 302–3, 368
Internet, 96–97
interstellar dust clouds, 29–31, 32, 322
iPhone, 44, 133, 299–300
  sales of, compared to bottled water sales,
    133, 342
Iraq, 197, 281, 365
Iredale, David, 147, 345
irrigation:
  in Australia, 171, 183–84, 186, 187, 188,
    191–92, 194, 214–16, 352
  efficiency in, 224, 232, 303–4, 369
  crops and, 119
  in Imperial Valley, 267–69, 331, 362–63
  in India, 247–48, 358
  in Las Vegas, 66–68
Israel, 47, 122
Italy, water utility loss in, 303, 368
iTunes
  as model for water-data management,
    130–31
  sales, compared to bottled water sales,
    133, 342

Jackson, Miss., 101, 102–3, 109, 337
Jacobsen, Steven, 33–34, 37, 38, 39
Jaitly, Ashok, 218, 224, 229, 231–32
Jamaica, 15

Jargali, India, 239–44, 245, 247, 248, 360
Jawaharlal Nehru University, 218, 221
Johnson, Harvey, 101, 337
Jones, James Earl, 49
Jones, Laurie, 160–61
Jurassic Park (rice paddock), 183–84, 185

Kampala, Uganda, 233, 358
kangaroos, 182, 183, 188, 213
Kennedy Space Center, 1, 316
Kent, Muhtar, 120, 340
Khan, Mehmood, 244–50
Kleinschmidt, Alan, 145, 162, 165, 351
Kroger, 103

laminar flow, of water, 296–97
language, water references in, 3, 23, 28,
    321–22
Lanier, Lake, 73, 74, 75–76, 77, 78, 79, 80,
    100, 332, 333
lanolin, 112, 113, 116
Las Vegas, Nev., 14, 51–72, 80, 81–87, 118,
    122, 142–43, 151, 176, 205, 298, 309–10,
    327
  average high temperature in, 66, 327
  desalination plant for, 84
  economic impact of, 268–69, 363
  fountains of, 52, 56–58, 69, 87, 295, 329
  front lawns forbidden in, 71
  golf courses in, 59, 66–69, 72, 330
  grass removal in, 68–69, 70, 86, 267, 330
  groundwater for, 60–61, 82, 86
  infrastructure costs in, 20, 82, 109
  Lake Mead allotment of, 54, 60, 61, 62, 66,
    67, 70, 328, 331
  Lake Mead intake pipes of, 81–82, 109,
    334
  laundry facilities in, 63–66, 330
  per-capita water use in, 58, 77, 83–84, 306,
    329, 334–35
  population growth in, 15–16, 54–55, 60,
    234, 328
  precipitation in, 54, 327
  recycled water in, 62–63, 64–66, 67,
    71–72, 86–87, 331
  the Strip of, 51–54, 57, 63, 69, 86, 197, 329
  total water use in, 20, 55, 58, 59–60,
    71–72, 73, 75, 77, 86–87, 339
  water culture in, 51–54, 56, 62–63, 70–71,
    73, 83
  water features of, 51–54
  water management in, 54–58, 59–60, 65,
    82
  water rates in, 60, 267–68, 277, 362
  water regulation in, 56–58, 69, 70–72,
    330, 332
  water tax in, 82, 83
Las Vegas Valley Water District (LVVWD),
    58, 59, 61, 372
Las Vegas Wash, 86–87
Latter, William, 29
lawn watering, 21, 59–60, 69, 70–71, 79, 149,
    155, 305, 330, 331
Lawrence, D. H., 1
LeBlanc, Steve, 104, 108
Levi Strauss, 117, 123, 341
lightbulbs, 19–20, 131
Little Bushkill Creek, 310–11
London, 228, 266, 279, 303, 365
Los Angeles, Calif., 75, 102, 138, 234, 304
Lowing, Nick, 182

McCormick, Gerald, 73, 332
Macfarlane, Ian, 148, 159, 161, 347
McKinsey & Company, 226, 355
McLeod, Robyn, 195–97, 199, 274, 354
magnetron, 40
Magnuson, Paul, 77–78, 79, 80, 333
Malta, 131
Mandalay Bay hotel, 51, 58, 63, 86
Manners, Snow, 160–61, 165, 181
Manu, 22
Marseille, France, 10
Mawson Lakes, Australia, 116–17
Mayas, 15
Mayer, Edgar, 148
Mead, Lake, 14, 16, 59, 64, 69, 74, 82, 84, 267,
    268
  drop in water level of, 55, 56, 81–82, 83,
    328–29
  evaporation from, 72
  Las Vegas allotment from, 54, 60, 61, 62,
    66, 67, 70, 328, 331
  recycled water returned to, 71–72, 86–87
  size of, 14, 54, 327–28
  third Las Vegas intake pipe from, 81–82,
    109, 334
Medicis, 278
Melbourne, Australia, 9, 166, 167, 169–70,
    172, 194, 202, 212, 214, 302, 349–350,
    353, 354
Melbourne *Age*, 192, 352
Melnick, Gary, 29–31
Memphis, Tenn., 99
men, water content of, 3, 316
Mewat, India, 245–50
Mexico, 84
Meyer, Wayne, 215

MGM Resorts International, 21, 117, 118, 122, 127, 142–44, 309; *see also* Aria hotel

Michell, David, 114, 116

Michell Wool, 112–17, 142, 338

Michigan, Lake, 277

microchips, 2, 23, 44–48, 120, 125, 126–27, 129, 326, 340, 341

micropollutants, 8, 84, 109, 160–61, 173–78, 304–5, 350–51
  as environmental problem, 178
  in recycled water, 162–64, 173

Microsoft, 229

microwave ovens, and water, 40–41, 325

Miller, Grant, 7, 317

Mirage hotel, 52–53, 58, 62

Mission Industries laundry, Las Vegas, 63–66, 67, 68, 330

Mississippi, 101, 102–103, 109, 337

Mississippi River, 99, 168, 170, 240, 336, 359
  canals to divert flooding of, 85

Molhaka, India, 240, 247

Monsanto, 117, 118–19, 300, 349

Moon, 109, 258
  water on, 34, 227–28, 235, 294, 356–57

Morley, Rosemary, 150–51, 152–54, 160–61, 165, 178, 181

Mother Dairy, India, 247

Moulamein, Australia, 183, 224

Mulroy, Patricia, 51, 56–63, 66, 69–70, 71, 72, 74, 80, 81–86, 176, 306, 329

Mumbai, India, 16, 226, 228, 261

Mundaring Weir, 197–98, 200, 353

municipal water systems, *see* water utilities

Murphy, Sue, 203, 210–12, 273–74, 354

Murray-Darling Basin, Australia, 170–71, 190–91, 214, 274, 350, 352, 353

Murray Irrigation, 191, 192, 352

Murray River, Australia, 154, 155, 167, 199, 216, 287
  description of, 170–71, 191, 194–5
  domestication of, 190–91, 195–96
  drought and, 115, 169, 170–71, 186, 187, 192, 196, 197, 280
  economic importance of, 169, 170–71, 190, 191, 194, 280
  irrigation from, 183, 184, 191, 194, 196, 215
  management of, 215, 280–81
  overallocation of, 196–97, 280
  total inflow to, 196, 283, 350

mussels, 74, 75, 76

Nahata, Vijay, 261, 262, 263

Nai Nangla, India, 244–50

Nalco, 123

Napoleon, Ohio, 267, 276

NASA, 1, 34, 227, 316, 323, 324, 356, 366, 368

Nassau, USS, 106, 107, 108

National Weather Service, "leave or die" hurricane warning of, 92, 336

natural gas, 24, 65, 94–96, 103, 298, 350–51, 367

Naumann, Bruce, 116

Navi Mumbai, India, 260–64

Navy, U.S., 106–7, 108

"nega-liters," 211

Nestlé, 134; *see also* Poland Spring

Nevada, 14, 54, 56, 80, 190, 328
  water rights in, 60–61
  see also Las Vegas, Nev.

NEWater program, Singapore, 158, 159, 346

Newman, Paul, 49

New York, N.Y., 6, 137, 259, 268, 279, 303, 310, 346

New York New York hotel, 57, 59, 329

New York Times, 60, 101, 351

Niagara Falls, 2, 205, 298, 353, 370

Noah, 22

North Carolina, 14, 28, 194

nuclear power plants, water used by, 5, 18–19, 74, 320

Nunes, Sharon, 130, 131, 132

"oarsman," 296

Obama, Barack, 84–85, 138, 167, 343

Occoquan Reservoir, Va., 158, 346

oceans and lakes, 30, 32, 33, 34, 37, 43, 204
  evaporation from, 35, 72
  tempering influence of, 42
  volume of, 32, 39, 323

Ogallala Aquifer, 20, 85

Okeechobee, Lake, 20, 140

Old Testament, water in, 22, 49, 321

olivine (mineral), 37

Orange County, Calif., water recycling in, 158, 162, 346

Orange County, Fla., water recycling in, 305–6, 369–70

Orion and Orion Molecular Cloud, 30–31, 323

Orlando, Fla., 69, 305–6, 330

Orme, Tenn., 11–12, 16, 20, 318

Ortega, Cindy, 118–19, 143–44, 309–10

Palmer, Arnold, 66, 68

parts per trillion, water contaminants, 163, 174–75, 177

Patton, Paul, 143, 345

Payne, Daren, 100

PCB, pollution by GE, 142, 344–45
Pennant Hills Golf Club, Australia, 140, 344
Pennsylvania, 100–101, 102, 139, 270–72, 298
Pentair, 251–52
Pepsi, 47, 134
Perdue, Sonny, 73, 78–79, 332, 334
Perrier, 138
Persian Gulf, 281, 365
Perth, Australia, 151, 166, 167, 168–69, 197–209, 210–12, 214, 215, 237, 273–74, 349
   climate of, 203
   desalination plant in, 203–8
   English-style gardens in, 168, 203
   rainwater inflows to, 168, 199–203, 349
   water conservation in, 204, 209, 211–12
   water culture in, 203–4, 208–9, 210–12, 302
   water politics in, 198, 203–7, 208–9
Peter Cooper Village, 279
pharmaceutical residue, in water, 84, 156, 160, 175, 177
Philadelphia, Pa., 310, 362
   early water system of, 6, 267, 310, 362
Phnom Penh, Cambodia, 233, 358
Phoenix, Ariz., 54, 184, 351
Planet Hollywood hotel, 53
plate tectonics, 2, 37, 39
Plato, on water, 275, 364
plumbing, indoor, 109, 337–38
Pocono Mountains, Pa., 310–11
Poland Spring water, 6, 19, 132–33, 136
Poland Spring, Me., 6
polarity and polar molecules, in water, 41–42, 49
politics, *see* water politics
pollution, 8, 16, 24, 43, 123, 140, 142, 174, 224, 230–31, 233, 257–58, 344, 350, 358
   bacteriological, 17, 238, 258, 358, 361
   see also micropollutants
"Poowoomba," Australia, 160, 164
population, world, 12–13, 14, 16–17, 319
population growth, 21, 289, 300–301
   in Atlanta, 74, 77, 303, 332
   in India, 16, 226–27, 261, 355
   in Las Vegas, 15–16, 54–55, 60, 328
poverty, 13, 20, 225, 226–27, 233, 245–46, 290–91; *see also* water poverty
power plants:
   nuclear, 18–19, 320
   water use of, 5, 18–19, 21, 76, 316
precipitation and rainfall, 14, 35, 54, 55, 152, 168, 201, 327, 349

proteins, shape, and water, 27–28
Prozac, 24, 174
purple bankclimber mussel, 74, 75
purple-pipe water systems, 115, 116, 117, 305–6
PUR water filters, 48–49, 177

Queensland, 148, 159, 179
   recycled water in, 163, 180, 348
Quicken Loans Arena, Cleveland, 287–89, 365–66
Quran, water in, 22, 320

Radarranges, 40
rainwater harvesting, 151, 253, 255–56
Ramia, Joe, 148
Rao, Hanumantha, 250–51
Raytheon Manufacturing, 40–41
Reames, Tony, 11–12
recycled water, 9, 17, 114–17, 158, 211, 305–6, 346
   for golf courses, 67, 68, 140
   on International Space Station, 302, 368
   in Las Vegas, 62–63, 64–66, 67, 71–72, 86–87
   public education about, 157–158, 162–165, 181
   science of, 162–64, 173–75, 348
   Toowoomba's planned use of, 150–51, 152–54, 155–66, 178, 179, 180
   yuck factor and, 145, 160, 164, 165, 172
recycling, of plastic water bottles, 137, 343
red blood cells, 43–44, 46, 326
redwood tree rings, 201
Reinhold, Robert, 60
residence time, water in the atmosphere, 324
Rethod, Bhimrao, 264
reuse water, *see* recycled water
reverse osmosis (RO), 47, 156, 163, 207–8, 237, 251–52
rice farming, 18, 171, 182, 183–89, 190–93, 214, 351
   water requirements for, 185–86, 192–93, 352
Riley, Bob, 76
Rio Grande, 281
rocks, age of, 28
Rohret, Bill, 67–69, 330
Roman Empire, aqueducts of, 6, 96
Ronak (Indian college student), 247–48
Royal Caribbean Cruises, 21, 117, 124–25, 127
runoff, 168, 349
Ruyters, Herman, 148

Sacramento Bee, 273
Safe Drinking Water Act, 176, 342, 367
St. Lucia, 15
Salisbury, Australia, 112, 113, 114–17
San Francisco, Calif., 137
San Luis Resort, 92–93, 94, 103, 104, 111, 336
Santa Fe, N.Mex., water rates in, 277, 278, 365
Saturn, water on moon of, 294, 366
Saudi Arabia, 131–32
SA Water, 114, 115, 116, 117, 349, 354
Schaefer, Vincent, 141
Schuyler, Ronald, 336–37
scum lines, 98
Seabees, 106–7
sea levels, 39
Seattle, Wash., 4, 137, 335, 343
semiconductors, *see* microchips
Senate, U.S., 84, 299
serpentine (mineral), 33, 37
sewage, 257
sewage systems, drinking water and, 6, 165, 237–38, 261–62
sewage treatment plants, *see* wastewater treatment plants
"sewer mining," 140
sewers, 160, 230, 262
Shafer, Dick, 73
Shakespeare, William, water imagery of, 321–22
Sharks, in Las Vegas, 51, 53, 86
Sharma, Jyoti, 252–54, 256
Shatt al Arab river, 281, 365
sheep, 112–13, 171, 188, 216, 338, 350
Shiklomanov, Igor, 36–37, 324
showerheads, 118, 122, 142–43, 204, 309, 339, 345
Sichem Defender, 10, 318
Siegfried and Roy, 53
Sinatra, Frank, 103, 296
Singapore, 132, 162
  NEWater program in, 158, 159, 346
smart water, 125, 130, 131–32, 303, 308–9
Smith, Adam, on value of water, 270, 275, 363
Smyth, Joseph, 37–38, 39
Snyder, Shane, 176, 177, 178, 351
Socrates, 275
solar energy, 131–32, 341
solar system, 28, 30, 322–23
Sonawane, Sharda, 263–64
Sonawane, Vandana, 263–64
Soni, Vikram, 221–23
Sonoran Desert, 184

South Australia, 113, 114, 116, 195–96, 274, 278, 354
  permanent water sprinkler ban in, 171–72
South East Queensland Water Grid (Seqwater), 180
Southern Nevada Water Authority (SNWA), 61–62, 69, 70–72, 84, 176, 328, 331, 332, 334, 363
  incentive program of, 66, 68–69, 70, 340
South Queensland, University of, 149, 164, 347
space shuttle, 1, 316, 368
Specter, Arlen, 298–99, 309, 367
Spencer, Percy, 40–41, 325
Stahl, Lindsey, 46, 47–48
Stark, Dennis, 94
Starwood hotels, 265, 361
steel mills, and water use, 18, 320
Steenrod, Scott, 124, 125, 341
storm water runoff, 114–16
Stulbach, Juju, 139
Stuyvesant Town, 279
Sun Rice, 214
sustainability, 17–19, 130; *see also* recycled water
Swannell, Peter, 164
swimming pools, 53, 72, 170, 203, 332, 349
Sydney, Australia, 140, 147, 166, 167, 212, 344, 362
Sydney Sun-Herald, 113
Syria, 15

Taihu, Lake, 140, 344
Tampa Bay, Fla., 4, 205, 207, 353
Tanami Desert, 145–46
tap water, 24, 28, 133, 134, 135, 137, 225, 235, 237
  concerns about safety of, 174–78, 298–99, 304–5, 309, 350
  cost of, 8, 135, 318, 361
  in-home filtering of, 177, 237
  regulation of, 176, 342–43, 367
Tarragona, Spain, 10
television, 2, 109, 275, 338
Tennessee, 11, 73, 332; *see also* Orme, Tenn.
Tennessee River, 11, 73
Texas, 86, 97, 111; *see also* Galveston, Tex.
Thames Water utility, London, 228, 266, 279, 357, 362, 365
thirst, 141, 145–47, 300
"This Land Is Your Land," 73
Thorley, Dianne "Mayor Di," 150, 152–54, 157–58, 159, 160, 161, 162, 165, 178, 180–81, 346

tiered pricing, of water, 277–78, 290
Tigris River, 197, 281
Times of India, 227, 356–57
Titanic, 2
toilet flushes, 4–5, 13, 72, 305, 362, 373
toilets, dual-flush, 204, 212, 309–10
"toilet to tap" water, 17, 157; *see also* recycled
    water; Toowoomba
Toowoomba, Australia, 148–66, 168, 178–
    181, 185, 202, 206, 309, 346
    Cooby Dam reservoir of, 156, 161
    Cressbrook reservoir of, 159, 179
    description of, 148–49, 345
    prayer services in, 148, 149, 154
    wastewater recycling plan for, 150–51,
        152–54, 155–66, 178, 179, 180, 187
    water infrastructure of, 154
    water politics in, 152–54, 158–66, 172–74,
        181, 186–87, 198, 203, 206
    water scarcity in, 148, 149–50, 154–55,
        166, 178, 187, 197, 210, 346
    Wivenhoe-Cressbrook pipeline to,
        178–80, 351
Toowoomba *Chronicle*, 153, 158
topless swimming pools, 53, 86
Traverse City, Mich., water rates in, 266, 277
Treasure Island hotel, 58, 62
turbulence, in water, 296
Turkey, 197
Tylenol, residue in water, 163, 174, 348
typhoid, 271

Uganda, 233
ultrafiltration, 156
    micropollutants and, 162–64
ultra-pure water (UPW), 44–48, 125, 127,
    128, 156
Unilever, 245, 246, 250
United Kingdom, 5, 228, 266
    water leakage in, 303, 368
    water use in, 5, 317
United States:
    agriculture in, 21, 69, 184, 197, 267–69,
        276, 277, 304, 320, 368–69
    average water bill in, 8, 109, 266, 267,
        276–78, 317, 338, 361–62
    bottled water consumption in, 133–135,
        136, 137, 138, 320, 342
    desalination plants in, 205, 207
    economic growth in, 226
    gross domestic product (GDP) of, 21
    health care crisis in, 302
    indoor plumbing in, 109, 337–38
    life expectancy in, 7, 299, 317

mortality rates in, 7
    new construction in, and water, 306
    per-capita water use in, 7–8, 13, 22, 90
    public schools in, 259
    total water use in, 4–5, 21–22, 35, 316, 323
    water infrastructure costs in, 109, 135,
        235, 343
    water leakage in, 5–6, 303, 317
    water politics in, 173
    water purity in, 174–78, 298–99, 304, 342,
        309, 350–51
urban runoff, 114–16
used water, *see* recycled water
USGS, *see* Geological Survey, U.S.
UV radiation, 47, 65, 126, 156

value of water, versus price, 144, 270, 275;
    see also economics of water
Van Staden, Jacques, 124
Vasudevan, Mukund, 251–52
Venetian hotel, 53, 57
Venezuela, 15
Victoria, Australia, 170, 350
Vietnam, 233
virtual water, 18, 67
volcanoes, water from, 37–38
Voyager 2 spacecraft, 322–23

Wade, Brandon, 93, 103
Wal-Mart, 40, 103, 130
washing machines, 63–64, 113, 123, 339
Washington, D.C., 273, 274, 364
Washington, University of, 201, 202
wastewater, 99, 305, 336–37
    micropollutants in, 175–76
    reused, 67, 71, 72, 86, 155–56, 180, 305
wastewater treatment plants, 89, 90, 99–100,
    111, 230, 343
    advanced (AWTP), 155–56, 161, 163, 180
    bacteria for, 99–100, 336–37
water:
    access to, worldwide, 12–13, 136, 138,
        223–26, 290–91, 300
    age of, 28–29
    as basic human right, 290
    beauty of, 1–2, 25–26, 49–50, 194–95,
        293–94, 310–11
    bottled, *see* bottled water
    consumption, *see* water statistics
    cost of, *see* water bills; economics of water
    creation of, 29–32, 38–39
    forms of, 32–35, 37–38
    as $400 billion-a-year business, 131
    as "genesis ingredient," 44

water (*cont.*)
golden age of, 3–6, 7–9, 14, 23–24, 131, 279, 289–90, 303, 304, 309
human daily requirements of, 13, 16–17, 243, 319, 359–60
human relationship to, 2, 3, 6, 8–9, 22–24, 25–27, 28, 50, 117, 133, 136, 143–44, 157, 160, 164, 165, 172, 174, 189–90, 192–93, 209–10, 230–32, 266–67, 286, 289–92, 297–298, 306–9, 310–13, 321–22; *see also* water consciousness; water culture
importance of, 2–3, 43–44, 291
insulating properties of, 42, 43
invisibility of, 3–4, 7–9, 24, 234
myths about, 26–27, 32, 309
natural purification of, 17–18
ostentatious display of, 53–54
permanence of, 17–19, 313
personality of, 16, 25–26, 41, 48–50, 293–94, 298, 313
as price inelastic, 274, 275–76, 364
price sensitivity and, 272–74, 275–77
pricing of, 8, 23, 136, 143–44,193, 210, 265–92, 361–62; *see also* water bills
proposed market system for, 280–87
purification of, 44–48; *see also* recycled water; water treatment plants
qualities of, 2, 16–19, 41–44, 47, 48–50, 296–97, 313
recycled, *see* recycled water
resilience of, 16–19
as sacred in India, 230–32, 233
Shakespeare references to, 321–22
as solvent, 43, 45, 47
temperature change and, 42
territorial protectiveness about, 85
total amount of, 32, 35–36, 322–23
trading in, 286–87
versatility of, 1–3, 19, 25–26, 41, 44, 48–50
walking to supply, 13, 20, 220, 225, 240–44, 246, 248, 290–91, 360
wasting of, 5–6, 20, 189, 224, 232, 303–4
Water: A Comprehensive Treatise (Franks, ed.), 307–8
water anxiety, *see* water insecurity
water bills, 8,109, 122, 126, 127, 129, 135, 210, 233, 262, 266–67, 271, 272–73, 276–77, 317, 338, 361–62
water complacency, 5–6, 9, 24, 74, 131, 203, 303
in Atlanta, 74–76, 79–80

water conflict, 20
urban vs. rural in, 169, 187–88, 192–93, 194
see also Toowoomba; water politics; water scarcity
water consciousness, 117, 121, 122–23, 125, 126, 127, 129, 141, 157, 189, 231
water conservation, 118, 121, 124–25, 130, 231, 302
in Atlanta, 78–80
at IBM Burlington, 126, 128–29
in Las Vegas, 63, 70–72, 83
in Perth, 204, 209, 211–12
see also recycled water
Water Corporation, Australia, 198–209, 210–12, 273–74
water culture, 306–9
in Australia, 166, 168–70, 171–72, 189, 192–93, 209, 210–12, 354
in India, 221–23, 228–29, 230–33, 239–44, 245–50, 251–52, 259, 263–64
in Las Vegas, 51–54, 56, 62–63, 70–71, 73, 83
in Perth, 203–4, 208–9, 210–12, 302
see also water, human relationship to
water cycle, 35–36, 37, 323
water envy, 189–90, 193–94, 215–16
water footprint, 18
water illiteracy, 9, 138, 300, 309
Water in Crisis (Gleick, ed.), 36–37
water infrastructure, 24, 84–85, 96–97, 109, 251–52, 290, 291–92
cost of maintaining, 20, 81–82, 109, 233, 235, 273, 274, 343
cost of rebuilding, 110–11, 166–67, 179–80
cross-contamination of sewage and, 237–38, 256, 261–62
failures of, 101–103
in India, 223–24, 233, 235, 259, 260–63
intelligent networks for, 130–31
lack of resilience of, 102–3, 209–10
leakage in, 5–6, 134, 233, 234, 237–38, 250, 270–71, 303, 317, 342, 368
and price differences, 277
water insecurity, 114, 141
corporate response to, 117
water management, 18, 23, 84, 142, 215, 285, 354
Atlanta's lack of, 74–75, 83, 102, 234, 303
cascading benefits of, 65, 68
corporate, 117–32
by IBM, 129–32

India's lack of, 224, 234, 250, 259
in Las Vegas, 56, 58, 59–60, 65, 82, 234, 306
water-mark, 5, 9
water molecules, 29, 31, 41, 322, 356
properties of, 41–42, 43, 49
size of, 31, 43–44, 46, 47, 326
water pipes, cost of replacing, 109
cross-contamination of, 237–38, 256, 261–62
water politics, 23, 85, 86, 173, 282, 286
in Atlanta, 73, 74, 75–76, 78–80
in Australia, 152–53, 161–62, 166, 273
and cost of, 272–73, 282
in India, 231–33
in Perth, 198, 203–7, 208–9
science vs. perception in, 152, 162–65, 173–74
in Toowoomba, 152–54, 158–66, 172–74, 181, 186–87, 198, 203, 206
water poverty, 13, 20, 218–21, 225–26, 233, 246–50, 290–91, 359
U.N. report on, 318–19, 359
water problems, as local, 19–20, 301–2
water productivity, 117, 120, 286, 303, 340–41
in agriculture, 232, 303–4, 369
of Coca-Cola, 120
of IBM, 129, 340–41
of Intel, 119–20, 340
in U.S. economy, 21–22, 303–4
water recycling, *see* recycled water
water regulation, 70–71, 142, 149–50, 168–70
water scarcity, 9, 18, 20, 23, 188, 189, 190, 193, 289, 301
in Australia, 114, 148, 149–52, 154, 166–174, 187, 192, 198, 212
economic impact of, 151, 166, 167, 171, 187
as public policy issue, 152, 153
water slavery, 221, 291
water statistics *(page numbers are to easiest reference; pages in source notes include note citation)*
access to water, people without, worldwide, 13
age of water on Earth, 28
Bellagio, Fountains of, water use, 69
bottled water, total spending on, U.S., 133
bottled water, total consumption, U.S., 134
Campbell Soup, water required per can of tomato soup, 306
carrots, water required to raise 3-pound bag, 268
children who die from water-related illnesses, 13
cost of water, U.S. home, 317n14
Coca-Cola, water required per liter of soda, 120
Coca-Cola, total water use, 120, 340n10
diarrhea, cost of to economy of India, 223
dinosaur, urine volume from, 17
Earth, total surface water volume of, 32, 322n2
electricity, water required to generate, per capita, U.S., 2
electricity, water required to generate, total, U.S., 5
evaporation, total into atmosphere, 35
food, water required to raise, per calorie, 366n23
golf, water required per 18-hole round, Las Vegas, 67
hotel room, water required to launder linens, Las Vegas, 64
human beings, daily minimum water required, 13
human beings, total water consumed in history, 16–17
human beings, total water contained inside all, 36
human beings, water content of, 3
indoor plumbing, percent of homes with, U.S., 109, 337n17
infrastructure spending, total U.S. on water, 135
living creatures, total water contained inside all on Earth, 36
men, water content of, 3
nuclear power plant, water required to cool, 18–19
rainfall, gallons of water in 1 inch of rain, 35
space shuttle, water required to launch, 1
steel, water required to manufacture, 18
toilet flushing, water use by, U.S., 4–5
volcanic eruption, percent of water in cloud, 38
walk, people who must to fetch water, worldwide, 240
women, water content of, 3
water loss by utilities, worldwide, 303
water use, Las Vegas, 58, 331n21, 334n45
water use, New York City, 346n13
water use, U.S. total, 21, 316n2
water use, U.S. per capita, 7
water use, U.S., water productivity doubled relative to GDP, 21

Water Stewardship Act (Georgia, 2010), 78–80

water tankers, 218–21, 239, 253, 262

water treatment plants, 6–7, 109; *see also* wastewater treatment plants

water utilities, 54, 74, 96–97, 228, 232–34, 238

  bills from, *see* water bills

  electricity use by, 276–77, 365

  lack of resilience in, 102–3, 209–10

  outsourcing for efficiency, 262–63

  water loss in, *see* water infrastructure, leakage in

  water treatment by, 6–7, 109, 177

water walk, *see* water, walking to supply

"Water Your Body," 299–300

Wealth of Nations, The (Smith), 270, 275

weather, 2, 32; *see also* climate change; drought

wells, 7, 61, 63, 201, 204, 224, 244, 247–48, 254, 255–56

Western Australia, 168, 198–203, 205, 206–7

  normal rainfall in, 199–203

WET Design, 52, 293, 294–98, 367

wetlands, water-filtering, 114, 115

Whole Foods, 133, 341

Williams, Sam, 78

Wilson, Eric, 88, 92–96, 98, 103, 104, 107–8, 110–11, 336

windshield washers, 143

Wipro, 228, 260, 357

Wivenhoe reservoir, Australia, 179, 180

Wolfenden, Richard, 25, 27–28

women:

  education of, 20, 225, 227, 240, 245, 246–50

  management of water systems by, 354

  water content of, 3, 316, 360

Wong, Penny, 166, 348, 354

Woods, Jarman, 146

wool, 112–13, 216, 338

  scouring of (cleaning), 112–14, 115, 116, 122

Wooten, James, 106–7

World Health Organization, 13, 355

Wynn, Steve, 58, 62, 70

xeriscaping, 67–68

Yamuna River, India, 230–31, 233, 256–58, 259, 358, 361

Yarragadee Aquifer, 204, 205–7

Young, Mike, 215, 265, 278–87, 290, 291, 365

Young, Ross, 172

YouTube, 297, 344, 367

yuck factor, recycled water and, 145, 160, 164, 165, 172

ABOUT THE AUTHOR

As a reporter, Charles Fishman has tried to get inside organizations, both familiar and secret, and explain how they work. Fishman's previous book, the *New York Times* bestseller *The Wal-Mart Effect*, was the first to crack open Wal-Mart's wall of secrecy, and has become the standard for understanding Wal-Mart's impact on our economy and on how we live. *The Economist* named it a Book of the Year.

In the course of reporting about water, Fishman has stood at the bottom of a half-million-gallon sewage tank, sampled water directly from the springs in San Pellegrino in Italy and Poland Spring in Maine, and carried water on his head for three kilometers (more than a mile and a half) with a group of Indian villagers.

Fishman is a former metro and national reporter for the *Washington Post*, and was a reporter and editor at the *Orlando Sentinel* and *The News & Observer* in Raleigh, North Carolina. Since 1996, he has worked for the innovative business magazine *Fast Company*. Fishman has won numerous awards, including three times receiving UCLA's Gerald Loeb Award, the most prestigious award in business journalism.

Fishman lives outside Philadelphia with his wife, also a journalist, their two children, and their yellow Labrador. He likes his water from the refrigerator spigot, with ice, or splashing across the bow of a Sunfish.

Learn more at www.charles-fishman.com.